Collinear Holography

# Collinear Holography

## Devices, Materials, Data Storage

*Xiaodi Tan*
*Hideyoshi Horimai*
*Tsutomu Shimura*
*Xiao Lin*

WILEY-VCH

**Authors**

***Prof. Xiaodi Tan***
Fujian Normal University
College of Photonic and Electronic Engineering
No.1 Keji Road, Shangjie
Minhou District
350117 Fuzhou
Fujian
China

***Dr. Hideyoshi Horimai***
HolyMine Corporation
2032-2-301 Ooka, Numazu
410-0022 Shizuoka
Japan

***Prof. Tsutomu Shimura***
The University of Tokyo
Institute of Industrial Science
4-6-1 Komaba, Meguro-ku
153-8505 Tokyo
Japan

***Prof. Xiao Lin***
Fujian Normal University
College of Photonic and Electronic Engineering
No.1 Keji Road, Shangjie
Minhou District
350117 Fuzhou
Fujian
China

**Cover Image:** © SergeyBitos/Shutterstock



**Library of Congress Card No.:** applied for

**British Library Cataloguing-in-Publication Data**
A catalogue record for this book is available from the British Library.

**Bibliographic information published by the Deutsche Nationalbibliothek**
The Deutsche Nationalbibliothek lists this publication in the Deutsche Nationalbibliografie; detailed bibliographic data are available on the Internet at <http://dnb.d-nb.de>.



**Print ISBN:** 978-3-527-41363-8
**ePDF ISBN:** 978-3-527-81045-1
**ePub ISBN:** 978-3-527-81047-5
**oBook ISBN:** 978-3-527-81048-2

**Typesetting** Straive, Chennai, India
**Printing and Binding** CPI Group (UK) Ltd, Croydon, CR0 4YY

C9783527413638_040624

# Contents

# Preface

In big data era, large capacity, high speed, long life, and low-cost data storage technology is urgently demanded. Collinear Holography is a promising candidate technology for mass data storage. The biggest feature of optical holographic storage is that it breaks through the two-dimensional surface storage mode of traditional optical disc storage and adopts the three-dimensional volume storage mode. As the data recording is increased by one dimension, the existing optical storage density can be increased by several orders than traditional optical disc. Data transfer is in two-dimensional data page, so the increase of data transfer rate is also more than several orders of magnitude.

Holographic data storage technology moved out of the laboratory and went to commercialization after the collinear holography was proposed. In 2005, the paper published in the *Journal of Applied Optics* (edited by the Optical Society of American) is the first report that introduced the details of collinear holography. The paper had been ranked eighth of "the top 15 highly cited articles published in the journal over the last 10 years" in 2014. Collinear holography has been applied and developed not only in the area of data storage system, but also expanded to other fields of information processing systems.

Collinear holography data storage technology has now grown to become an exciting interdisciplinary field of research and development, involving device configuration, materials, opt-electronics, and data processing. To meet the rapid development of this technology, we summarized the research results of authors in the book and emphasized the fundamental principles that can be employed for understanding the limitation of various collinear holographic systems, and to analyze their performance of data storage systems. A significant effort is made to bridge the gap between theory and practice through the use of numerical examples based on real system, materials, and page information. The book covers a very wide range of topics, including the basic physical properties of holography, the propagation of plane waves, the Fourier optics, the concept of data format, key devices of system, recording materials, system margins compared with other method, characteristics of data writing and reading. At the end of the book, the applications that use collinear technology are introduced to treat the optical transmission of information systems. In writing this book, we have assumed that the reader has been introduced to basic electromagnetic plane waves in an undergraduate course in electricity and magnetism. It is further expected that the reader has some background in elementary Fourier optics and data processing.

In addition to our four authors, many other authors have also participated in the writing of individual chapter of the book. For the editing format reason, we cannot list their names on the cover and at every chapter. The book is organized in such a way that Chapter 1 introduces the background of big data era, history of holographic data storage, and present problems. Chapter 2 introduces the collinear holographic data storage systems, in which the system control and optical standardization tester are contributed by Masaki Igarashi (New Optware) and Masao Endo (The University of Tokyo). Chapter 3 summarizes the theoretical research results of collinear holography, including the simulation results of Shuhei Yoshida (Tokyo University of Science), Jianhua Li (Beijing Institute of Space Long March Vehicle) and Liangcai Cao (Tsinghua University). Chapter 4 summarizes the key equipment and devices related to holographic data storage, including the contributions of Jinpeng Liu (Xidian University), Hiroyuki Matsuda (New Optware), Tadashi Sasada (New Optware), Yawara Kaneko (New Optware) and Masaharu Kinoshita (New Optware). Chapter 5 introduces the materials, and Yasuo Tomita (The University of Electro-Communications), Jinxin Guo (Beijing University of Technology), Lin Cao (Beijing University of Technology), Jialing Jian (Beijing University of Technology), Dayong Wang (Beijing University of Technology), Xinping Zhang (Beijing University of Technology), Fenglan Fan (Hebei Normal University for Nationalities) and Ying Liu (Beihang University) contributed all of this chapter. In Chapter 6, the data format quotes the author's research results for many years. In Chapter 7, the reconstructed characteristics of reading and writing from collinear holographic data storage system are given by the selectivity, related parameters, and symbol error. In Chapter 8, various system margins are discussed, and Jianhua Li (Beijing Institute of Space Long March Vehicle), Yeh-Wei Yu (National Central University), Liangcai Cao (Tsinghua University) and Jinpeng Liu (Xidian University) have contributed a part of them. In Chapter 9, the related applications of collinear holography are reviewed, while Kanami Ikeda (The University of Electro-Communications) and Eriko Watanabe (The University of Electro-Communications) described their correlation system in detail. All of the chapters contain the research results of authors' research and development in data storage and applications for more than 20 years.

With the continuous growth of storage market demand in the era of big data, holographic data storage will receive extensive attention from the industry and commercialization. Although the collinear holography has shown its infinite charm, it still needs to further improve its performance and stability, and strive to reduce costs. We hope this book can attract more readers to participate in the research and development of collinear holography to seek greater progress in collinear holography.

*Xiaodi Tan*
*Hideyoshi Horimai*
*Tsutomu Shimura*
*Xiao Lin*
*Fuzhou, China*
*2020 August*

# Foreword

Collinear holography was originally proposed by the holograph inventor Dennis·Gabor in 1948; to avoid the interference of zero-order light, off-axis holography was adopted to realize the recording and reconstruction of three-dimensional images until the 1960s, after laser was invented. Holographic data storage was also proposed at the same time, due to its large storage capacities and high transfer rates. Therefore, two-beam interference holography has become a common method of traditional holography. Because two beams are easily affected by environmental vibration, holography is usually inseparable from the vibration isolator. To eliminate the commercial limitation caused by the problem of off-axis holography, collinear holography has gained a new life and is perceived as a major breakthrough in the field of holographic data storage.

Collinear holography, in this book, is a new form of collinear holography, which combines the advantages of D. Gabor's holography and traditional off-axis holography. The information and reference beams are divided co-axially by a special light modulator. With this unique configuration it can produce a small, practical holographic data storage system more easily than traditional two-axis holography. The advantages of the collinear holography have been summarized by the authors with their theoretical study to experimental results for a long term.

This book covers the data storage system, principle, basic theoretical models, key devices and components, materials, and other applications. The book includes many technical challenging works of the authors in detail. This book suggests that the collinear holography is very effective in increasing recording density and data transfer rates.

What will be the next generation of data storage technology? We do not know yet. However, I think collinear holography will increase in importance over the coming decade with the development of related technologies and devices. Collinear holography will enable us to expand its applications into other optical information systems.

*Xiangang Luo*
Academician of Chinese Engineering
Beijing
China

# 1 Introduction

## 1.1 Big Data Era

Information has long been regarded as an important resource, and with the continuous development of information technology, information protection, mining, and data storage have become particularly important. The year 2012 was declared as the Big Data Meta Year with the sudden explosion in the amount of global data generated and the advent of the big data era has forced us to prioritize data storage. According to the research results of International Data Corporation (IDC), the total amount of data generated worldwide in 2010 was 1.2 ZB (1 ZB = $10^3$ EB = $10^6$ PB = $10^9$ TB), the data volume in 2013 was 4 ZB, and that in 2018 was 33 ZB. The latest forecast of IDC predicts that by 2025, the total global data will exceed 175 ZB.

Today's data storage technology is still dominated by magnetic storage devices, such as hard disk storage and magnetic tape storage. Although this technology has matured, its data storage capacity has only increased by about 20% each year, which is far from keeping up with the current rate of data growth. Moreover, its limitation of two-dimensional storage still persists. For further increasing the magnetic storage density, the size of the recording magnetic particles needs to be continuously reduced. However, when the magnetic particles are small enough, they are affected by the superparamagnetic effect, making it difficult for the recording particles to maintain magnetic stability even at normal temperatures. The traditional magnetic storage density has approached its theoretical limits. In practice, to ensure data redundancy, three hard disks are generally required to back up one copy of the data simultaneously, and the data in each hard disk need to be transferred to a new hard discovery four to five years; otherwise, the information will likely be lost forever. Professionally stored tape storage also needs to be transferred every 10 years or so to avoid data loss. Therefore, companies with large data centers, such as Google, require a large server scale, and the annual cost of data transfer is huge, accounting for about one-third of the total cost of data storage. Additionally, during magnetic data writing and reading, the driver radiates a large amount of heat, and an exorbitant amount of power is consumed to cool the server. Accordingly, the power cost also accounts for about one-third of the total cost. Therefore, the data

*Collinear Holography: Devices, Materials, Data Storage*, First Edition.
Xiaodi Tan, Hideyoshi Horimai, Tsutomu Shimura, and Xiao Lin.

storage cost of the magnetic storage technology is in sync with the amount of data growth, and both are exploding, and as a result, the pressure on magnetic storage technology is increasing.

Optical storage is another important data storage technology. At present, the most common optical storage technology is optical disks. If light with a wavelength of $\lambda$ is used as a light source for data storage, its theoretical storage area density is nearly $1/\lambda^2$. From the earliest compact disc (CDs) to video compact (VCDs), digital video discs (DVDs), and now Blu-ray discs, the density of disc storage is also increasing. However, the optical discs are still two-dimensional surface storage devices based on bit storage (although some optical discs can achieve multilayer storage, but the number of layers is limited); each recording bit represents only a 0 or 1 state, and its storage density is affected by the record. Moreover, although a smaller-sized recorded bit results in a greater data storage density, the size of the bit is limited. Each recording bit is etched by converging the energy of an incident laser light. To obtain a smaller recording bit size, the numerical aperture of the recording objective lens must be increased, and a shorter wavelength laser should be used as the recording light source. At present, the numerical aperture of the data recording objective lens in the most advanced Blu-ray disc has reached about 0.85, and the laser wavelength has also been shortened to 405 nm. Now a single-sided single-layer Blu-ray disc can have a storage capacity of 25 GB, while a double-sided four-layer Blu-ray disc can exhibit a storage capacity of 200 GB. Therefore, the famous data company Facebook started to build a Blu-ray disc-based database with a total capacity of 1000 PB in 2014. Its data access energy consumption is 80% lower than that of the magnetic hard disks. The Blu-ray disc can be stored for 30–50 years, and the cost of data transfer is greatly reduced. This Facebook storage system uses a very cost-effective redundant backup method, which can achieve a data backup redundancy with a coefficient lower than two to ensure data security and not cause performance degradation due to scale expansion.

Despite the rapid development in the optical disc storage, it is still a two-dimensional surface bitwise storage device. To further increase the storage density, the numerical aperture of the recording objective lens needs to be increased further, and the recording light wavelength should be further shortened. However, the theoretical aperture of the objective lens in air is less than or equal to one. To develop a numerical aperture greater than one, only the method of immersion can be used. The application environment for such a storage device is limited, and the implementation cost will be exorbitant. With the increase in the numerical aperture, the thickness of the protective layer on the surface of the optical disc needs to be reduced excessively, which in turn will eventually result in the loss of the protective effect. The wavelength of the recording light is currently close to the range of invisible light, and the cost of using violet or extreme ultraviolet lasers is also very high. Therefore, further increasing the storage density of the traditional optical disc is an arduous task. According to the current 200 GB data storage capacity of a Blu-ray disc, by 2020, the number of optical discs required to store the global data is expected to exceed 500 billion, and their combined thickness will exceed the distance between the Earth and the Moon. Therefore, traditional optical disc storage is unable to meet the growing requirements of big data storage, and the discovery and development of new technologies is imminent.

Optical holographic data storage is an optical storage technology that follows the principle of holography for data storage and reproduction. The most noticeable feature of the optical holographic storage is that it breaks through the two-dimensional surface storage mode of the traditional optical disc storage and adopts the three-dimensional volume storage mode. Further, its theoretical storage density is $1/\lambda^3$. By increasing the storage density by one dimension, the existing optical storage density can be increased by several orders of magnitude.

The optical holographic data storage technology records the amplitude and phase information of objects in the form of holograms in holographic materials. As shown in Figure 1.1, during the recording process, the light passes through a spatial light modulator (a two-dimensional optical element that can display the two-dimensional pattern that you want to upload), and the information carrying the two-dimensional pattern is called the object light. This object light interferes with another beam of the known light field (reference light) in the holographic material to form a complex light field distribution. The holographic material responds to the light fields of different intensities accordingly, resulting in changes in the material and finally forming some kind of stable structure, namely a hologram. This hologram records the object light information in the holographic material. During the reading process, only the hologram in the material needs to be irradiated with the same reference beam in the same state as that used during the recording process, and the light energy is coupled to the object light through the coupling effect of the holographic structure on the reference light. In the field, diffraction of light occurs, and the diffracted light is also

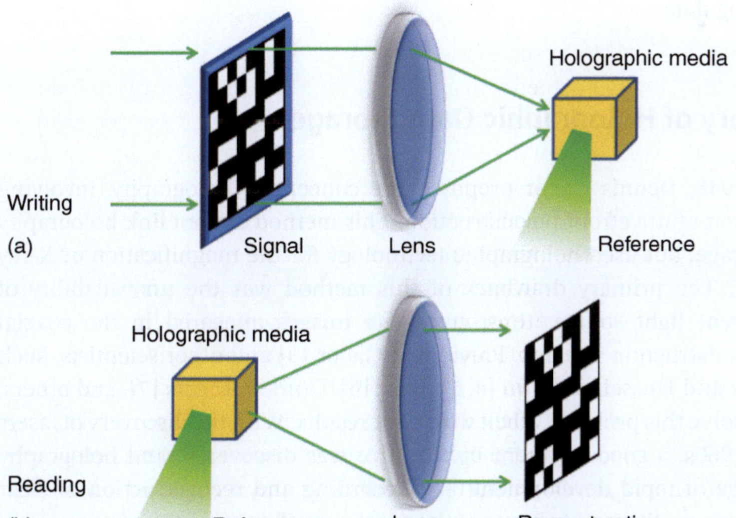


**Figure 1.1** Schematic illustration of holographic data storage: (a) writing process and (b) reading process. Source: Xiao Lin.

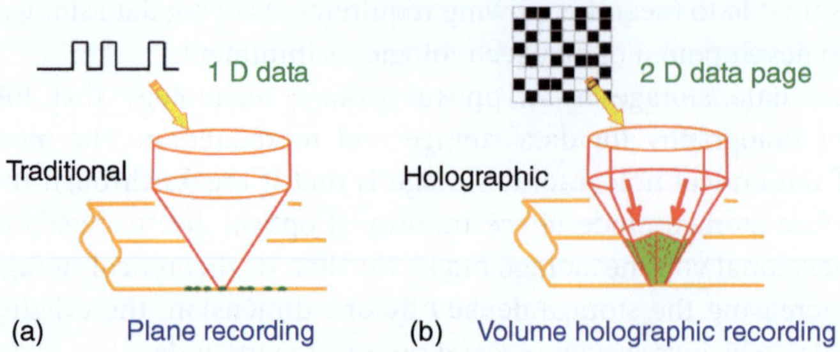


**Figure 1.2** Comparison between traditional data storage and holographic data storage. Source: Xiao Lin.

called as the reconstructed light, which is actually consistent with the distribution of the original light field, enabling readout of the object light information.

Optical holographic storage can get rid of the one-dimensional data storage limitation of bit storage. Each recording position represents a two-dimensional coding pattern, as shown in Figure 1.2. Further, because the holographic storage has the characteristics of reusable recording, the recording bit and its area can be superimposed on each other, which leads to a large recording holographic storage density of the order of $TB/in^2$. Simultaneously, the data reading conversion rate can also reach the order of 10 GB/s, because each recording position readout is a two-dimensional encoding pattern. Additionally, the performance of the materials used for holographic storage is significantly improved. The materials, viz. the early photorefractive crystals to the current photopolymers, are easy to prepare and cost effective, and the life of data can be as long as 50 years. Therefore, holographic storage is considered to be the most promising and a key optical storage technology in the era of big data.

## 1.2 History of Holographic Data Storage

As early as 1948, Dennis Gabor proposed the concept of holography, involving only the method of wavefront reconstruction. This method did not link holography with data storage, but used holographic technology for the magnification of X-ray images [1, 2]. The primary drawback of this method was the unavailability of a good coherent light source; thus, conjugate images appeared in the coaxial wavefront reconstruction systems. Previously, Gabor [3] and other scientists, such as Kirkpatrick and Hussein El-Sum [4, 5], Baez [6], Gordon Rogers [7], and others, attempted to solve this problem, albeit with poor results. With the discovery of lasers in the early 1960s, a good coherent light source was discovered, and holography entered a stage of rapid development. The recording and reconstruction of clear images became a reality, and because of the excellent coherence of the light source, E.N. Leith and J. Upatnieks proposed a reference light off-axis holographic recording system [8, 9] that overcame the drawback of Gabor hologram, in which the reconstructed conjugate images are aliased in one place. In 1963, van Heerden formally proposed the concept of holographic data storage [10]. He attributed holographic data storage to three-dimensional solid-state optical information storage and estimated that its storage density limit was $V/\lambda^3$ ($V$ is the volume of the recording material, and $\lambda$ is the wavelength of the recording light). Subsequently, van Heerden also discussed the possibility of multiplexing with the reference light angle and the wavelength in holographic data storage. Leith et al. also proposed and validated the technique of recording disk rotation multiplexing in the early developmental stages of holographic data storage [11]. At this time, the holographic technology had been quite embryonic, but owing to the lack of effective recording materials, most of the research on holographic data storage stayed was confined only to the discussion of methods. In 1966, Ashkin et al. of Bell Labs unexpectedly discovered the photorefractive effect while performing frequency doubling experiments with lithium niobate crystals [12]. These crystals have been extensively studied as photorefractive crystals and applied to holographic data storage [13–16], making three-dimensional volume holographic data storage once again a research hotspot. Due to the development of materials, a series of verification experiments on holographic data storage have been published, and more reuse technologies have been studied.

In 1973, researchers at the American RCA Corporation designed and verified a holographic memory that could read and write $10^6$ bit data [17]. In 1973, the Japanese company NEC proposed a holographic data storage system (HDSS) with read-only holographic encoder disks, but stored in two dimensions, on a 128 mm × 128 mm surface, with a storage capacity of $2.5 \times 10^5$ bits at a resolution of 1 line/mm [18]. In 1974, the American 3M company, Strehlow, and others designed a 7 MB holographic memory [19]. In 1973, Huignard and coworkers first demonstrated the law of angular multiplexing, and proposed that recording 100 holograms in one location can reach a total storage capacity of $10^{11}$ bits [20], and then experimentally demonstrated the angular multiplexing scheme in iron-doped niobium. Ten holograms were reused in the lithium acid crystals [21]. In 1975, Amodei and Staebler of RCA Corporation, for the first time, recorded 500 holograms in 1 cm$^3$ of iron-doped lithium niobate crystals [14]. In 1978, Andrei Mikaelianc described a rewritable holographic storage system based on iron-doped lithium niobate crystals [22]. In 1976, Hitachi, Japan reported a holographic video disc, which stored a 30 minutes color video in a disc of 300 mm diameter; a total of 54 000 holograms were recorded on the disc [23]. In 1980, Kubota et al. of NEC Company developed a one-dimensional Fourier transform hologram optical disc for recording audio information; this disc could achieve a data conversion rate of 256 Mb/s [24].

Although there was a rapid development of the holographic storage theory in the 1970s, it was still limited by the backward effects of recording materials, modulators, and detectors at that time, and a satisfactory storage density could not be achieved in the early works.

Until the early 1990s, with the development of recording materials, spatial light modulators, charge-coupled device detectors, and other key materials and devices, the storage density of the HDSSs could be greatly increased, and the optical holographic data storage entered a period of accelerated development. Overall, the materials used and the reuse theory have been comprehensively improved.

In 1991, Mok et al. stored 500 holograms of military vehicle shapes in 1 $cm^3$ of iron-doped lithium niobate crystals [25], and only two years later, F.H. Mok achieved the storage of 5000 images in 1 $cm^3$ of iron-doped lithium niobate crystals, indicating a 10-times-increased storage capacity [26]. In 1994, Hesselink and coworkers from Stanford University showed a full digital holographic storage system that converts images and videos into multiple data pages, and the bit error rate can reach $10^{-6}$ [27]. In 1994, IBM and Stanford University and other seven companies and university research groups formed a joint agency under the partial sponsorship of the United States Defense Advanced Research Projects Agency (DARPA). They expect to develop a HDSS with $10^{13}$ bit storage capacity and 1 Gb/s data conversion rate within five years. The material test system is provided by IBM. This system can not only store and reconstruct holograms of large data pages, but also perform bit error rate analysis on the reconstruction results [28]. The world's first complete HDSS was jointly established by Stanford University, Siros, IBM [29] and Rockwell, Thousand Oaks [30], and other companies. Researchers at Caltech and Lucent also completed similar system demonstrations [31].

In 1997, CIT's Allen Pu and D. Psaltis used a spherical reference light to obtain a volume holographic storage with an area density of 100 bits/$\mu m^2$ on a 1-mm-thick iron-doped lithium niobate crystal via shift multiplexing [32]. In 1998, Bell Labs' K. Curtis and coworkers used a similar multiplexing technology to store the areal density in iron-doped lithium niobate crystals at a rate exceeding 350 bits/$\mu m^2$ [33].

In terms of recording materials, in addition to photorefractive materials that have been studied since the 1970s [21, 22, 34], in 1994, the United States DuPont company developed a radical polymerizable photopolymer [35]. Pu et al. conducted an in-depth study on the holographic storage characteristics of this photopolymer material product, and performed holographic storage experiments using cyclic and angular multiplexing techniques. Subsequently, Allen Pu and Demetri Psaltis converted this material into a holographic disc, and multiplexed and stored 32 holograms in each flat cell area, and obtained a storage area density of $10^9$ bit/$cm^2$ [36]. Even though the sensitivity of the photopolymers is one to two orders of magnitude higher than that of the photorefractive crystals, the problem of material shrinkage is also severe. Therefore, the dynamic response range and shrinkage of the material must be balanced, as explained in detail by T. Bieringer [37]. The researchers at the Polaroid and Bell Labs took different approaches to materials research. Polaroid used experimental cationic ring-opening materials for the experimental verification [38, 39], whereas Bell Labs and Lucent used free radiating media [33]. Later, Stanford University and Aprilis Company (continued to develop after Polaroid) used a cationic open-loop material to record 250 GB of information on a DVD-sized disc with a data transmission rate of 10 Gb/s. In 2003, Waldman et al. used a photopolymer material, HMD-050-G-C-400 manufactured by Aprilis Company, with a thickness of 400 μm to achieve a holographic storage area density greater than 100 bits/$\mu m^2$ [40].

In 2002, Suzuki et al. first doped $TiO_2$ nanoparticles into methacrylate photopolymer films and found that the volume shrinkage of the materials during the holographic exposure was suppressed [41]. Since then, the method of nanoparticle doping to modify materials has been the focus of many studies. In 2008, Goldenberg et al. studied the doping of metal nanoparticles and modified gold nanoparticles. It has been proven that nanoparticles, when mixed with carboxyl functional acrylate monomers, can affect the structure of the material, inhibit shrinkage, and increase the stability of the material [42]. In 2010, Koji Omura and Yasuo Tomita studied the properties of $ZrO_2$ nanoparticle polymer composite films under a 404 nm laser and proposed that the shrinkage of the material was suppressed owing to the increase in the gel point [43]. In 2011, Hata et al. introduced thioolefin monomers into silica nanoparticle polymer composites, investigated their photopolymerization kinetics and volume holographic recording characteristics, and found that the material shrinkage and thermal stability were greatly improved [44]. In 2014, the research group of Chengming Yue and others proposed a holographic kinetic model to represent the kinetics of the hybrid grating in bulk polymer doped with photopolymers of gold nanoparticles quantitatively and described the polymerization of the gold nanoparticles. Further, because of the multicomponent diffusion behaviors, a mixed circular polarization-angle volume holographic multiplexing recording was then achieved in the prepared gold nanoparticle-doped phenanthrenequinone/polymethyl methacrylate (PQ/PMMA) photopolymer [45, 46]. In 2016, Tomita et al. introduced hyperbranched polymers, with ultrahigh refractive indices, as organic nanoparticles, and prepared nanoparticle–polymer composite holographic gratings, achieving a significant increase in the refractive index modulation and diffraction efficiency (close to 100%) at a wavelength of 532 nm [47]. In 2018, Liu et al. prepared a new material by dispersing silver nanopillars in a photopolymerizable mixture, and demonstrated the grating of this nanopillar-doped polymer composite under the exposure of an ultrafast nanosecond laser. To study the formation, they analyzed the causes of reciprocity failure and the improvement in the polymer holographic properties by the doped silver nanoparticles; the diffraction efficiency of the optimized polymer was as high as 51.4% [48].

In addition to incorporating nanoparticles into photopolymers, many optimization attempts have been made on photopolymer systems. In 2012, Ortuño et al. developed a new dry photopolymer called bio-photoelectronics, which has a low toxicity. This photopolymer was developed with the aim to counter the toxicity and poor environment compatibility exhibited by most of the photopolymer components. Additionally, its high thickness makes it ideal for holographic data storage applications [49]. Subsequently, in 2013, Ortuño et al. also proposed a new chain transfer agent 4,4′azo-bis-(4-cyanovaleric acid) (ACPA) with a polyvinyl alcohol (PVA)/acrylamide (AA) photopolymer, whose performance can be improved by the initiator ACPA [50]. In 2016, Cody et al. studied a novel composition of a low-toxic water-soluble holographic photopolymer and obtained a bright reflection grating with a recording diffraction efficiency of up to 50% [51]. Later, the research team of Fan Fenglan and others proposed to modify the material components via chemical methods to prepare PQ-supported dimonomer photopolymer materials, which improved the solubility of the photoinitiator in the photopolymer, thereby achieving material optics, as an improvement of features [52]. In 2018, Liu et al. proposed a new type of photosensitizer-doped photopolymer and proposed an optimized three-step thermal polymerization preparation method. The thermal initiators with different concentrations and the photoinitiator characteristics of the photopolymers [53] were studied in detail.

By doping nanoparticles and optimizing the polymerization methods, the current recording materials have been significantly improved to better meet the needs of optical holographic data storage. The next step is to steadily increase the diffraction efficiency while further enhancing the material stability and corresponding rate as well as improving the data storage density.

The supreme advantage of volume holographic storage is its ability to use a variety of multiplexing technologies to increase the storage density. The multiplexing technology of volume holographic storage was well developed in the 1990s [27, 54, 55].

Angular multiplexing is one of the more widely used off-axis HDSSs. It uses plane waves of the same wavelength as the reference light and forms multiple holograms at different angles of incidence in the same area. With a 1 cm thick lithium niobate crystal as the recording material and a 500 nm wavelength laser as the recording and reading light, the angular selectivity is approximately 50 μrad. Thus, 10 000 holograms can be recorded within 30° of the reference light angle multiplex.

Wavelength multiplexing uses plane waves of different wavelengths as the reference light, which is incident at the same angle, to form multiple holograms [56]. This multiplexing technique is usually more suitable for inorganic photorefractive crystal materials, as organic photopolymer materials have different sensitivity to different wavelengths and do not exhibit a simple linear response.

Phase code multiplexing uses phase coding of the reference light of the multiplexed hologram to prevent crosstalk with each other to the greatest extent. In particular, the orthogonal phase multiplexing of the reference light can effectively suppress the cross-page crosstalk. Moreover, the introduction of phase modulation improves the signal-to-noise ratio [57, 58].

Displacement multiplexing involves the separation and multiplexing of different holograms in space; that is, after recording a hologram somewhere, a certain distance is moved, and then the next hologram is recorded. The sharp drop in the diffraction efficiency after the Bragg mismatch is used so that one hologram does not affect the other when it is read. The displacement multiplexing technology is mostly used for organic photopolymer materials, which can be well matched with optical disc-type HDSSs, and is highly compatible. Moreover, the smaller the shift distance, the higher the multiplexing number [59–61]. The aforementioned methods are the main multiplexing methods used for optical holographic storage, while some systems use random phase multiplexing and other hybrid multiplexing methods [62–64].

Optical holographic data storage technology moved out of the laboratory after the year 2000 and became commercialized. In 2000, Stanford University developed the HDSS [59] with a storage area density of 70 bits/μm$^2$, and a continuously improved data transmission rate (up to 10 Gb/s) [65]. The Aprilis Company in the United States of America intensively researched materials for such a storage device. In 2002, it developed a photopolymer disc with a storage capacity of 200 GB, while in 2003, it launched a Vulcan™ drive hearing device prototype, but the transmission rate was only 75 Mb/s [40]. In 2005, Japanese Optware proposed a coaxial holographic data storage scheme [66–69], in which green light, which is sensitive to the recording material, is used as the recording read light, and red light, which is insensitive to the material, is used as the address light [70]. The optical path for reading and writing was improved, and a servo system was introduced to generate the entire holographic data. The storage system is more compatible with the widely accepted CD-ROM systems. The system stored 100 GB of data on a similar-sized polymer disc. In 2006, the American company InPhase, separated from Bell Labs, launched Tapestry™ HDS-300R, which was the first commercial HDSS [71, 72]. It matched with three types of photopolymer holographic versatile disc (HVD) [73, 74] with different capacities and different data transmission rates. The device featured a storage capacity of 300 GB, transmission rate of 20 MB/s, capacity of 800 GB, transmission rate of 80 MB/s, capacity of 1.6 TB, and transfer rate of 120 MB/s. For storage, an off-axis original "polytopic" method was adopted [75]. Since then, a parallel market exploration pattern has been formed for the coaxial HDSS represented by Optware and the off-axis HDSS represented by InPhase.

However, the research on optical holographic storage encountered a low tide after the successive collapse of both InPhase and Optware, further indicating that the technology at that time was not mature enough, especially in terms of materials. Although photorefractive crystals have the characteristics of recording and erasing, the data cannot be recorded and stored for a long time. Moreover, the problem of shrinkage and low efficiency of organic photopolymers need to be solved as well, and the storage method is still based on the amplitude and does not fully utilize the advantages of holographic storage. Another critical issue was that the data storage requirements at that time were not very large, and mature technologies, such as hard disks, could cover the storage requirements of that time.

After 2012, the era of big data arrived, and the market for data storage grew drastically. Additionally, the synthesis of organic photopolymer materials has progressed considerably in recent years. In later years, Tan Xiaodi's team proposed and demonstrated a phase-type holographic data storage method [76–79], which is fast becoming a research hotspot because of its high storage capacity and higher signal-to-noise ratio.

## 1.3 Present Problems

### 1.3.1 Recording Medium

The recording medium may be the most important factor. In the past, it was thought that photorefractive crystals were the main materials for holographic data storage recording, because a large change in the refractive index could be obtained with a weakly intense light. However, the photorefractive crystal exhibits a severely deteriorated reproduction, indicating that the reading light makes the recording hologram gradually disappear. Therefore, although the recording speed of the photorefractive crystal is very fast, the data storage life is short. In addition, the photorefractive crystal is not suitable for a rotating disk structure, and thus impedes its application. In recent years, the research on holographic materials has gradually shifted to photopolymers, whose refractive indices change due to irreversible chemical changes. Therefore, once the holographic structure is formed using such a material, its structural stability can be maintained for a long time, and it is less affected by the reading light. However, the photopolymers are characterized by a low refractive-index modulation as well as a low photosensitivity, which result in a long recording time. Additionally, the volume of the photopolymer material shrinks after light exposure, which in turn changes the recorded diffraction grating pitch, resulting in a significant deviation from the Bragg condition. Consequently, the reading light fails to reconstruct the original information correctly. Although the shrinkage can be compensated by adjusting the wavelength of the reading light and the angle of the beam incident on the material, this compensation is limited for more complex structures. In general, a material shrinkage below 0.5% is acceptable, and the higher the refractive index modulation of the photopolymer material, the higher the shrinkage rate. Therefore, disentangling this contradictory behavior is a major challenge for future of materials research.

### 1.3.2 Multiplexing Diffraction Efficiency

Theoretically, multiple multiplexed recording is expected to reduce the diffraction efficiency of holographic data storage as the diffraction efficiency is inversely proportional to the square of the number of multiplex recordings. A reduction in the diffraction efficiency leads to a reduction in the intensity of the light reaching the photodetector, and even a reduction in the signal-to-noise ratio. Because the intensity of the reading light is limited, the required diffraction efficiency also has a lower limit, and consequently, the number of multiplexing has an upper limit.

### 1.3.3 Media Duplication

Another problem with multiplexing holograms is media duplication. Flat optical storage devices, such as CDs, can be copied in large quantities at high speed through injection molding. For multiple recording media, sequential writing is inevitable, and thus, a very high writing speed is required for holographic recording.

### 1.3.4 Pixel Matching

Pixel matching arises while reading data. A photodetector pixel matching of $1:1$ is required for ideally reproducing a two-dimensional data page. However, realizing this condition in actual optical systems is difficult because of aberrations. Theoretically, according to the sampling theorem, if the pixel pitch of the photodetector is smaller than that of the reproduced image, then the signal can be reproduced by image processing even if the pixels are not completely matched. However, this approach is anticipated to increase the amount of calculation. Thus, the situation of the optimal pixel matching is determined by the performance of the optical system and the computer.

### 1.3.5 High-Speed Reading

Even though the reading speed of light is sufficiently fast enough, the data reading speed of the actual system depends only on the refresh speed of the connected electrical devices, such as spatial light modulators and photodetectors. Currently, a digital micromirror device (DMD) is the only spatial light modulator that can achieve high-speed modulation. A DMD can reach a refresh rate above 20 kHz, but it can only modulate the amplitude. To modulate information such as phase, usually liquid crystal on silicon (LCOS) is used as the spatial light modulator, whose refresh rate can reach only up to 180 Hz; this rate is exceedingly below the speed required to match the amplitude. In addition, high-speed photodetectors have been extensively developed, exhibiting a data throughput of 128 Gb/s. Nevertheless, it is necessary to further reduce the volume and cost of the high-speed cameras. In addition, a camera with a higher refresh rate usually requires a larger luminous flux to ensure sufficient illumination, enforcing higher requirements on the diffraction efficiency of the material and the laser power.

In general, fulfilling the recording material requirements is still the last hurdle in the direction of efficient big data storage. To mitigate the material-related deficiencies, an amalgamation of materials research and optics is indispensable. Optics expert should focus on improving the optical storage systems and provide clear guidelines, which will aid the material specialists engaged in holographic data storage research in realizing practical applications of this technology.

## References


1 Gabor, D. (1948). A new microscopic principle. *Nature* 161: 777–779.

2 Gabor, D. (1949). Microscopy by reconstructed wave-fronts. *R. Soc. London Proc.* 197 (1051): 454–487.

3 Gabor, D. Microscopy by reconstructed wave-fronts: II. *Proc. Phys. Soc.* 64 (6): 449–469.

4 Kirkpatrick, P. and El-Sum, H.M.A. (1956). Image formation by reconstructed wave fronts. I. Physical principles and methods of refinement. *J. Opt. Soc. Am.* 46 (10): 825–830.

5 El-Sum, H.M.A. (1953). *Reconstructed Wave-Front Microscopy*. Stanford: Stanford University.

6 Baez, A.V. (1952). Resolving power in diffraction microscopy with special reference to X-rays. *Nature* 169 (4310): 963–964.

7 Rogers, G.L. (1950). Gabor diffraction microscopy: the hologram as a generalized zone-plate. *Nature* 166 (4214): 237–237.

8 Leith, E.N. and Upatnieks, J. (1962). Reconstructed wavefronts and communication theory. *J. Opt. Soc. Am.* 52 (10): 1123–1130.

9 Leith, E.N. and Upatnieks, J. (1964). Wavefront reconstruction with diffused illumination and three-dimensional objects. *J. Opt. Soc. Am.* 54 (11): 1295–1301.

10 van Heerden, P.J. (1963). Theory of optical information storage in solids. *Appl. Opt.* 2 (4): 393–400.
11 Leith, E.N., Kozma, A., Upatnieks, J. et al. (1966). Holographic data storage in the three-dimensional media. *Appl. Opt.* 5 (8): 1303–1311.
12 Ashkin, A., Boyd, G.D., Dziedzic, J.M. et al. (1966). Optically-induced refractive index inhomogeneities in $LiNbO_3$ and $LiTaO_3$. *Appl. Phys. Lett.* 9 (1): 72–74.
13 Staebler, D.L. and Amodei, J.J. (1972). Coupled-wave analysis of holographic storage in $LiNbO_3$. *J. Appl. Phys.* 43 (3): 1042–1049.
14 Staebler, D.L., Burke, W.J., Phillips, W. et al. (1975). Multiple storage and erasure of fixed holograms in Fe-doped $LiNbO_3$. *Appl. Phys. Lett.* 26 (4): 182–184.
15 Ishida, A., Mikami, O., Miyazawa, S. et al. (1972). Rh-doped $LiNbO_3$ as an improved new material for reversible holographic storage. *Appl. Phys. Lett.* 21 (5): 192–193.
16 Shah, P., Rabson, T.A., Tittel, F.K. et al. (1974). Volume holographic recording and storage in Fe-doped $LiNbO_3$ using optical pulses. *Appl. Phys. Lett.* 24 (3): 130–131.
17 Stewart, W.C., Mezrich, R.S., Cosentino, L.S. et al. (1973). An experimental read–write holographic memory. *Radio Corporat. Am. Rev.* 34: 3–44.
18 Nishida, N., Sakaguchi, M., and Saito, F. (1973). Holographic coding plate: a new application of holographic memory. *Appl. Opt.* 12 (7): 1663–1674.
19 Strehlow, W.C., Dennison, R.L., and Packard, J.R. (1974). Holographic data store. *J. Opt. Soc. Am.* 64: 543–544.
20 D'Auria, L., Huignard, J.P., and Spitz, E. (1973). Holographic read–write memory and capacity enhancement by 3-D storage. *IEEE Trans. Magn.* 9 (2): 83–94.
21 D'Auria, L., Huignard, J.P., Slezak, V.C. et al. (1974). Experimental holographic read–write memory using 3-D storage. *Appl. Opt.* 13 (4): 808–818.
22 Mikaeliane, A. (1978). Holographic bulk memories using lithium niobate crystals for data recording. In: *Optical Information Recording*, 217–233.
23 Tsunoda, Y., Tatsuno, K., Kataoka, K. et al. (1976). Holographic video disk: an alternative approach to optical video disks. *Appl. Opt.* 15 (6): 1398–1403.
24 Kubota, K., Ono, Y., Kondo, M. et al. (1980). Holographic disk with high data transfer rate: its application to an audio response memory. *Appl. Opt.* 19 (6): 944–951.
25 Mok, F.H., Tackitt, M.C., and Stoll, H.M. (1991). Storage of 500 high-resolution holograms in a $LiNbO_3$ crystal. *Opt. Lett.* 16 (8): 605–607.
26 Mok, F.H. (1993). Angle-multiplexed storage of 5000 holograms in lithium niobate. *Opt. Lett.* 18 (11): 915–917.
27 Heanue, J., Bashaw, M., and Hesselink, L. (1994). Volume holographic storage and retrieval of digital data. *Science* 265: 749–752.
28 Bernal, M.P., Coufal, H., Grygier, R.K. et al. (1996). A precision tester for studies of holographic optical storage materials and recording physics. *Appl. Opt.* 35 (14): 2360–2374.
29 Shelby, R.M., Hoffnagle, J.A., Burr, G.W. et al. (1997). Pixel-matched holographic data storage with megabit pages. *Opt. Lett.* 22 (19): 1509–1511.

30 Paek, E.G. (1995). Volume holographic memory systems: techniques and architectures. *Opt. Eng.* 34 (8): 2193–2203.

31 Curtis, K. (2000). Digital holographic data storage prototype. *Optical Data Storage. Conference Digest*, pp. 164–166.

32 Pu, A. and Psaltis, D. (1997). Holographic data storage with 100 bits/$\mu m^2$ density. *Optical Data Storage Topical Meeting. Conference Digest*, pp. 48–49.

33 Dhar, L., Curtis, K., Tackitt, M. et al. (1998). Holographic storage of multiple high-capacity digital data pages in thick photopolymer systems. *Opt. Lett.* 23 (21): 1710–1722.

34 Thaxter, J.B. and Kestigian, M. (1974). Unique properties of SBN and their use in a layered optical memory. *Appl. Opt.* 13 (4): 913–924.

35 Zhou, H.J., Morozov, V., and Neff, J. (1995). Characterization of DuPont photopolymers in infrared light for free-space optical interconnects. *Appl. Opt.* 34 (32): 7457–7459.

36 Pu, A. and Psaltis, D. (1996). High-density recording in photopolymer-based holographic three-dimensional disks. *Appl. Opt.* 35 (14): 2389–2398.

37 Bieringer, T. (2000). Photoaddressable polymers. In: *Holographic Data Storage*, Springer Series in Optical Sciences (ed. H.J. Coufal, D. Psaltis and G.T. Sincerbox), 209–230. Berlin Heidelberg: Springer-Verlag.

38 Orlov, S.S., Bjornson, E., Phillips, W. et al. (2000). High transfer rate (1 Gbit/sec) high capacity holographic disk digital data storage system. *Lasers* and *Electro-Optics,* CLEO. OSA Publishing, pp. 190–191.

39 Waldman, D.A., Li, H.Y.S., and Horner, M.G. (1997). Volume shrinkage in slant fringe gratings of a cationic ring-opening volume hologram recording material. *J. Imaging Sci. Technol.* 41 (5): 497–514.

40 Waldman, D.A., Butler, C.J., and Raguin, D.H. (2003). CROP holographic storage media for optical data storage greater than 100 bits/$\mu m^2$. *The International Symposium* on *Optical Science & Technology. Proceedings of SPIE,* pp. 5–20.

41 Suzuki, N., Tomita, Y., and Kojima, T. (2002). Holographic recording in $TiO_2$ nanoparticle-dispersed methacrylate photopolymer films. *Appl Phys. Lett.* 81 (22): 4121–4123.

42 Goldenberg, L.M., Sakhno, O.V., Smirnova, T.N. et al. (2008). Holographic composites with gold nanoparticles: nanoparticles promote polymer segregation. *Chem. Mater.* 20 (14): 4619–4627.

43 Omura, K. and Tomita, Y. (2010). Photopolymerization kinetics and volume holographic recording in $ZrO_2$ nanoparticle-polymer composites at 404 nm. *J. Appl. Phys.* 107: 023107.

44 Hata, E., Mitsube, K., Momose, K. et al. (2011). Holographic nanoparticle polymer composites based on step-growth thiolene photopolymerization. *Opt. Mater. Express* 1 (2): 207–222.

45 Li, C., Cao, L., He, Q. et al. (2014). Holographic kinetics for mixed volume gratings in gold nanoparticles doped photopolymer. *Opt. Express* 22 (5): 5017–5028.

46 Li, C., Cao, L., Wang, Z. et al. (2014). Hybrid polarization-angle multiplexing for volume holography in gold nanoparticle-doped photopolymer. *Opt. Lett.* 39 (24): 6891–6894.

**47** Tomita, Y., Urano, H., Fukamizu, T. et al. (2016). Nanoparticle-polymer composite volume holographic gratings dispersed with ultrahigh-refractive-index hyperbranched polymer as organic nanoparticles. *Opt. Lett.* 41 (6): 1281–1284.

**48** Liu, P., Zhao, Y., Li, Z. et al. (2018). Improvement of ultrafast holographic performance in silver nanoprisms dispersed photopolymer. *Opt. Express* 26 (6): 6993–7004.

**49** Ortuño, M., Gallego, S., Márquez, A. et al. (2012). Biophotopol: a sustainable photopolymer for holographic data storage applications. *Materials* 5 (5): 772–783.

**50** Ortuño, M., Fernández, E., Fuentes, R. et al. (2013). Improving the performance of PVA/AA photopolymers for holographic recording. *Opt. Mater.* 35 (3): 668–673.

**51** Cody, D., Gribbin, S., Mihaylova, E. et al. (2016). Low-toxicity photopolymer for reflection holography. *ACS Appl. Mater. Interfaces* 8 (28): 18481–18487.

**52** Fan, F., Liu, Y., Hong, Y. et al. (2018). Improving the polarization-holography performance of PQ/PMMA photopolymer by doping with THMFA. *Opt. Express* 26 (14): 17794–17803.

**53** Liu, P., Wang, L., Zhao, Y. et al. (2018). Holographic memory performances of titanocene dispersed poly(methyl methacrylate) photopolymer with different preparation conditions. *Opt. Mater. Express* 8 (6): 1441–1453.

**54** Mok, F.H., Psaltis, D., and Burr, G. (1993). Spatially- and angle-multiplexed holographic random access memory. *The International Society for Optical Engineering. Proceedings of SPIE*, pp. 334–345.

**55** Orlov, S.S., Phillips, W., Bjornson, E. et al. (2002). High data rate (10 Gbit/sec) demonstration in holographic disk digital data storage system. *Conference on Lasers and Electro-Optics*. IEEE, pp. 70–71.

**56** Rakuljic, G.A., Leyva, V., and Yariv, A. (1992). Optical data storage by using orthogonal wavelength-multiplexed volume holograms. *Opt. Lett.* 17 (20): 1471–1473.

**57** Denz, C., Tschudi, T., Pauliat, G. et al. (1991). Volume hologram multiplexing using a deterministic phase encoding method. *Opt. Commun.* 85 (2): 171–176.

**58** John, R., Joseph, J., and Singh, K. (2005). Holographic digital data storage using phase-modulated pixels. *Opt. Lasers Eng.* 43 (2): 183–194.

**59** Hesselink, L. (2000). Digital holographic demonstration systems by Stanford University and Siros Technologies. In: *Holographic Data Storage*, 383–397.

**60** Steckman, G.J., Pu, A., and Psaltis, D. (2001). Storage density of shift-multiplexed holographic memory. *Appl. Opt.* 40 (20): 3387–3394.

**61** Pu, A. and Psaltis, D. (1996). Holographic 3-D disks using shift multiplexing. *Lasers and Electro-Optics*. IEEE, p. 165.

**62** Darskii, A.M. and Markov, V.B. (1991). Angular sensitivity of holograms with a reference speckle wave. *Three-Dimensional Holography: Science, Culture, Education. Proceedings of SPIE*, pp. 392–395.

**63** Barbastathis, G., Levene, M., and Psaltis, D. (1996). Shift multiplexing with spherical reference waves. *Appl. Opt.* 35 (14): 2403–2417.

**64** Markov, V., Millerd, J., Trolinger, J. et al. (1999). Multilayer volume holographicoptical memory. *Opt. Lett.* 24 (4): 265–267.

**65** Orlov, S.S., Phillips, W., Bjornson, E. et al. (2004). High-transfer-rate high-capacity holographic disk data-storage system. *Appl. Opt.* 43 (25): 4902–4914.

**66** Horimai, H. and Tan, X. (2006). Collinear technology for a holographic versatile disk. *Appl. Opt.* 45 (5): 910–914.

**67** Horimai, H., Tan, X., and Li, J. (2005). Collinear holography. *Appl. Opt.* 44 (13): 2575–2579.

**68** Horimai, H. and Tan, X. (2005). Advanced collinear holography. *Opt. Rev.* 12 (2): 90–92.

**69** Horimai, H. and Tan, X. (2007). Holographic information storage system: today and future. *IEEE Trans. Magn.* 43 (2): 943–947.

**70** Shih, H.F. (2007). Integrated optical unit design for the collinear holographic storage system. *IEEE Trans. Magn.* 43 (2): 948–950.

**71** Wilson, W.L., Curtis, K.R., Anderson, K.E. et al. (2003). Realization of high performance holographic data storage: the InPhase technologies demonstration platform. *Organic Holographic Materials and Applications. Proceedings of SPIE*, pp. 178–191.

**72** Dhar, L., Curtis, K., and Fäcke, T. (2008). Holographic data storage: coming of age. *Nat. Photonics* 2: 403–405.

**73** Wilson, W.L. (2004). Toward the commercial realization of high performance holographic data storage. *Lasers and Electro-Optics*. IEEE, pp. 29–37.

**74** Schnoes, M., Dhar, L., Schomberger, G.L. et al. (2003). Photopolymer use for holographic data storage. *Advanced Optical Data Storage. Proceedings of SPIE*, pp. 68–76.

**75** Anderson, K. and Curtis, K. (2004). Polytopic multiplexing. *Opt. Lett.* 29 (12): 1402–1404.

**76** Lin, X., Huang, Y., Shimura, T. et al. (2017). Fast non-interferometric iterative phase retrieval for holographic data storage. *Opt. Express* 25 (25): 30905–30915.

**77** Lin, X., Huang, Y., Li, Y. et al. (2018). Four level phase pair encoding and decoding with single interferometric phase retrieval for holographic data storage. *Chin. Opt. Lett.* 16 (3): 032101.

**78** Tan, X., Horimai, H., Arai, R. et al. (2016). Phase modulated collinear holographic data storage system. *International Workshop on Holography and Related Technologies* (IWH 2016).

**79** Lin, X., Huang, Y., Cheng, Y. et al. (2016). Inter-page-crosstalk reduction in holographic data storage system through phase modulation in signal region. *Jpn. J. Appl. Phys.* 55 (9S): 09SA07-1-3.

# 2

# System of Collinear Holography

## 2.1 Polarized Collinear Holography

### 2.1.1 Introduction

Though volume holography is suitable for data storage, there are some remaining issues to realize practical ultra-high-speed and large-capacity optical data storage. It has become clear that we need to get a breakthrough. If the hologram can be recorded/reproduced using only one objective lens, like an optical disk [1], then the optical system becomes extremely simple, and the optical disk technology can be integrated with the hologram technology. However, despite travelling parallel to each other, the reference and information light form an interference fringe on the recording medium after their coaxial incidence on the optical axis of the objective lens. This necessitated the development of a holographic optical system in which the reference light and the reproduced information light are optically and efficiently separated.

This chapter describes the polarized collinear holography, which was the first collinear method devised based on the aforementioned principle. Since the origin of the idea of polarization collinear holography goes back to the research on "magneto-optical holography technology" from 1982 to 1984, the principle of polarization-separated hologram reproduction is outlined here using the magneto-optical effect. Next, the polarization collinear holography technique is described with a focus on the structure of the optical system and control of polarization. This technique optically realizes the principle of polarization separation and reproduction using an optical element, namely a composite optical rotation plate. Finally, the experimental results verifying the basic principle and the corresponding inferences are elucidated.

### 2.1.2 Magneto-Optical Holography and Principle of Polarization Separation and Reproduction

It is well known that when the principle of holography was invented by Dennis Gabor in 1948, the arrangement of the recording and reproducing system was an in-line optical arrangement with the same optical axis [2]. However, in such an

*Collinear Holography: Devices, Materials, Data Storage*, First Edition.
Xiaodi Tan, Hideyoshi Horimai, Tsutomu Shimura, and Xiao Lin.

arrangement, the reproduction reference light (probe light) overlaps the reproduction image, and the signal-to-noise ratio (SNR) of the reproduction information is significantly reduced. An optical system that solves this problem is the "two-beam interferometry" proposed by Leith and Upatnieks in 1962 [3]. However, the configuration of this optical system is the last major obstacle to the practical use of holographic data storage. The reference beam for reproduction (probe light) and the corresponding reproduced image are separated. We know a CD or DVD is a simple one-beam optical system, therefore an inexpensive optical pickup can be manufactured. However, an off-axis spatial separation method necessarily results in a complicated optical system. Therefore, the dominant breakthrough to the practical use of holographic data storage is that the hologram can be recorded using a coaxial optical arrangement, in which the optical axes of the reference and information light are parallel to each other. Further, the reproduced hologram can be recorded with the coaxially traveling reference light. The idea is to create a hologram recording/reproducing optical system that is capable of separating reproduced light based on some properties, e.g., polarization. The key to this approach is magneto-optical holography, which was studied in the early 1980s [4].

In magneto-optical holography, as shown in Figure 2.1, the reproduction of a hologram image is reproduced from a recorded magnetic grating via interference between the components in the rotation direction of a polarization plane caused by a magneto-optical effect (Kerr effect or Faraday effect). Therefore, the polarization plane of the reproduced image is completely orthogonal to that of the irradiated reference light. Therefore, in hologram recording/reproduction on a magneto-optical

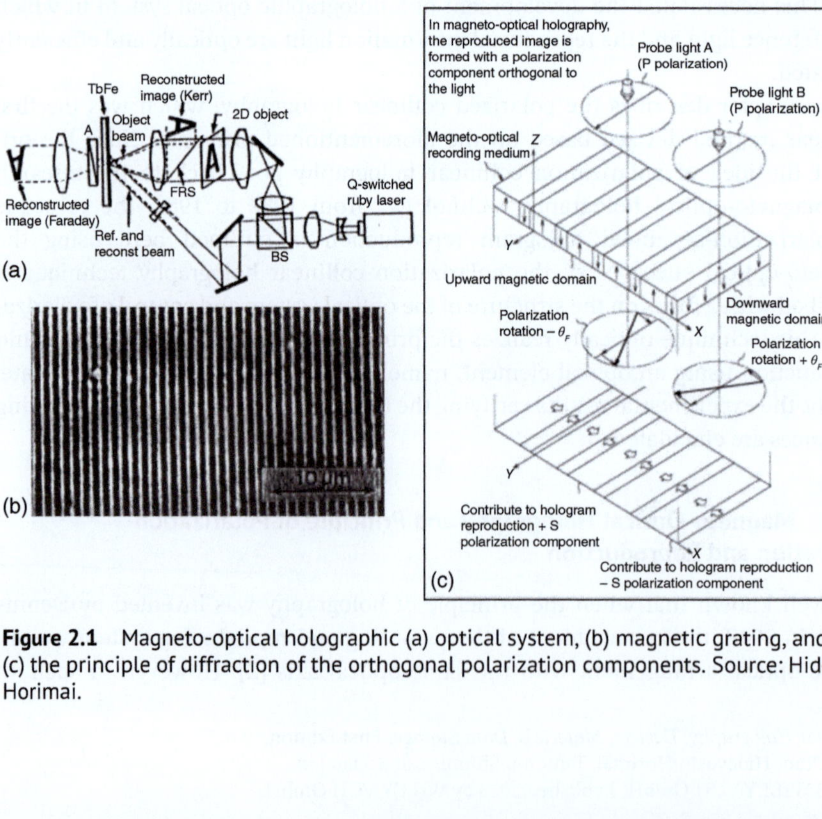


**Figure 2.1** Magneto-optical holographic (a) optical system, (b) magnetic grating, and (c) the principle of diffraction of the orthogonal polarization components. Source: Hideyoshi Horimai.

material, since the reference light and the reproduction light can be easily separated by polarized light, an in-line type optic distribution is obtained that advances the angle between the reference light and the information light in exactly the same optical axis direction. Using this arrangement, an extremely high SNR can be obtained.

### 2.1.3 Basic Principle of Polarization Collinear Holography

In magneto-optical holography, the recording medium is a ferromagnetic material. Therefore, the magneto-optical effect (Kerr rotation angle) of the recording medium itself allows the object light and the reference light to be easily separated by polarization. No rotation of polarization occurs in the photopolymer material used for recording. Therefore, we analyzed whether a similar phenomenon could be created by inserting some optical element into the optical path. As a result of trial and error, we found a method of realizing the same principle as orthogonal polarization separation and reproduction in magneto-optical holography by splitting a light beam into two with respect to the optical axis and giving each light beam an opposite optical rotation. This is a polarization collinear holography recording/reproducing method. An optical rotation plate [5] is used as an element to give optical rotation, but to give the opposite optical rotation to each of two divided areas symmetrical to the optical axis, a composite optical rotation plate (gyrator) was used, in which right and left optical rotation plates cut out from the right and left crystals, respectively, were polished and adhered. This composite optical rotation plate was previously proposed by the author when he studied a new method for detecting a magneto-optical signal on a magneto-optical disc (e.g. a MiniDisc) [6]. For hologram recording/reproduction, the thickness of the crystal was adjusted by polishing so that each optical rotation angle was ±45°. This is sandwiched between two pieces of glass with an anti-reflective (AR) coating on one side to suppress surface reflection. Figure 2.2 shows an outline and a photograph of the appearance of the prototype composite optical rotation plate.

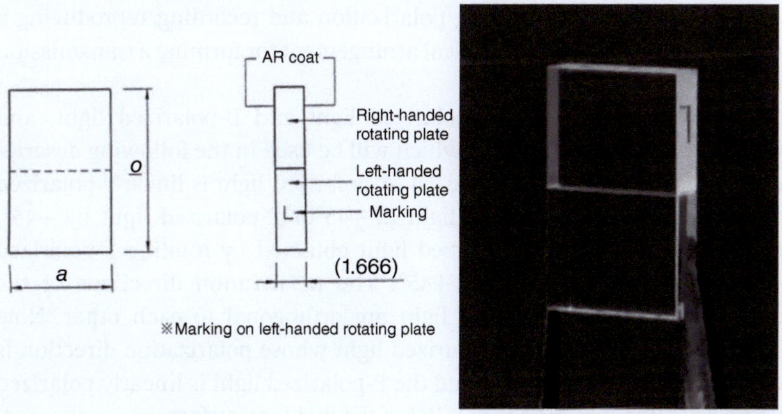


**Figure 2.2** Composite optical rotation plate (gyrator) for polarization collinear holography. Source: Hideyoshi Horimai.

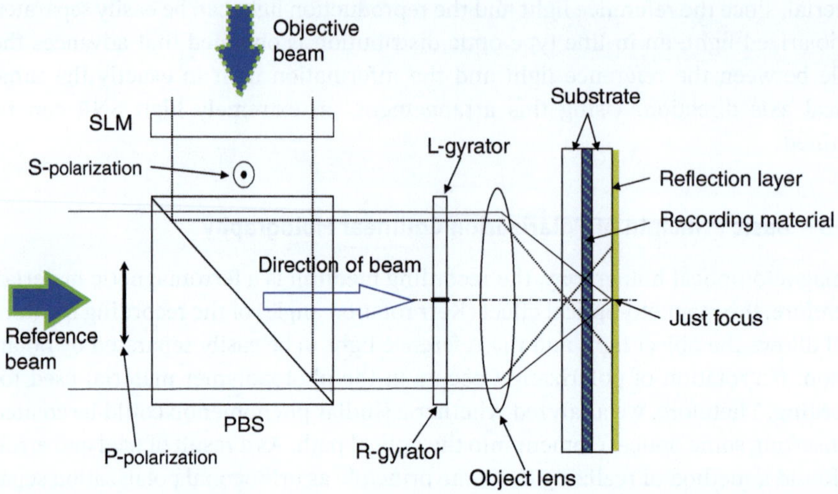


**Figure 2.3** Relationship between main part of optical system and light ray in polarization collinear holography using composite optical rotation plate. Source: Hideyoshi Horimai.

Figure 2.3 schematically shows the relationship between the arrangement of main optical components and the optical path of polarization collinear holography. Other optical components such as a laser light source and a beam splitter are omitted. The compound optical rotation plate is disposed near the entrance pupil plane of the objective lens. When the condensing function of a compound optical rotator and an objective lens is applied to holography, a non-reciprocal polarization rotation function such as the Faraday effect is realized in a pseudo manner, and a collinear hologram in which the reference light and the information light are coaxial and have no angle. A recording/reproducing optical arrangement is realized. Notably, a transmission hologram is formed when the object beam (objective beam) is designed to be focused before the recording medium, and a reflection hologram is formed when the object beam is designed to have the same focus as the reference light. In the present optical system, a method of converting polarization and recording/reproducing a hologram will be described using an optical arrangement for forming a transmission hologram.

First, the relationship between A-polarized light and B-polarized light, and P-polarized light and S-polarized light, which will be used in the following description, are defined as follows using Figure 2.4. A-polarized light is linearly polarized light obtained by rotating S-polarized light at +45°or P-polarized light by −45°, and B-polarized light is linearly polarized light obtained by rotating S-polarized light at −45° or P-polarized light by +45°. The polarization directions of the A-polarized light and the B-polarized light are orthogonal to each other. Note that the S-polarized light is linearly polarized light whose polarization direction is perpendicular to the incident surface, and the P-polarized light is linearly polarized light whose polarization direction is parallel to the incident surface.

Next, returning to Figure 2.3, how the polarized light is converted by the combination of the compound optical rotation plate and the objective lens will be described.

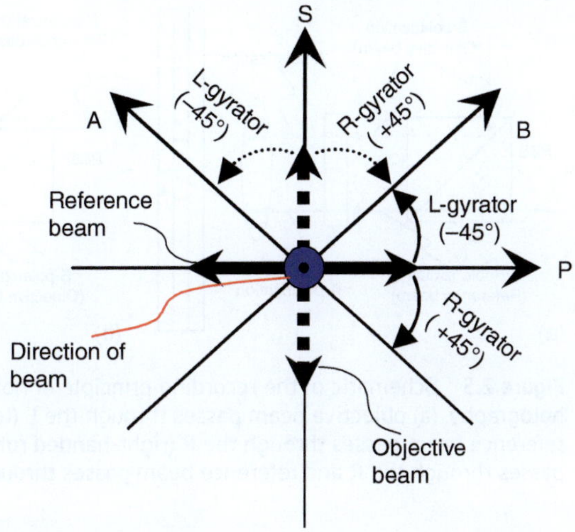


**Figure 2.4** Polarization conversion in polarization collinear holography optical system. Source: Hideyoshi Horimai.

First, in the drawing, the reference beam (Reference Beam) is P-polarized light, so that 100% of the beam passes through the polarizing beam splitter (PBS) and is incident on a right-handed rotating plate (hereafter, abbreviated as R) and a left-handed rotating plate (hereinafter abbreviated as L). The polarization planes after transmission by R and L are converted into mutually different polarizations. Explaining one by one, as shown in Figure 2.4, the P-polarized light rotates through +45° after transmitting R and becomes A-polarized light. This light is reflected by the objective lens after being condensed on the reflection surface of the disk, becomes parallel light again by the objective lens, and is incident from the surface opposite to L this time. Therefore, the A-polarized light is converted to the S-polarized light. As a result, 100% is reflected by the PBS, and the return light is changed to another optical path from the incident light. The other reference light is converted into B-polarized light by first transmitting L while being the same P-polarized reference light. However, when the light is reflected by the reflection surface of the disk and transmitted through R in the opposite direction, the light is converted into S-polarized light. The polarization of the object light is also converted according to the same rule. As can be seen from the polarization conversion method described above, the optical rotation (rotation of polarized light) of the optical rotation plate itself is relative, but a composite optical rotation plate in which a right optical rotation plate and a left optical rotation plate are bonded around the optical axis. By using the objective lens and the reflecting surface, it is possible to perform non-reciprocal polarization control like a Faraday element in a pseudo manner.

According to the above polarization state, the principle of recording a hologram in polarization collinear holography will be described with reference to Figure 2.5. As already shown in Figure 2.4, the reference light and the object light do not form interference fringes because their polarization planes are orthogonal to each other. However, the A-polarized light and the B-polarized light of the object light and the reference light form interference fringes. Both transmission holograms are formed from the positional relationship between the light traveling direction and

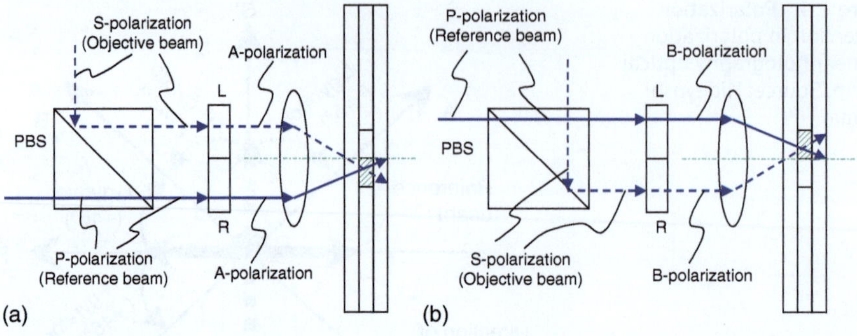


**Figure 2.5** Schematic of the recording principle of hologram in polarization collinear holography. (a) objective beam passes through the L (left-handed rotating plate) and reference beam passes through the R (right-handed rotating plate), (b) objective beam passes through the R and reference beam passes through the L. Source: Hideyoshi Horimai.

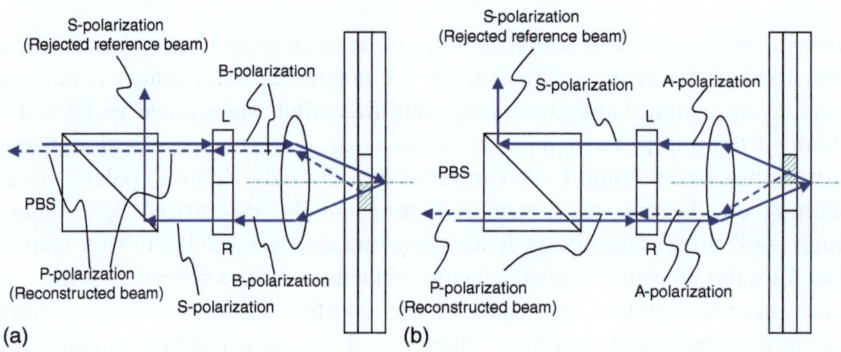


**Figure 2.6** Schematic of the reproduction of information from hologram and separation of reference light by polarization. (a) reference beam passes through the L, (b) reference beam passes through the R. Source: Hideyoshi Horimai.

the recording medium. In addition, two half-moon-shaped holograms face each other and are combined like a full moon.

Finally, a method of reproducing information from the hologram recorded as described above and a method of separating the reference light for reproduction using polarized light will be described with reference to Figure 2.6. First, a method of reproducing an image from the formed half-moon hologram will be described with reference to Figure 2.6a. The reproduction of the image is performed by the reference light traveling in the recording and reflection directions, that is, the light once reflected by the reflection surface of the disk. Since the hologram formed at this time is a transmission hologram, the reconstructed object light travels in the opposite direction toward the incident direction. The reproduction light by the reference light that has passed through the left (L) is changed in direction by diffraction and transmits through the same left (L) again, so that the reproduced image becomes P-polarized light again, so that it passes through the PBS. On the other hand, the reference light (zero-order light) for reproducing the hologram is transmitted to the right (R), is converted to S-polarized light, and is reflected by the PBS. The situation of Figure 2.6b is the same. As described above, the referencc light and the reproduced object light have polarization planes orthogonal to each other, and an effect equivalent to the principle of orthogonal separation and reproduction in magneto-optical holography can be expected.

### 2.1.4 Results of Experiments to Verify the Basic Principle of Polarization Collinear Holography

Figure 2.7 shows a schematic of the main part of the experimental optical system of the polarization collinear holography and the results of the polarization separation/reproduction experiment. Although not shown, the laser light source used for recording/reproducing the hologram was 532 nm wavelength Spectra-Physics MILLENNIA VS, and the spatial light modulator (SLM) for displaying two-dimensional images as recorded information was CRL MINATURE LCD (small liquid crystal display device) was used. The complex optical rotation plate developed and developed by custom design has a vertical × horizontal × thickness = 10 × 6 × 3.5 mm, and the optical rotation angle of the optical rotation plate divided into two is +45° and −45° respectively (both values at a wavelength of 532 nm). As a recording medium, a photopolymer developed for trial production for hologram recording was used.

Since the purpose is to demonstrate the basic principle, the recording medium was a transmission type without a reflective film, and the same optical system as the incident side was installed on the opposite side (right side in the figure) across the recording medium.

During hologram reproduction, if PBS is not inserted on the right side of the recording medium, it is difficult to recognize the reproduced image because the reference light for reproduction is mixed coaxially. However, when PBS is inserted, the coaxial optical arrangement is used. The reference light was polarized and separated, and a reproduced image with a high SNR was successfully obtained. Thereby, the basic operation of the polarization collinear holography was proved in principle.

## 2.2 Principle of Collinear Holography

### 2.2.1 Introduction

The technology that enables hologram recording and reading with same objective lens is called "collinear holography" or simply "collinear." However, at the initial stage of research, there are many skeptical opinions about the possibility to form a volume hologram (interference fringe) in the recording medium simply by focusing the reference beam and information beam traveling in the same direction by the same lens without a predetermined angle.

In this chapter, the basic technology of holographic optical memory will be introduced. The configuration of collinear optical system will be described.

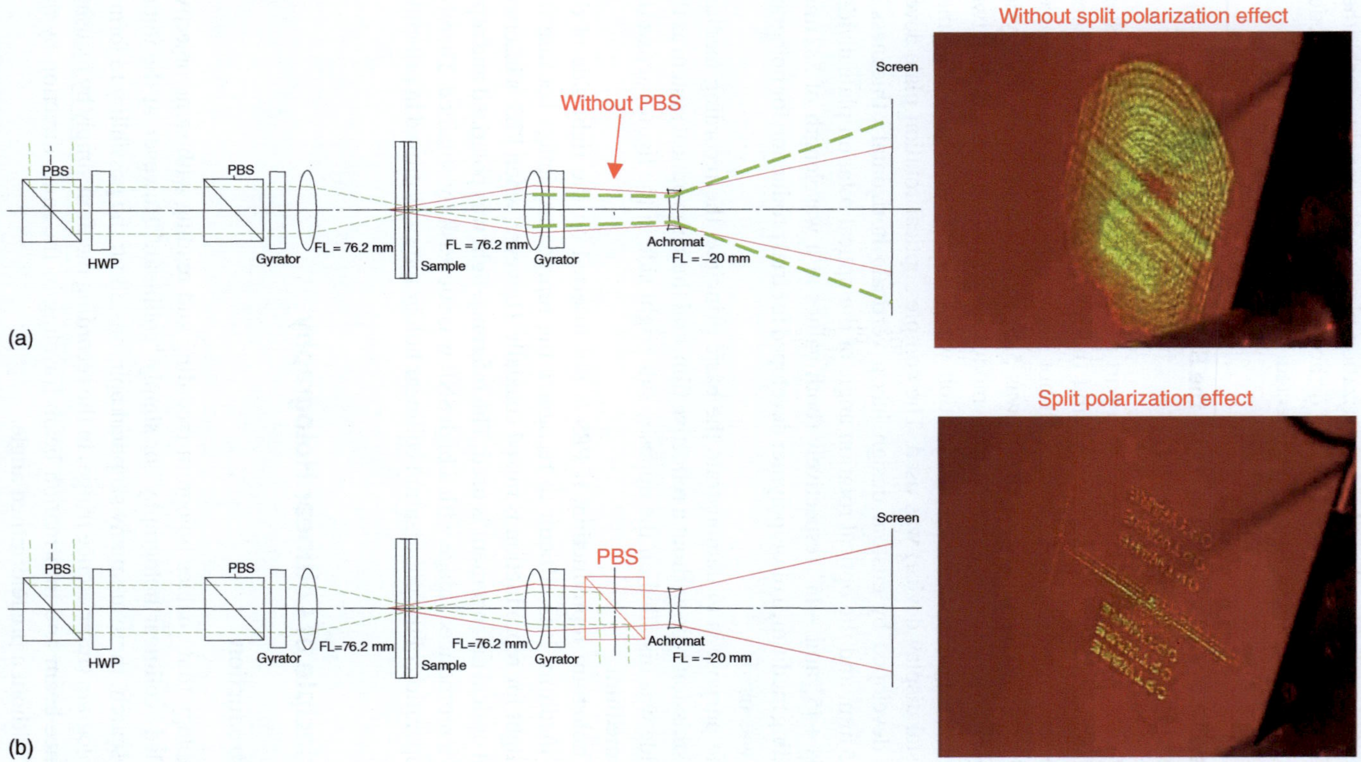


**Figure 2.7** (a) Principal parts of the experimental optical system of polarization collinear holography and (b) the results of proof of principle of polarization separation and reproduction. Source: Hideyoshi Horimai.

### 2.2.2 Basic Technology of Holographic Optical Memory

#### 2.2.2.1 Interference and Diffraction

The core of holographic storage technology is holography, which can be divided into two parts, interference and diffraction. Interference is used when recording information, and diffraction is used when reading information. Interference is the result of superimposing two light beams of the same frequency (one beam is object light or information light and the other is reference light) in a certain spatial area. Since the interference result is the distribution of the light field and contains all the information of the two beams of light, the pattern recorded by the interference is also called a hologram. Because the recording material of holographic storage often has a certain thickness, it cannot be ignored to define it as a thin hologram. Therefore, it is regarded as a thick hologram in the analysis, which is also called a volume hologram. The interference pattern of reference light and object light in a thick recording medium is a three-dimensional volume grating, and its schematic is shown in Figure 2.8.

In Figure 2.8, $d$ is the thickness of the recording medium, $\mathbf{K}$ represents the grating vector, $\varphi$ represents the angle between the grating vector and the $z$-axis, and $\theta$ represents the angle between the reference light and the interference fringe in the medium. $\Lambda$ stands for fringe pitch and is also called grating period. Reference light and object light here are the simplest form of plane wave. If the reference or object light is a complex wavefront, the resulting three-dimensional volume grating will become more complicated.

The angle $\theta$ between the reference light in the recording process and the recording fringes in the recording medium is also called the Bragg angle. During the reading process, a beam of reference light is incident on the three-dimensional grating, that is, diffracted light. When the incident direction of the reference light happens to meet the Bragg angle, the direction of the diffracted light will be exactly the same

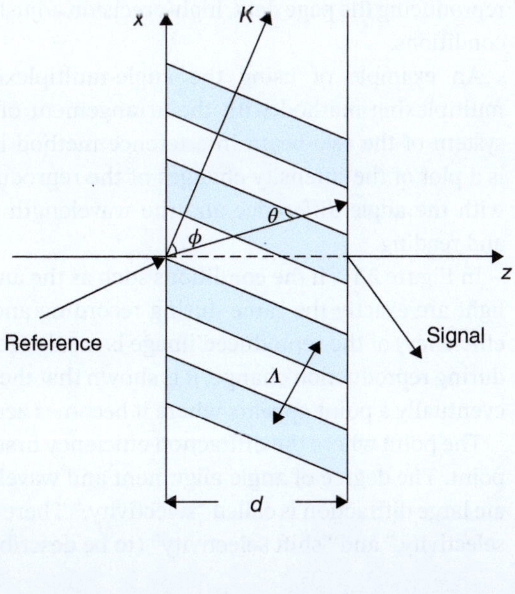


**Figure 2.8** Schematic of three-dimensional volume grating diffraction. Source: Hideyoshi Horimai.

as the original light, and the diffraction efficiency at this time is the highest. In this case, we call it satisfying the Bragg condition, as shown in Eq. (2.1),

$$2n\Lambda \sin\theta = \lambda \tag{2.1}$$

In Eq. (2.1), $n$ represents the refractive index of the recording medium, and $\lambda$ represents the wavelength of the incident reference light. Since the refractive index of the recording medium can be regarded as a fixed value, once the wavelength and direction of the recording reference and object light are constant, a fixed Bragg angle will be generated. When reading, when the direction and wavelength of the incident reference light meet the Bragg condition, we call it Bragg matching, and the intensity of the diffracted light is the highest at this time. When the direction or wavelength of the incident reference light does not meet the Bragg condition, we call it the Bragg mismatch. At this time, diffracted light is still generated, but its intensity will decrease sharply as the Bragg mismatch becomes larger. For this phenomenon, it can be seen that volume holography has wavelength and angle selectivity, which is a huge advantage for holographic storage, because this selectivity creates the possibility of multiplexed recording.

#### 2.2.2.2 Multiplexing Technology

Holographic optical memory, also known as volume hologram memory, increases recording density by recording three-dimensional page data one after another on a thick volume recording material. The structure of the interference fringes in the material with a large number of page data recorded is not only corresponding to each recorded page in a spatially independent area, but mixed (multiplexed) in the recording material. In the volume hologram memory, multiple page data are recorded in the recording material, and take advantage of Bragg diffraction. Then every page can be reproduced from mixed fringe with low cross talk. Therefore, in multiplexing, it is necessary to record the Bragg conditions for each page data separately, and when reproducing the page data, high-precision adjustment is required to satisfy the Bragg conditions.

An example of using the angle-multiplexing method and the wavelength-multiplexing method with the arrangement of the recording/reproducing optical system of the two-beam interference method is shown in Figure 2.9. Figure 2.10 is a plot of the intensity changes of the reproduced image due to Bragg diffraction with the angle difference and the wavelength difference at the time of recording and reading.

In Figure 2.10, if the conditions such as the angle and wavelength of the reference light are exactly the same during recording and reading, the intensity (diffraction efficiency) of the reproduced image is maximized. And if the angle and wavelength during reproduction change, it is shown that the diffraction efficiency decreases and eventually a point appears where it becomes zero (arrow in the figure).

The point where the diffraction efficiency first becomes zero is called the first null point. The degree of angle alignment and wavelength alignment required to generate large diffraction is called "selectivity." There are "angle selectivity," "wavelength selectivity," and "shift selectivity" (to be described later) depending on the physical

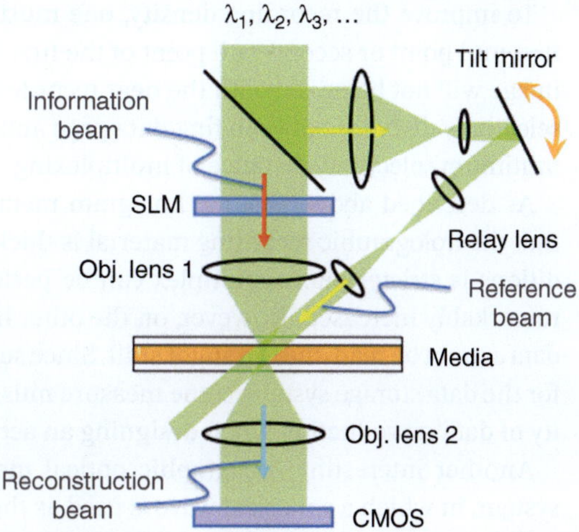


**Figure 2.9** Recording/reproducing optical system arrangement for angle multiplexing and wavelength multiplexing (two-beam interferometry). Source: Hideyoshi Horimai.

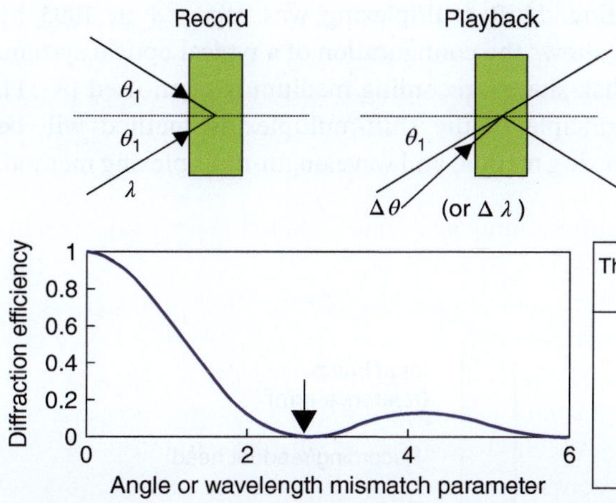


| Thickness (μm) | Angle (°) | Wavelength (nm) |
|---|---|---|
| 100 | 0.25 | 9.5 |
| 500 | 0.05 | 1.9 |
| 1 000 | 0.025 | 0.95 |
| 5 000 | 0.005 | 0.19 |
| 10 000 | 0.0025 | 0.095 |

**Figure 2.10** Change in intensity of reconstructed image due to Bragg diffraction: difference in angle and wavelength of reference light during recording and reading. Source: Courfal et al. [7]/Springer Nature.

quantity as a parameter. In particular, the above-mentioned case where the Bragg condition is directly effective is also called Bragg selectivity.

Further, the selectivity of the reference light with respect to the change in angle or wavelength is related to the thickness of the recording material as a parameter. The table on the right of Figure 2.10 shows the angle difference and wavelength difference values at which the first null point appears with respect to the thickness of the recording material. Thus, the thicker the recording material, the smaller the angle difference and the wavelength difference at which the first null point appears, and the Bragg condition becomes strict. For example, when the thickness of the recording medium is 500 μm, the angle difference to the first null point is about 0.05°, and the wavelength difference is about 1.9 nm. But when the thickness of the recording medium is 5000 μm, which is 10 times that of before, the angle difference to the first null point is about 0.005°, and the wavelength difference is about 0.19 nm.

To improve the recording density, one must record the second data page at the first null point or second null point of the first data page. The reproduced data page image will not be mixed with the near recorded data page. This also means that the selectivity distance between first data page and second data page without mix is the minimum selectivity distance of multiplexing.

As described above, volume hologram memory multiplexes using selectivity, so that the holographic recording material is thicker and the adjustment to Bragg conditions is stricter, more multiplex can be performed, and the recording density is remarkably increased. However, on the other hand, there is a risk that desired page data cannot be read due to a slight shift. Since such a situation leads to a fatal problem for the data storage system, some measure must be taken to ensure the reproducibility of data reproduction when designing an actual system.

Another interesting holographic optical memory system is the shift multiplex system, in which a spherical wave is used as the reference light, and the parameters are changed by shifting the position of the recording material itself to perform multiplex recording/reproduction. Shift multiplexing was reported in 1995 by D. Psaltis et al. [8]. Figure 2.11 shows the configuration of a typical optical system. As shown in the drawing, a disk-shaped recording medium is often used [8–11]. Hereinafter, the operation principle of the shift-multiplexing method will be described. In the angle-multiplexing method and wavelength-multiplexing method,

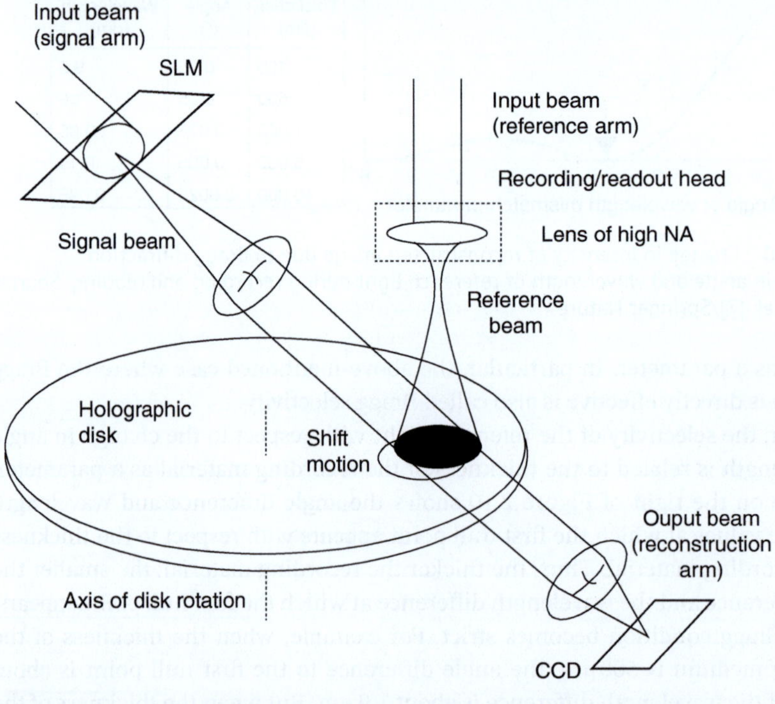


**Figure 2.11** Configuration example of optical system in shift multiplex recording/reproducing. Source: Based on Barbastathis et al. [9]/with permission of Optical Society of America.

the reference light is generally a plane wave, but in the shift-multiplexing method, a spherical wave reference light is used, so when focusing on one point of the recording material, it is irradiated by shifting the media. The wavefront direction of the reference light will change. Using this phenomenon, the shift-multiplexing method shifts the media, changes the effective angle of the reference beam, and multiplexes different records in spatially overlapping areas.

Although it is common to the above multiplexing system, a problem in practical use is that there is no standard for recording media. For this reason, when the hologram recording medium is removed and then mounted again, it is difficult to accurately access the recorded location and irradiate the light beam with the parameter corresponding to the desired data in one shot. In the shift multiplex method, surface rotation or eccentricity (called "wobble" in the paper) occurs due to the rotation of the disk, and the playback position shifts in the optical axis direction or in-plane direction. It may be deteriorated [7]. In addition, since the arrangement of the optical system is asymmetric, it is still a problem that sufficient shift selectivity cannot be obtained depending on the shift direction

### 2.2.3 Recording and Playback Method of Collinear Holography and Its Proof of Principle

Although not so serious in general holography, an important technical issue is how to separate the reference light for reproduction from the reproduced information light, especially for the purpose of realizing a holographic optical memory. In collinear holography, a spatial separation method was adopted. Hereinafter, a method for recording and reproducing information by the collinear method will be described.

First, the recording method will be described. As shown in Figure 2.12, an information light pattern and an annular reference light pattern surrounding the information light pattern are simultaneously displayed on the same SLM. When this SLM is irradiated with laser light as shown in Figure 2.13, the light of the information light pattern and the reference light pattern travels as one coaxial light beam and is collected by the objective lens. As a result, the information light and the reference light are automatically mixed on the recording medium to form a hologram (interference fringe).

Next, a reproduction method will be described. At the time of reproduction, only the reference light pattern is displayed on the SLM as shown in Figure 2.14. When this is irradiated with laser light, the light of the reference light pattern proceeds as a donut-shaped light beam and is irradiated onto the hologram recorded by the objective lens. As a result, the information light pattern is reproduced in the blank portion at the center of the reference light pattern for reproduction. When used as a holographic optical memory, a reflective film is provided on the recording medium. Then, since the reproduced image is returned to the objective lens again, the reproduced image can be detected by the complementary metal-oxide semiconductor (CMOS) sensor by being reflected by the beam splitter. On the CMOS sensor, the reference light pattern is imaged together with the information light pattern. However, the information light and the reference light are spatially separated

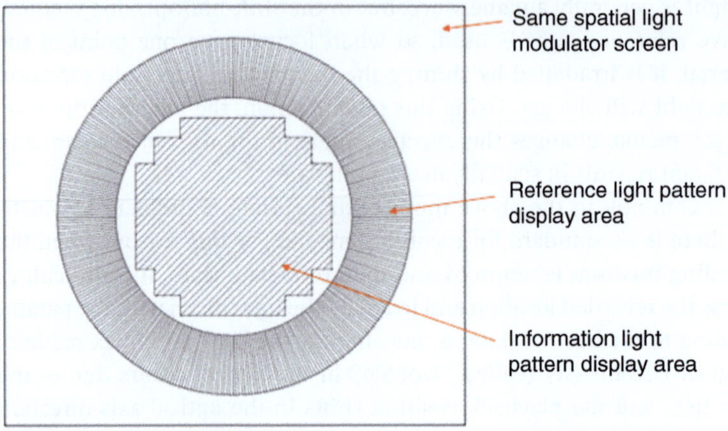


**Figure 2.12** Information beam and reference beam on spatial light modulator in collinear holography. Source: Hideyoshi Horimai.

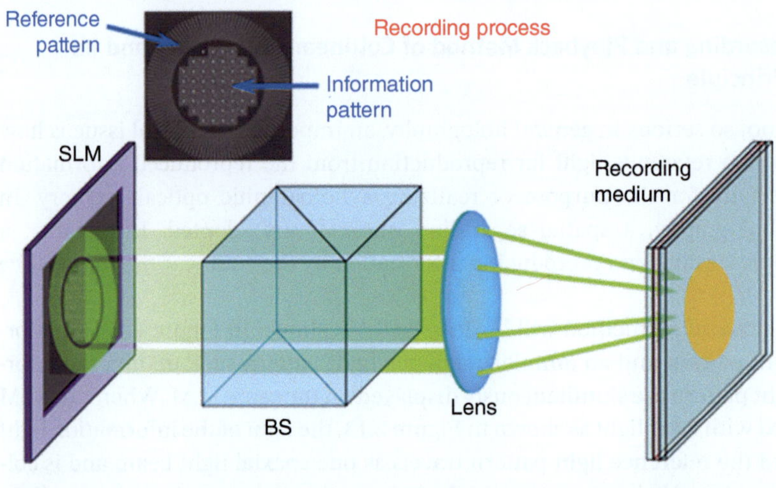


**Figure 2.13** Recording method by collinear holography. Source: Hideyoshi Horimai.

because they only need to be shielded by an aperture or only the information light is electrically detected. This is the recording/reproducing principle of collinear holography.

Based on this recording/reproducing principle, an attempt was made to actually record/reproduce the hologram. The experimental results are shown in Figure 2.15. Figure 2.15a schematically shows the information and the reference light pattern displayed on the SLM, and the black pixel becomes an ON pixel and emits light. Figure 2.15b is the reproduced image of the information light pattern that has been recorded and reproduced and detected by the CMOS sensor. The reference light portion was removed with an aperture placed in front of the CMOS sensor. Although some noise was observed at the center of the reproduced image, a reproduced image with uniform brightness was obtained over the entire page.

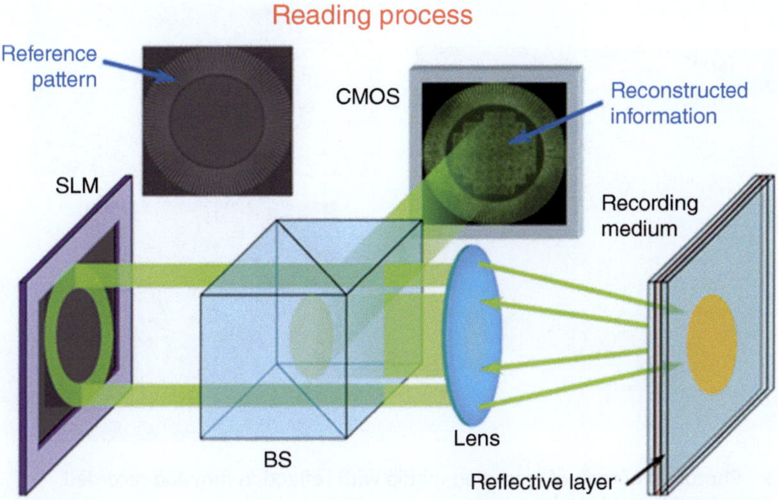


**Figure 2.14** Reading method using collinear holography. Source: Hideyoshi Horimai.

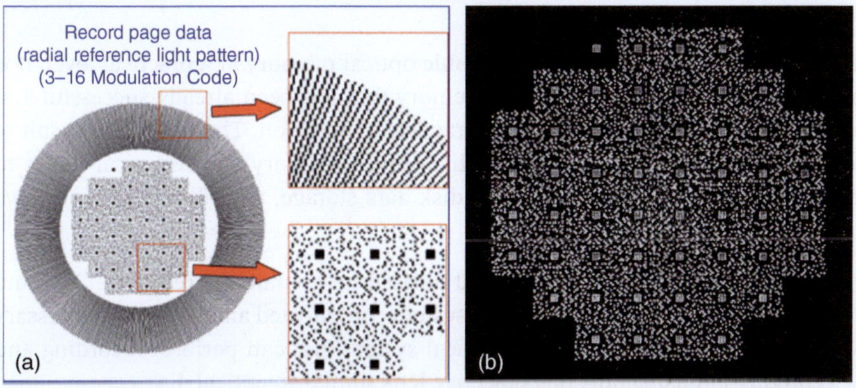


**Figure 2.15** (a) The information and the reference light pattern displayed on the SLM, (b) the reproduced image of the information light pattern captured by the CMOS sensor. Source: Hideyoshi Horimai.

Figure 2.16 is an appearance photograph of a recording medium (with a reflective film) after a recording/reproducing experiment. By devising the illumination method, several holograms recorded under different conditions can be confirmed in green. When one of them was observed with a microscope, it was confirmed that the size of the hologram was about 200 μm in diameter.

From the above results, a hologram (interference fringes) is recorded in the recording medium by simply squeezing the reference light pattern and the information light pattern that travel with the same coaxial light in the same direction without an angle. The basic principle was demonstrated by experiments for the first time in the world that an information light pattern can be reproduced by irradiation with the same reference light pattern.

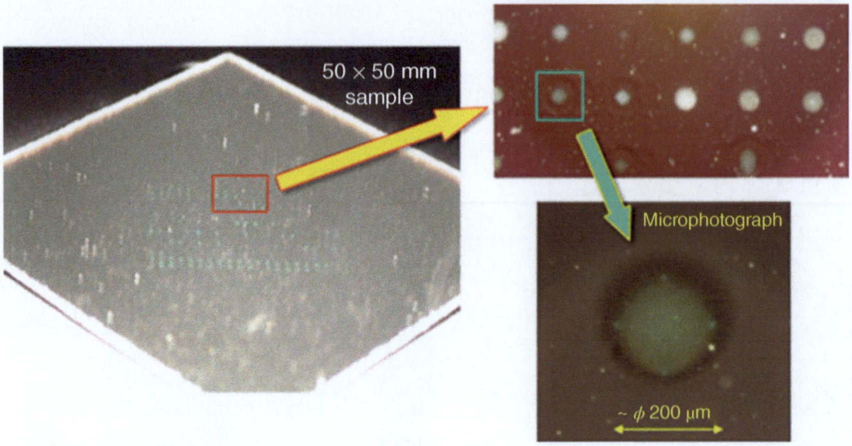


**Figure 2.16** Photomicrograph of recording media with reflection film and recorded hologram. Source: Hideyoshi Horimai.

## 2.2.4 Practical Collinear Method of Optical Memory

### 2.2.4.1 Basic Concept

To realize the next-generation holographic optical memory at a practical level, it is needed to compare with the optical memory that has been already successful and highlight the particular hindrance to practical application. Therefore, as a result of considering the problems of holographic optical memory and technical issues to be cleared, compared with the optical disk data storage, the following points have emerged.

(1) The two-beam interference method devised to separate the reference light and information light made the optical system complicated and large. It is necessary to create a simple holographic optical system that can perform recording and reproduction using only one objective lens similar to optical disk system.
(2) To achieve the high storage density, multiplexing recording is used such as small change in angle of reference beam and small shift of recording position. The reproducibility of the accuracy when the recording medium is set again in the apparatus cannot be maintained, in another words, there is a problem in the removable property.
(3) Even with the same type of recording device and recording medium, interchangeability (compatibility) between devices is not ensured, such as recording on one device and reproducing on another device. It is necessary to secure a wide system margin for absorbing variations between apparatuses and between recording media, and to construct a recording medium and a recording/reproducing system having a common standardized physical address and standard.
(4) In particular, when recording a hologram, the vibration is extremely disliked. However, even if the apparatus is made robust, the disc-shaped recording medium has low rigidity, and thus a large vibration occurs due to flapping of

the media. This cannot be removed with a vibration isolation table. Therefore, an application such as an optical servo that dynamically follows the movement of the recording medium is essential.

(5) In hologram recording and reproduction, a single-mode laser light source with high coherency and high quality has been used. However, to reduce the cost of the drive and the media for commercialization, it is ultimately necessary to consider the use of semiconductor lasers. Therefore, it is necessary to establish a recording/reproducing method with a wide wavelength margin that can realize recording/reproducing with a laser light source having low coherency.

Based on the above considerations, we ignored the specific implementation technologies for the time being and envisioned the ideal global form of the next-generation holographic optical memory. As a result, the basic concept is shown in Figure 2.17.

The biggest feature is that hologram recording/reproduction is performed using only one objective lens, which is the basis of the collinear method. As a result, the optical system becomes simple and the optical head can be arranged only on one side of the recording medium, so that there is a possibility that the entire system can be designed compactly. In addition, the reference light and the information light are common paths that follow the same optical path, and the optical path difference between the reference light and the information light is minimized. For this reason, we thought that hologram recording and reproduction could be realized even with a low-coherency semiconductor laser light source or Q-switched laser.

The recording medium is coated with a reflective film, and information light reproduced thereby is reproduced in a phase-conjugate manner through the same objective lens as that for recording. A pre-formatted address is also formed on the reference reflecting surface. This makes it possible to apply the optical servo that

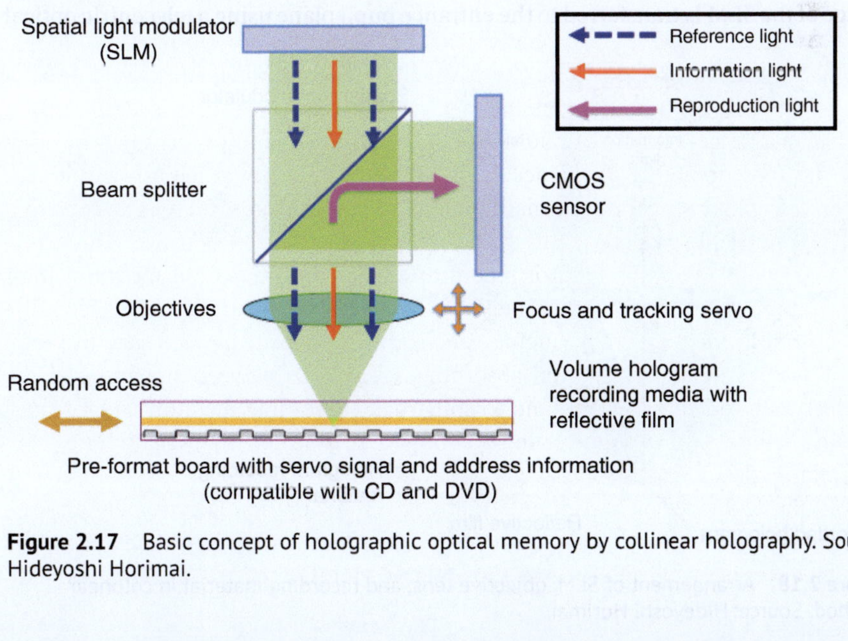


**Figure 2.17** Basic concept of holographic optical memory by collinear holography. Source: Hideyoshi Horimai.

has been cultivated for optical discs, and to allow the optical system to follow the recording medium on the nanometer order and to perform recording/reproducing with suppressed vibration. In addition, it is possible to realize exchangeability and compatibility of the recording medium, and random access for instantly searching for desired data. Furthermore, since the configuration of the optical system is arranged perpendicular to the recording medium and is symmetrical, there is a possibility that a wide optical and mechanical system margin can be secured.

#### 2.2.4.2 Basic Collinear Optical System

As schematically shown in Figure 2.18, in the collinear method, the page data pixels are arranged on the entrance pupil plane of the objective lens. Each pixel is regarded as a small light aperture and travels light toward the objective lens like a spherical wave emanating from a point light source. After passing through the objective lens, the small spherical wave from each pixel becomes a small parallel light and is concentrated on the focal plane of the objective lens. The concentrated and traveling light is incident on the hologram recording medium, but is focused on the opposite end surface of the hologram recording medium. A reflection film is provided on the opposite end surface of the recording medium, and focus servo is applied so that the reflection film surface is focused, and the relative position between the objective lens and the recording medium is controlled to be always constant.

Ideally, if the SLM can be arranged at the position of the entrance pupil plane of the objective lens, the optical system becomes extremely compact. However, it is difficult to secure the arrangement space in the optical dimension of the objective lens. Even if such an arrangement is realized, there is a problem when excessive exposure is performed because high-order diffracted light by Fourier transform of information light and reference light is irradiated into the recording medium. Therefore, in the actual recording/reproducing optical system, as shown in Figure 2.19, the display image of the SLM is transferred to the entrance pupil plane using a telecentric optical

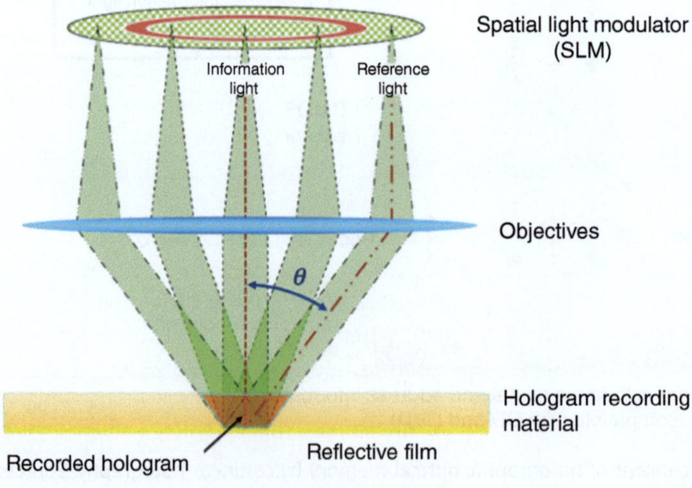


**Figure 2.18** Arrangement of SLM, objective lens, and recording material in collinear method. Source: Hideyoshi Horimai.

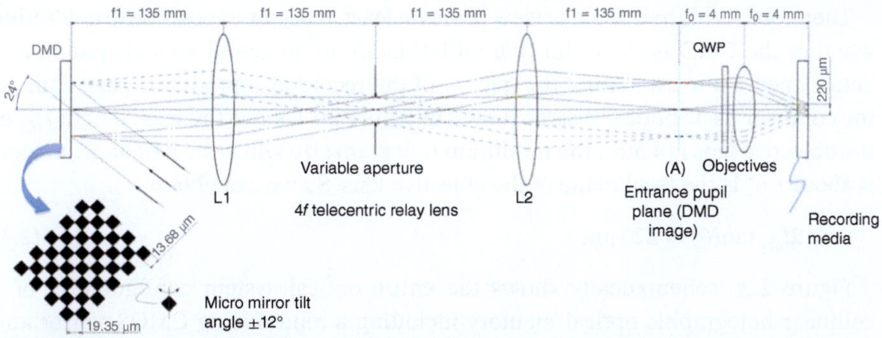


**Figure 2.19** Basic optical system configuration of collinear using DMD. Source: Hideyoshi Horimai.

system (so-called 4*f* system) in which the principal ray is parallel to the optical axis. Such an optical system was constructed. In addition, an aperture was placed in the center of the telecentric relay lenses $L_1$ and $L_2$ to cut higher-order diffracted light, and the hologram size was minimized to decrease material consumption. To simplify the description, the return light optical system including the reproducing CMOS sensor and the servo optical system for focus adjustment are omitted.

As a recording/reproducing SLM, a digital micromirror device (DMD) manufactured by Texas Instruments was used to realize a high frame rate. The page data displayed on the DMD are relayed at the same magnification by telecentric relay lenses $L_1$ and $L_2$, and re-imaged on the entrance pupil plane (position A) of the objective lens. This re-imaged DMD pattern is Fourier-transformed by the objective lens and recorded in recording medium.

The following is basic optical system design using actual device parameters. The DMD pixel is a small square mirror on micro electro mechanical systems (MEMS), and controls the ON/OFF of the light by changing the reflected mirror angle to a binary value of ±12°. Since the pixel rotation axis of the DMD is diagonal, it is difficult to align the optical system. To simply configure the optical axis of the entire optical system in the same horizontal plane, this rotation axis must be set to the vertical direction. Therefore, we decided to install the DMD unit itself in a state rotated 45° in the plane. When laser light is incident on the DMD arranged with 24°, the light beam reflected by the ON pixel is emitted from the DMD device in the vertical direction (the optical axis direction of the optical system toward the objective lens).

A light beam from the DMD generated in this way is incident on a telecentric relay lens composed of camera lenses $L_1$ and $L_2$ having a focal length of 135 mm, and a variable aperture is disposed at the focal position of the relay lens. The diameter of the aperture is set from the highest spatial frequency when DMD is modulated on/off in chess board pattern for each pixel. When a DMD pixel having a side size of 13.68 μm is regarded as a diffraction grating, the minimum grating pitch $d$ is 19.35 μm when tilted by 45°. When recording light with wavelength of $\lambda = 532$ nm is diffracted by this diffraction grating, the diffraction angle (half angle of view) of ±1st order light, which is the lowest order, is

$$\theta = \arcsin\left(\frac{\lambda}{d}\right) = \arcsin\left(\frac{0.532}{19.35}\right) = 1.575° \tag{2.2}$$

Therefore, if this half angle of view passes at least, it becomes possible to determine whether the DMD is ON (white) or OFF (black). In an actual optical system, it is determined while evaluating the quality of the recorded/reproduced page data. In the collinear basic optical system shown in Figure 2.17, since the focal length $f_{\text{obj}}$ of the objective lens is 4 mm, the minimum hologram size when the half angle of view is about 1.6° is the focal plane of the objective lens. So we can obtain,

$$2f_{\text{obj}}\tan(\theta) = 220\,\mu\text{m} \tag{2.3}$$

Figure 2.20 schematically shows the entire optical system configuration of a collinear holographic optical memory including a reproducing CMOS sensor and a servo system for focus adjustment. A PBS is placed on the optical path closest to the DMD, and in combination with a quarter wave plate (QWP) placed in front of the objective lens, the reproduction light returning from the recording medium is led to a CMOS sensor for reading page data. The ring reference light pattern is cut by the ring mask, and only the information light pattern is incident on the CMOS sensor. The magnification is finely adjusted so that 1 pixel of the DMD is detected by $3\times3=9$ pixels (three times oversampling) of the CMOS sensor with a lens arranged in front. A dichroic beam splitter (DBS) is arranged in front of the objective lens and is configured to guide red laser light for optical servo to the objective lens coaxially with the recording/reproducing light. Although not shown, the recording medium used in the experiment has a cross-sectional structure in which a photopolymer

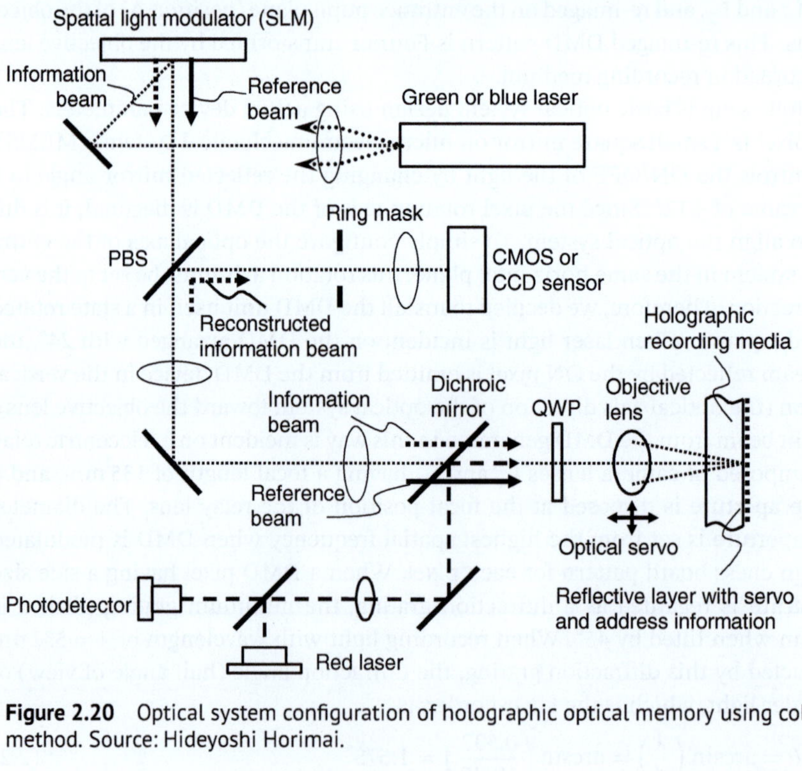


**Figure 2.20** Optical system configuration of holographic optical memory using collinear method. Source: Hideyoshi Horimai.

as a recording material is sandwiched between a transparent glass substrate with a thickness of 600 μm and a glass substrate coated with an Al reflective film on the surface. When the transparent substrate and the photopolymer are included in material, the thickness to the Al reflective film is 1.1 mm to 1.2 mm. For this reason, the plate thickness is corrected using a spherical aberration correction ring provided in the objective lens. The focus position is controlled by an autofocus control so that the Fourier transform surface of the page data comes to the Al reflective film by a focus servo mechanism based on the astigmatism method used in optical disks. Here, a 532 nm laser is used for recording. However, the servo is realized by combining a recording material (photopolymer) with a non-photosensitive 650 nm semiconductor laser.

The basic principle proof experiment of recording/reproducing by the collinear method and optical configuration are already shown in Figures 2.13 and 2.18. The details of parameters are described below,

Hologram recording material: photopolymer
Recording material thickness: 500 μm
Pre-exposure with green LED: 150 seconds
Hologram recording/reproducing wavelength: 532 nm
Recording power: average 3.5 mW at 20 kHz, 10 ns pulse × 20 pulse = 3.5 μJ
Reading at 60 seconds after recording ends
Reading power: same as recording average 3.5 mW at 20 kHz pulse × 20 pulse = 3.5 μJ
CMOS sensor (Photron): detection gain 6 dB
Light intensity ratio between information light and reference light: approximately doubled density.

## 2.3 Collinear Recording and Reproduction Characteristic Evaluation Device: S-VRD

To evaluate the recording and reproducing characteristics using the collinear method in detail, an experimental apparatus was used that was launched using the same optical parameters as the holographic optical memory system. The official name is the collinear hologram recording media evaluation device "Static Tester: S-VRD." Figure 2.21 shows an external view of the overall configuration of the system, and Figure 2.22 shows a photograph of the internal structure of the opto-mechanical unit. The configuration of the optical system is substantially the same as that shown in Figure 2.20. This apparatus is the first collinear-type recording and reproducing experiment equipment developed and commercialized at a practical level. It has a mechanism for positioning the recording medium with high precision using a six-axis stage control function, and an objective lens autofocus adjustment mechanism that links the piezo actuator and optical servo function. It also has a software servo function that corrects and detects the deviation of the detected reproduced image at high speed.

In static tester for volumetric hologram recording disc (S-VRD), recording and reproducing characteristics evaluation using collinear digital page data, recording

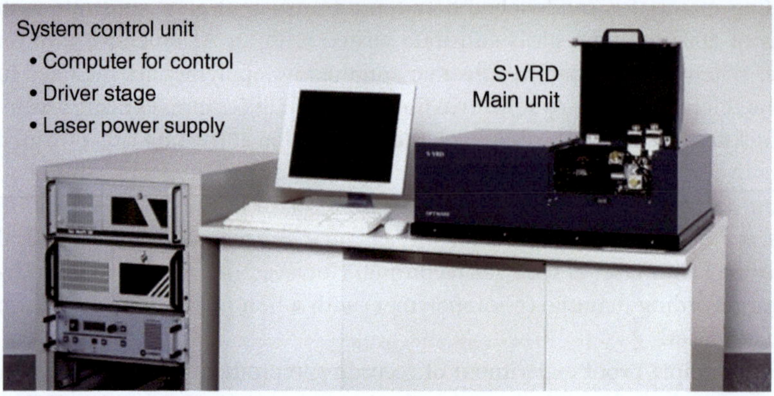


**Figure 2.21** System appearance of collinear hologram recording media evaluation device "S-VRD." Source: Hideyoshi Horimai.

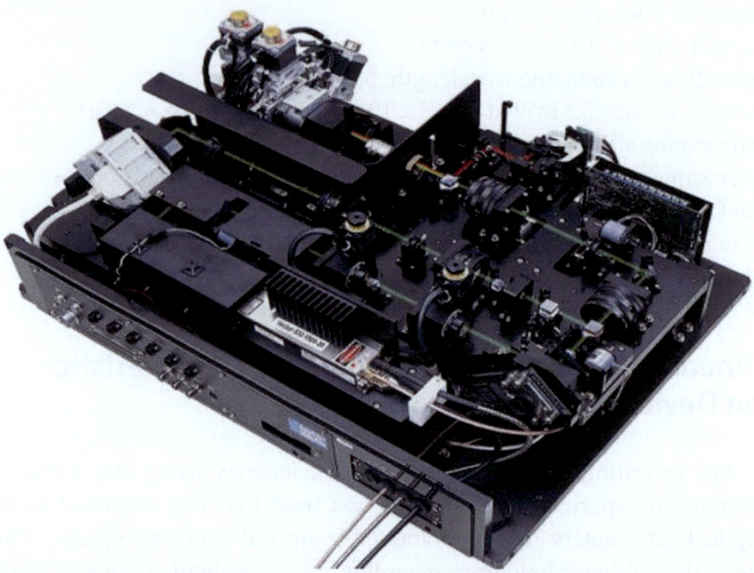

**Figure 2.22** S-VRD opto-mechanical unit (inside the main unit cover removed). Source: Hideyoshi Horimai.

and reproducing are compared on a pixel basis [12], playback signal quality evaluation based on SNR, and error rate. In addition to quantitative evaluation, it is designed to perform shift multiple recording characteristics evaluation and mechanical margin evaluation. In addition, a photodetector that is automatically inserted into the optical path is incorporated, and the power of the laser beam can be arbitrarily set by a combination of an automatic rotation stage of a half-wave plate and PBS. Thereby, it is possible to evaluate the multiplex recording and reproducing characteristics using scheduling for the hologram recording material having nonlinear sensitivity. Further, an LED light source with low coherence is incorporated, and it can be automatically inserted into the optical path before or after recording of the hologram to irradiate the hologram recording medium locally through the objective lens. As a result, pre-exposure for the purpose of removing the non-photosensitive energy region in the initial stage of exposure in the hologram recording material and bleaching processing for removing the photosensitivity of the recording material after exposure can be performed.

As a feature of the recording and reproducing optical system, a QWP is disposed in front of the objective lens to reduce interference dependency of the polarization component by the light condensing angle of the high magnification objective lens (NA = 0.55), and an interference optical system by circularly polarized light is adopted. This also contributes to maximizing hologram reproduction light to a high-speed CMOS sensor and is effective in evaluating weak diffraction efficiency. In addition to a Q-switched green laser (wavelength of 532 nm) for recording and reproducing a hologram, a nano semiconductor laser (wavelength: 630 nm) having no photosensitive property is used as a recording medium, and a nano control is performed by an optical servomechanism. Even if the optical servo is always operated, the recording material is not deteriorated by excessive exposure. For quantitative evaluation of recording or reproducing characteristics, recording and reproducing is performed using digital page data by 3 : 16 modulation (3 : 16 Modulation Coding), evaluation of recording sensitivity characteristics and diffraction efficiency, SNR, error rate, etc. Application software has been developed so that it can be evaluated on a sub-page basis. Figure 2.23 shows an example of a basic characteristic analysis screen of the recording and reproducing page data of S-VRD.

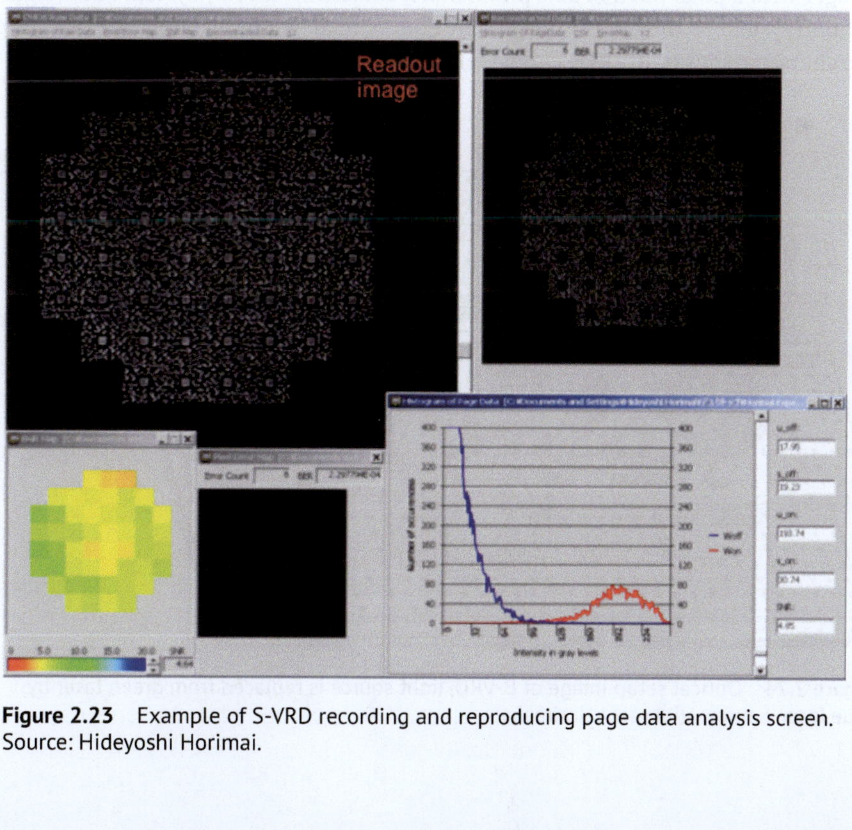


**Figure 2.23** Example of S-VRD recording and reproducing page data analysis screen. Source: Hideyoshi Horimai.

## 2.4 B-VRD

Blue light static writing and reading for volumetric hologram recording disc (B-VRD) is transformed from S-VRD. As shown in Figure 2.24, the main change is that the light source of S-VRD holographic recording system is replaced from green laser by blue laser. We call this static tester for volumetric hologram recording disc using blue laser as B-VRD. It is also a collinear holographic optical storage performance evaluation system, which is customized for Sony Corporation. The laser is a newly developed blue external cavity laser diode (ECLD) unit, which is manufactured by Sony [13].

It is known that the information density of the holographic recording is inversely proportional to the wavelength of the laser. That is to say, the shorter the wavelength, the higher the recording density, and the shorter the wavelength of the blue laser is than that of the green laser. Therefore, the main purpose of B-VRD is to select the blue laser.

Figure 2.25 shows the optical setup and optical path of B-VRD, which is based on the collinear hologram recording system S-VRD developed by Optware [14]. A blue laser of ECLD with a wavelength of 410 nm, developed at Sony, is implemented in the system. The maximum power of the blue laser of ECLD is 60 mW. Other constituent elements of the system are a mechanical shutter with minimum exposure time of 4 ms, a beam-expander with the scaling power of 3.94, DMD as an SLM with a pixel pitch of 13.7 μm, an objective lens with NA = 0.55 and $f$ = 4 mm, and CMOS imager with a pixel pitch of 12.0 μm. Lenses, consisting of relay system and magnification system before CMOS imager, are commercial enlarging ones. The details of the characteristics of B-VRD$_1$ can be referred to S-VRD section.

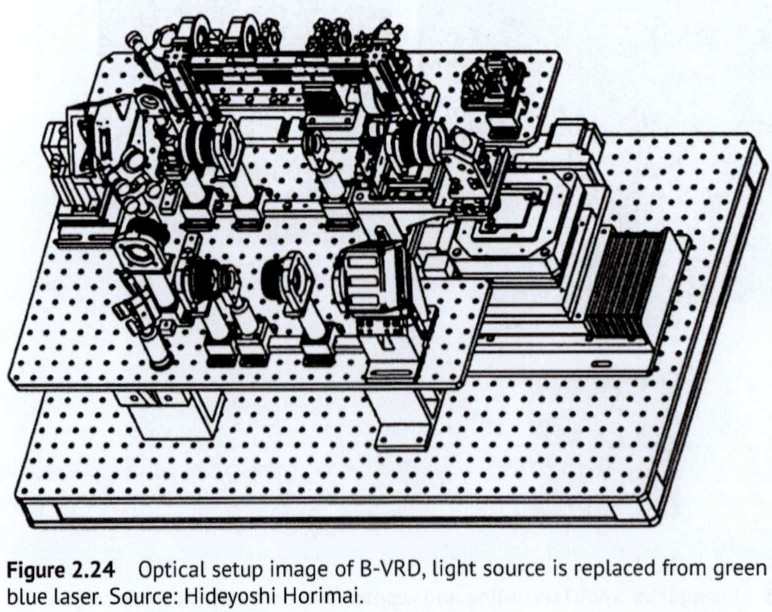

**Figure 2.24** Optical setup image of B-VRD, light source is replaced from green laser by blue laser. Source: Hideyoshi Horimai.

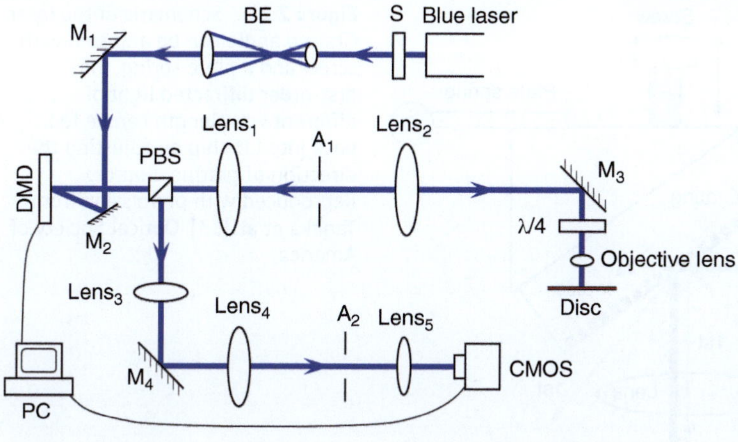


**Figure 2.25** Optical path of B-VRD. S, mechanical shutter; BE, beam expander; DMD, digital micromirror device; M, mirrors; PBS, polarization beam splitter; Lens, commercial lenses; A, apertures; Objective lens, objective lens; and complementary metal oxide semiconductor (CMOS), imager sensor. Source: Hideyoshi Horimai.

**Figure 2.26** Front view of the blue laser of ECLD. Include heatsink, Peltier element, and base plate. Source: [13] Tanaka et al. (2007), Optical Society of America.

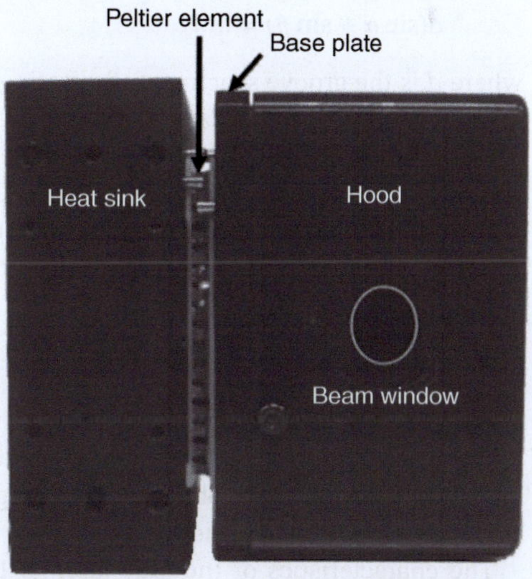


The blue laser of ECLD is developed by Sony Corporation. The laser has been embedded in the line-segment hologram writer already [15]. Figure 2.26 shows the front view of the blue laser ECLD. Figure 2.27 shows the principle of how multi-mode light can be converted into single-mode light. The multimode laser emitted from laser diode (LD) is collimated by a lens and then projected onto a grating, which diffracts the incident light into the first and zero-order beams. The first-order diffracted light of different wavelength can be fed back into LD chip by adjusting the direction of grating. When the intensity of the first-order diffraction light is enough, the feedback wavelength is dominant in the chip, so that the LD emits single-mode light. At this time, the beam of zero order can be used as a laser source.

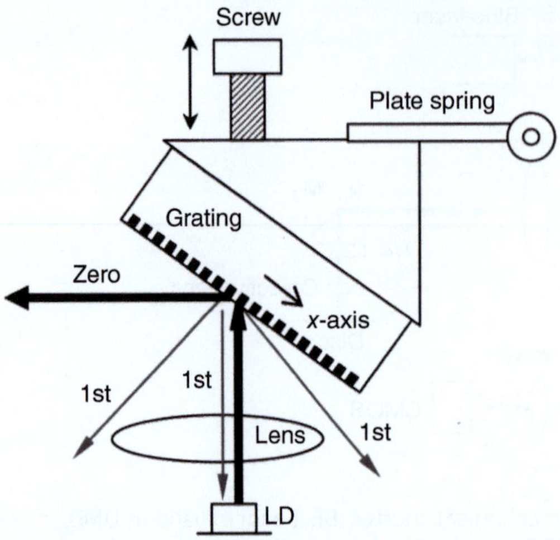


**Figure 2.27** Schematic of the laser. Grating angle can be adjusted with a screw and a plate spring, the first-order diffracted light of different wavelength can be fed back into LD chip by adjusting the direction of grating. Source: Reproduced with permission from Tanaka et al. [13], Optical Society of America.

The relationship between the incident light and the diffracted rays on the grating can be explained. The grating equation is written as

$$d(\sin\alpha + \sin\beta) = m\lambda \tag{2.4}$$

where $d$ is the groove spacing of the grating, $\alpha$ is the incident angle on the grating normal, $\beta$ is the diffracted angle, $m$ is the diffraction order, and $\lambda$ is the wavelength. In our blue laser of ECLD, $d$ is 278 nm and $\lambda$ is 405–410 nm in the laser. In Eq. (2.4) there is not solution for $m = -1, \pm 2, \pm 3, \ldots$ Only the 0 order and +1 order diffracted light exist. In the Littrow configuration, in Figure 2.27, if the +1 order diffracted light returns to the chip, the incident angle and +1 order diffracted angle are same ($\alpha = \beta$). Equation (2.5) can be rewritten as

$$2d\ \sin\alpha = \lambda \tag{2.5}$$

For example, if we need to produce a single-mode laser with a wavelength of 407 nm, we have to adjust the angle of incidence to 47° by adjusting the screw shown in Figure 2.27. To obtain a stable single mode, we must also pay attention to the angle of rotation around the $x$-axis.

The characteristics of the blue laser of ECLD have been investigated by Sony Corporation, e.g. the threshold current and slope efficiency of the ECLD laser are lower than those of the conventional LD. The power of the laser is confirmed when the laser can emit a single-mode ray up to 80 mW by measuring with an optical spectrum analyzer. The coherent length is about 14 m at ambient temperature. The stability of the laser is so high that the deviation of wavelength is less than 5 pm over 10 hours of continuous operation, at 50 mW output and 30 °C [13]. The spectrum of the ECLD blue laser is shown in Figure 2.28.

In the collinear holography, because the image of SLM is located at the front focal plane of the objective lens and the recording medium is located at the back focal plane of the objective lens, the actual recording medium records the SLM Fourier transform image. In B-VRD, as the DMD represents the amplitude-modulated

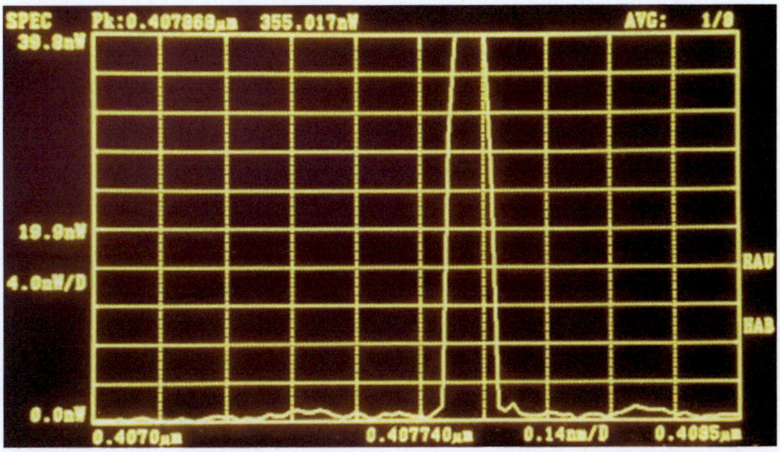


**Figure 2.28** The spectrum of the ECLD blue laser, the coherent length is about 14 m. Source: Hideyoshi Horimai.

data page, a strong DC component will be generated on the Fourier plane in the recording medium, which directly affects the effective dynamic range of the recording medium. To suppress the DC component of the Fourier transform and reduce the influence of the dynamic range of the recording medium, the random-phase mask plate is introduced while DMD modulates the amplitude signal. Therefore, Sony Corporation reformed the original B-VRD developed by Optware Corporation. In the modified B-VRD, as shown in Figure 2.29, the original position of the DMD is replaced by a random-phase mask plate, and the new position of the DMD is

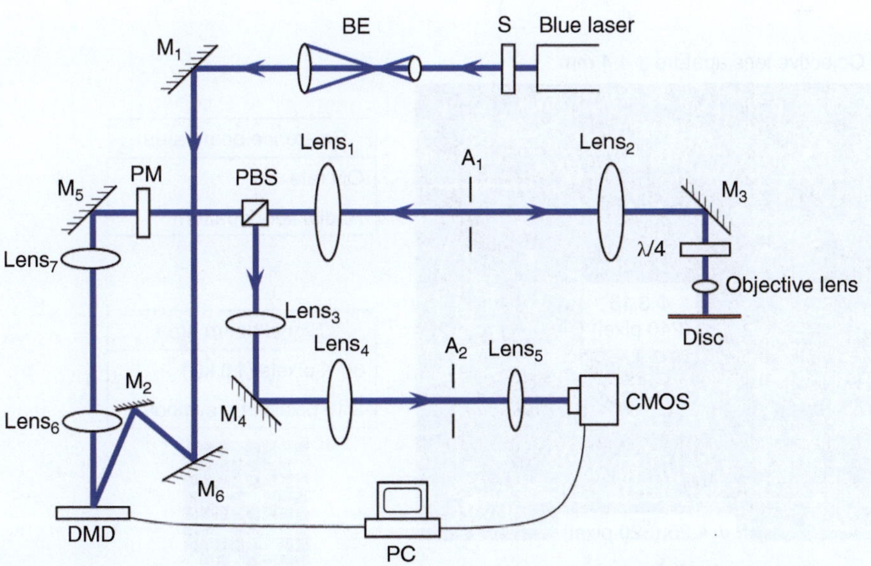


**Figure 2.29** Optical path of modified B-VRD. PM, phase mask. Source: Hideyoshi Horimai.

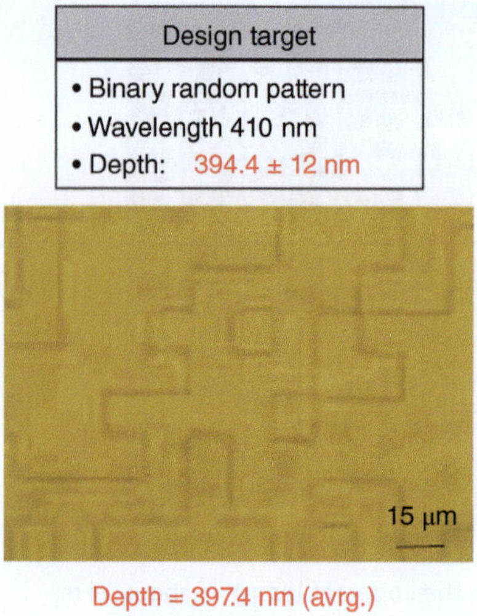


**Figure 2.30** Phase mask's structure and its parameters. Source: Hideyoshi Horimai.

moved backward to the light source, in the expanded optical path. Two lenses as a 4$f$ system form the real image of the DMD on the random-phase mask plate [16].

A random-phase mask plate is implemented at the conjugate plane to the DMD reflecting plane. The random-phase mask plate has a binary depth of $\lambda/2$ ($\lambda$: wavelength) as shown in Figure 2.30 attached by the measurement result of the step depth. The mask is fabricated by ion-milling technique.

Data pattern used in the B-VRD system, the page data are modulated by amplitude only, is shown in Figure 2.31. One page signal pattern consists of 52 sub-pages. A sub-page consists of 36 symbols and a symbol consists of 16 pixels. Three pixels out of 16 in each symbol turn ON at a same time. In each sub-page, centered four

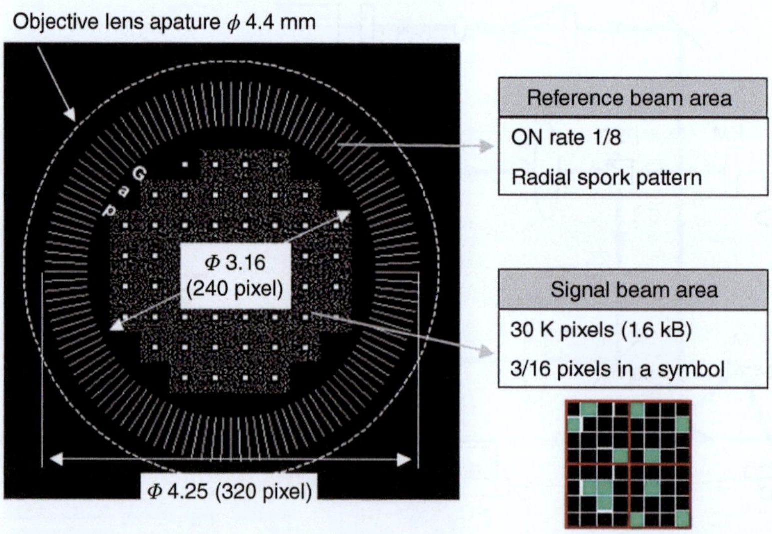


**Figure 2.31** Page data are modulated by amplitude only. Source: Hideyoshi Horimai.

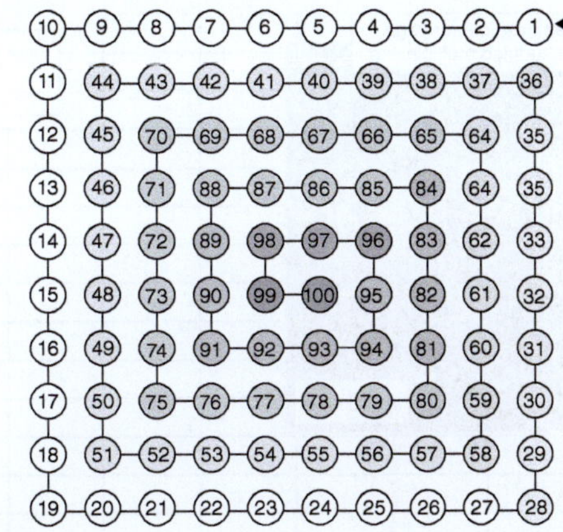


**Figure 2.32** Page arrangement and recording order – 10 × 10 pages multiplexing. The holograms are recorded one by one in the order from the periphery to the center and reconstructed in reverse order. Source: Hideyoshi Horimai.

symbols are sub-page sync mark, and the remainder symbols are effective for signal modulation. A single sub-page, located at the upper left of the pattern, includes only sync mark. This sub-page is processed to synchronize whole hologram page. The amount of the data size of a hologram is 1.6 kB. A reference part of a page shaped in a radial spoke pattern, and the ON rate in a reference beam area is 1/8.

In the verification experiment, the arrangement and recording sequence of 10 × 10 pages multiplexing are adopted as shown in Figure 2.32. The holograms are recorded one by one from the top right corner of the outermost layer in the order from the periphery to the center. Then the holograms are reconstructed one by one in reverse order of recording. The distance between adjacent holograms is a constant.

In their report [16], the B-VRD system has been tried to record three cases of multiplexing density experiments: 10 × 10 pages at 20 μm intervals, 12 × 12 pages at 16 μm intervals and 14 × 14 pages at 14 μm intervals. 12 × 12 multiplexing is almost equivalent to the recording density of 33 Gb/in$^2$ and 14 × 14 multiplexing is to that of 42 GB/in$^2$. To equalize the diffraction efficiency of each page and effectively utilize the media dynamic range, recording energy scheduling was applied by controlling the exposure time. Exposure time for each hologram was determined based on the results of the constant-energy multiplexing. The recording energy at non-scheduled multiplexing was 4.7 μJ for 10 × 10, 2.2 μJ for 12 × 12 and 14 × 14.

The transition of the number of errors and a retrieved center page with scheduled multiplexing is shown in Figures 2.33–2.35. A value of diffraction efficiency, $\eta$, is attached to each page image. In these figures, page number is assigned in recording order, that is, the maximum page number corresponds to the center position of multiplexing. The discriminated value of multiplex recording assumes that the error numbers of all pages are around 100.

In the result of 10 × 10, and 12 × 12, the number of errors was suppressed to around 100 by introducing the scheduled recording, while, in the result of 14 × 14 multiplexing, 400–500 errors remain at maximum. In all three cases, the effect to suppress the error number at the multiplex center is observed. And especially in the case of 14 × 14 multiplexing, fluctuation of error number is revealed. This fluctuation has some correlation to the positional difference of page overlapping. A scheduling algorithm

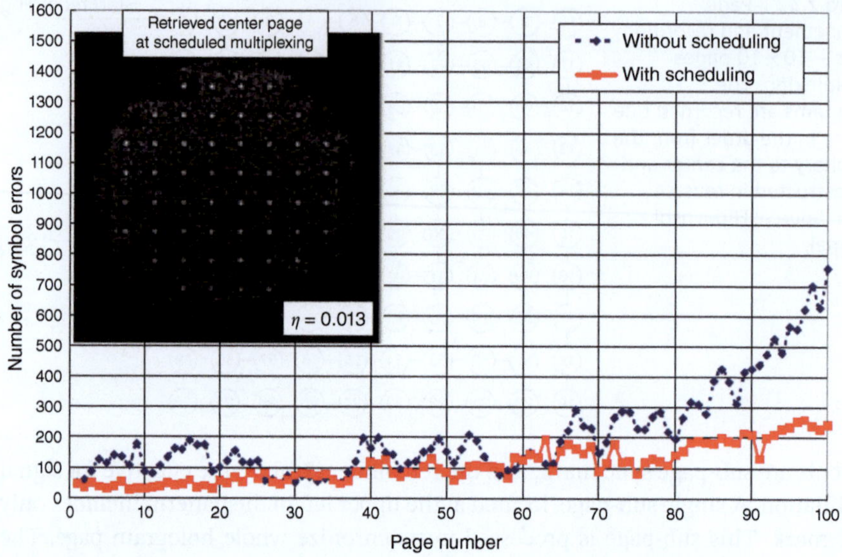


**Figure 2.33** Result: 10 × 10 pages at 20 μm pitch. Source: Hideyoshi Horimai.

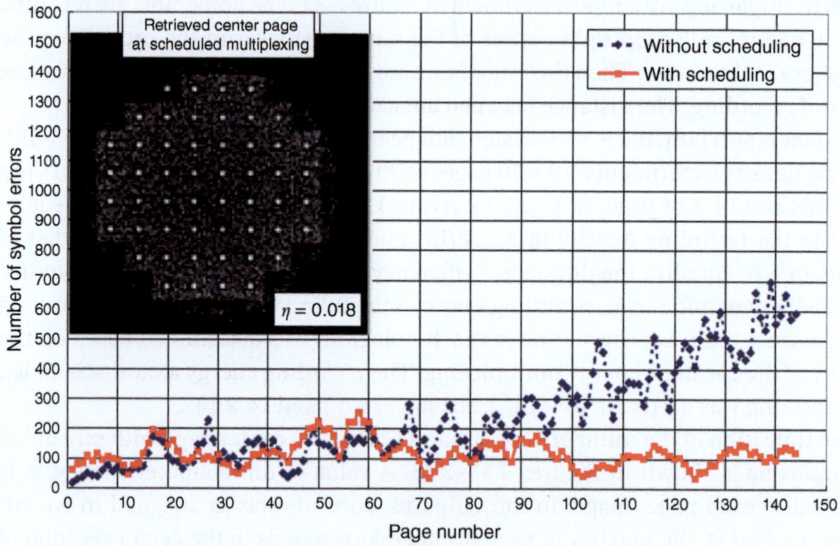


**Figure 2.34** Result: 12 × 12 pages at 16 μm pitch. Source: Hideyoshi Horimai.

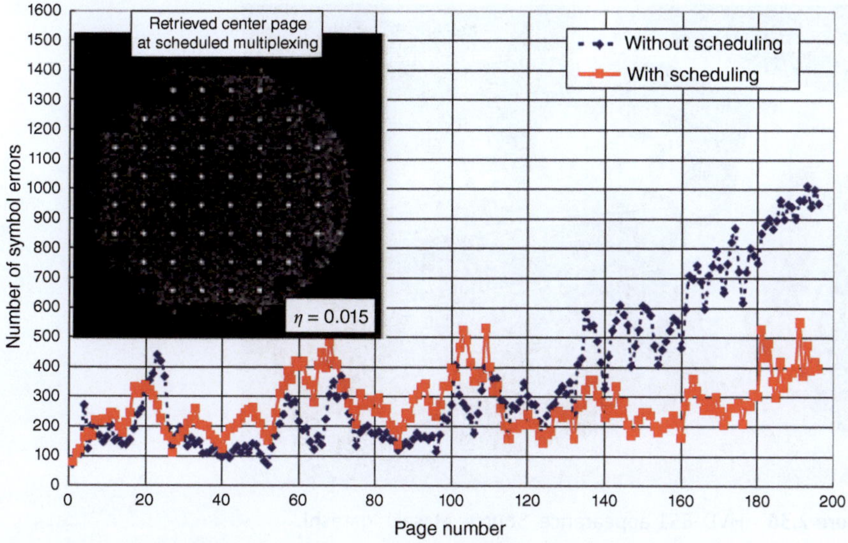


**Figure 2.35** Result: 14 × 14 pages at 14 μm pitch. Source: Hideyoshi Horimai.

considering geometrical nonuniformity of multiplexing leads to further suppression of error number and increase of multiplexing density.

In the conclusions, a new technique of high-density recording confirmation for blue laser collinear holographic system B-VRD is introduced. A blue ECLD and random-phase mask were first implemented and were experimentally evaluated. The result shows that 10 × 10 pages at 20 μm pitch and 12 × 12 at 16 μm pitch multiplexing were attained. And a possibility to 14 × 14 at 14 μm pitch multiplexing was implied. 12 × 12 multiplexing is almost equivalent to the density of 33 GB/in$^2$ and 14 × 14 multiplexing is to that of 42 GB/in$^2$, the density of these is higher than that of the Blu-ray Disc.

## 2.5 HVD-ES1

### 2.5.1 System Overview and Background of Development

HDV-ES1 is a system that implements dynamic recording and reproduction by applying the fundamental of conventional static recording and reproduction. The HDV-ES1 is capable of recording and reproducing hologram discs formed in a disk shape.

This system not only realizes the function of merely recording and reproducing a hologram disc but also it is possible to evaluate each margin, etc. at the time of designing a drive and saving a reproduced hologram as an image at the time of reproduction. It is possible. Based on these facts, considering practical application as an information storage system, it is designed with a view to the future.

**Figure 2.36** HVD-ES1 appearance. Source: Masaki Igarashi.

As described above, HDV-ES1 is a system developed for the purpose of demonstration of dynamic recording and reproduction of holograms and acceleration of research and development of hologram media (Figure 2.36).

### 2.5.2 Overall System Configuration

The appearance of this system is shown in Figure 2.37.

| | Name | Model number |
|---|---|---|
| ① | HVD-ES1 Unit | EV1500D-ES01 ver.2.5 |
| ② | PC Controller and R/W Channel Unit | EV1500D-PC01 |
| | R/W Channel Firmware | HVD R/W-Enc/Dec Software ver.1.6 |
| ③ | Servo and Timing Controller Unit | EV1500D-DR01 |
| ④ | Laser Power Supply Unit | EV1500D-LP01 |
| ⑤ | (Half wave-plate) HWP Rotary Driver Unit | EV1500D-AZ01 |
| ⑥ | Application Software | HVD ES1 PC APL ver.1.0 |
| | Error evaluation Offline Software | HVD Evaluation Software ver.1.0 |
| ⑦ | HVD-R Cartridge Sample | HVD-R Cartridge Sample ver.1.0 |
| ⑧ | HVD-R Loading Mechanism | HVR-R Loading Sample ver.1.0 |
| ⑨ | DMD Device Software | DMD 1100 Software ver.1.0 |
| ⑩ | Disc | |

The functional block of this system is shown in Figure 2.38.

The HDV-ES1 controls each function in the PC shown in Figure 2.38. The PC is equipped with a custom board of EB 4 PCI (peripheral component interconnec), which is connected with a monitor, keyboard and mouse and becomes a PC

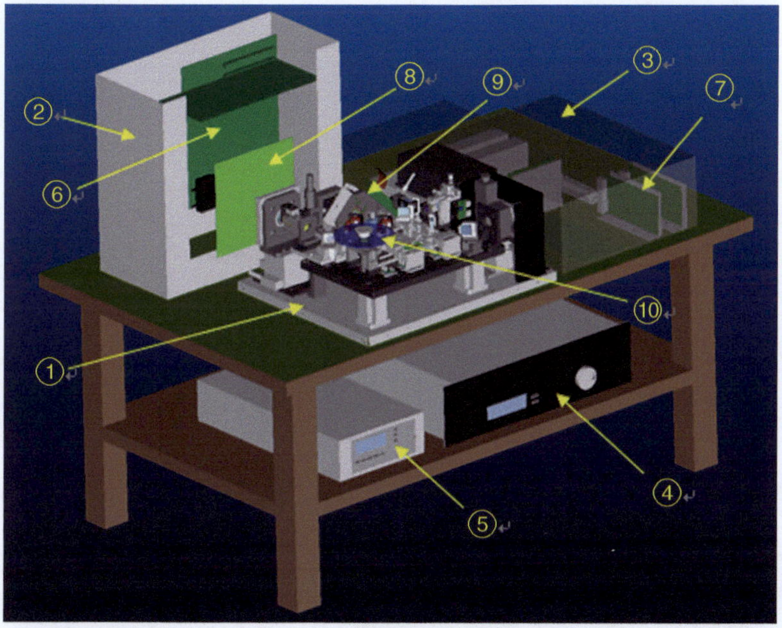


**Figure 2.37** System appearance. Source: Masaki Igarashi.

controller and R/W channel unit, two controller boards of RS-232C (commercially available), a digital input/output (DIO) controller board (commercially available) are built in.

As shown in the red line in Figure 2.38, the controller board of the RS-232C is connected to the CMOS, DMD, servo, and HWP rotation drivers and controls them. In Figure 2.38, error correction code (ECC) was scheduled to be performed by hardware, but in ES 1, ECC processing is performed in PC.

The DIO controller board is connected to the servo and timing controller unit and performs timing control such as recording and reproduction.

### 2.5.3 Main Unit

#### 2.5.3.1 Hardware and Parameters

Separable optical system is adopted based on two-wavelength-type collinear optical system (Tables 2.1–2.6).

HWP rotary driver unit used Precision Rotation Stage Driver Unit (SC-200) manufactured by Suruga Seiki Co. Ltd.

#### 2.5.3.2 System Controller Section

The system controller part is an application that controls the entire system created in Visual Studio, controls each unit as described below, and performs recording

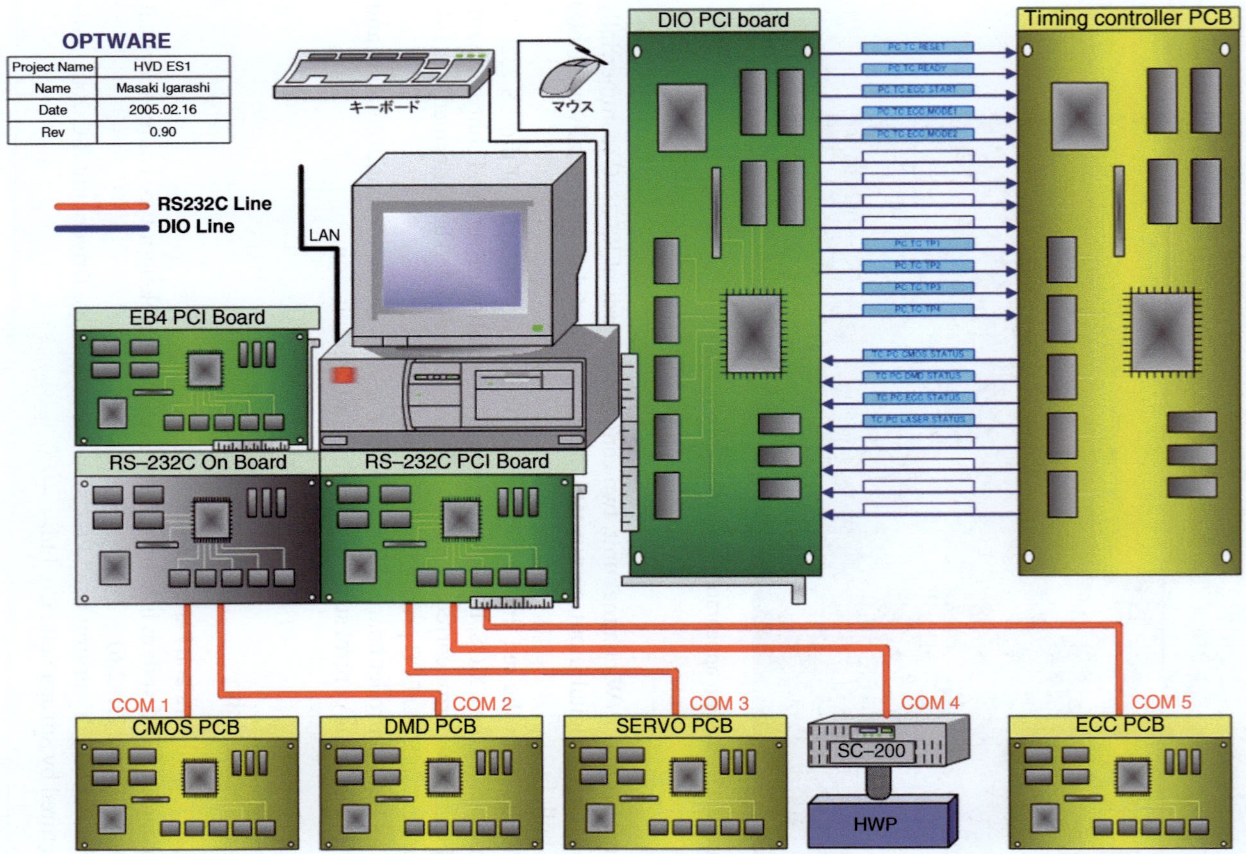


**Figure 2.38** Functional block diagram. Source: Masaki Igarashi.

**Table 2.1** Optical system.

| |
|---|
| Mechanism unit ES 1 mechanism |
| Green optical system ES 1 optical system |
| Red optical system ES 1 optical system |
| Actuator ES 1 actuator |
| Mechanical ES 1 |
| Optical alignment application CMOS raw data, video graphics array (VGA) display |
| Optical correction moving mirror relay lens correction |

**Table 2.2** Laser light source for hologram recording/reproducing.

| |
|---|
| LASER unit Photonics Company DC 150-532 |
| Power 3000 mW |
| Pulse energy (Q-switch method) 300 μJ/pulse |
| Based on power control method HWP |
| Pulse width 30 to 40 ns |
| Operating frequency 10 kHz |
| Drive control method external pulse input |

**Table 2.3** DMD (SLM) device part.

| |
|---|
| DMD board D1100+field programmable gate array (FPGA) evaluation board (Texas Instruments [TI]) |
| DMD device DMD 0.7 extended graphics array (XGA)- Double Data Rate SDRAM (DDR) |
| Signal processing chip XILINX Corporation FPGA: SP3 |
| Function Optware encoding |
| Page data recording frame rate 3.5 kfps (256 × 256) |
| Page data transfer rate 2.5 Gbps (83MHz × 32 bits) |

and reproduction of the hologram. It has an ECC function, it can encode/decode to collinear page data, and it is possible to record and reproduce still images, movie files, etc.

The recorded information, so-called TOC information, is managed by the PC, and even if the disk is mounted or demounted, the desired recorded data can be accessed again. By doing this, we have a function to evaluate the storability of data and perform reliability test. In addition, in recording and reproduction, to fully demonstrate

**Table 2.4** High-speed CMOS sensor part.

| |
|---|
| CMOS board ES1 custom board |
| CMOS device MI-MV 02 (μm) |
| Signal processing chip XILINX Corporation FPGA: V2P |
| Function image buffer, offset correction |
| Page data playback frame rate 2.4 kfps |
| Page data transfer rate 10 Gbps (166 MHz × 32 bits × 2) |
| Sampling × 2 sampling |
| Pixel gradation 8bit gradation output |

**Table 2.5** R/W channel signal processing unit.

| |
|---|
| R/W channel Processor board IPFLEX Corporation Evaluation Board: EB 4 |
| Signal processing chip DAPDNA-2 × 2 (parallel processing not possible) |
| System-to-computer I/F PCI (33 MHz × 32 bits) |
| Page data transfer rate 10 Gbps (166 MHz × 32 bits × 2) |
| Recording processing function data transfer only (no encoding) |
| Playback processing function sync detection, downsampling |
| Optware decode |

**Table 2.6** Focus servo/tracking servo optics.

| |
|---|
| SERVO board ES1 board |
| SERVO chip SANYO |
| Signal processing chip ALTERA cyclone II |
| Seek function track search F/W (TBD) |
| Timing control function External timing controller board |
| Rotation speed 300 rpm |
| Minimum pulse interval clock reference up to eight times |

the performance of the hologram material, it is possible to designate the shift amount of the recording hologram position and the number of holograms to be recorded on the track.

R/W channel processor board control
CMOS settings
DMD settings
Disk operation control by SERVO PCB (it is possible to move the pickup to an arbitrary position on the disc)
HWP control (laser power adjustment for recording and reproduction)

ECC setting (setting of redundancy for error correction, etc.)
Recording function
Reproduction function (data can be reproduced even as a hologram image)
Disk information management function (with recorded data)
Script function (setting of hologram recording order, etc.)
Verify function of data (it is possible to compare playback data and record source data)
Replay file combining function (restoration from reproduced hologram data to original data)
Record file division function (record file data divided into hologram data)
Loading control (TBD)

| System controller board | PC base |
|---|---|
| Processor | PC processor (Pentium 2) |

### 2.5.4 Recording Playback Condition and Corresponding Media

The recording playback conditions are shown in Table 2.7. Pre-formatted address information and reflective film are attached to the disc itself, and a hologram material is applied on top of it. Since the recording material has high photosensitivity, recording is performed in a dark room; but by exposing the medium to a medium, reproduction can be performed even in a bright room. Therefore, at the time of commercialization, the disc will be put in a light-tight cartridge. Recording media is called holographic versatile disc (HVD) and was approved as an international standard at ECMA International in May 2007. The diagram of HVD is shown in Figure 2.39.

### 2.5.5 Specifications

see Table 2.8.

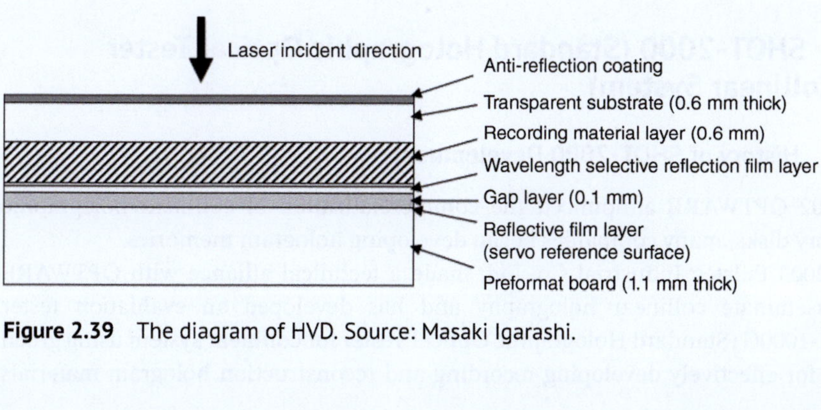


**Figure 2.39** The diagram of HVD. Source: Masaki Igarashi.

**Table 2.7** Recording playback condition.

| **Hologram recording/reproduction laser wavelength** | **532 nm (DPSS Nd:YAG)** |
|---|---|
| Hologram recording/reproducing laser power | Laser emission energy: maximum 1 (μJ/pulse) (tentative) (specified by the objective lens output when displaying recorded page data with power ratio objective/Reference = 1.0) |
| Hologram recording/playback mode | Q-switch pulse mode (synchronized with external trigger) |
| Pulse width | 20–30 ns (tentative) |
| Laser wavelength for focus servo | 650 nm (semiconductor laser) |
| Hologram data page display size (not including reference light) | Area of 192 × 192 pixels |
| Recorded data size per hologram | 1.632 kB (3–16 coding, code rate = 0.5) |
| Recording media | Media conforming to HVD standard |

**Table 2.8** Specifications.

| **Power supply voltage frequency** | **AC100 (V)±5 (%), 50/60 (Hz), earth essential** |
|---|---|
| Power consumption | Maximum time 1000 (W) (total value of all units) |
| Equipment base | Optical isolation table unnecessary (however, depending on the environment, the necessary optical anti-vibration table) |
| Installation environment | Darkroom, dustproof environment recommended (depending on the recording medium, red fluorescent light can be used) |
| Operating temperature range | 15–30 °C (depending on the recording medium, it depends on environmental temperature) |
| Operating humidity range | 20–60% (no condensation, etc.) |

## 2.6 SHOT-2000 (Standard Holographic Optical Tester for Collinear System)

### 2.6.1 History of SHOT-2000 Development

In 2002 OPTWARE announced the commercialization of collinear holographic memory disks, many companies began developing hologram memories.

In 2003 Pulstec Industrial Co. Ltd. made a technical alliance with OPTWARE to disseminate collinear holography and has developed an evaluation tester SHOT-1000G (Standard Holographic Optical Tester for collinear system using green laser) for effectively developing recording and reconstruction hologram materials

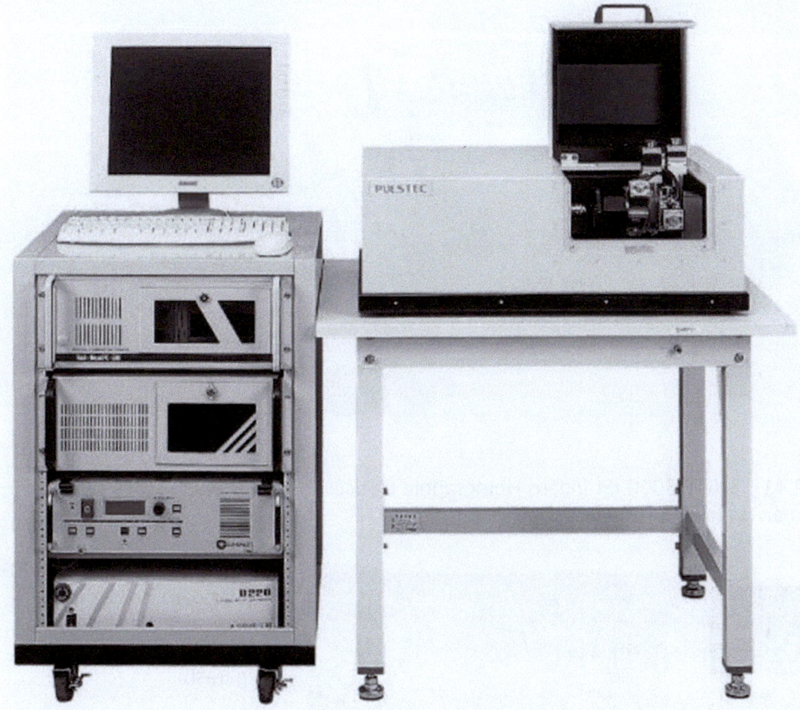

**Figure 2.40** SHOT-1000G (Standard Holographic Optical Tester for collinear system using green laser). Source: Masaki Igarashi.

compatible with collinear hologram system (Figure 2.40) and started supplying them to each company.

In 2004, Pulstec Industrial Co., Ltd. collaborated with Toyohashi University of Technology and developed a two-beam plane wave tester SHOT-500G (Standard Holographic Optical Tester for plane-wave system using green laser) (Figure 2.41) as an evaluation tester capable of basic performance evaluation of recording and reconstruction hologram materials, and started supplying them to each company.

In 2005, OPTWARE founded a group HVD alliance for international standardization of HVD.

In May 2007, HVD alliance succeeded in acquiring HVD standards at ECMA International., an international standardization organization.

To acquire international standards, it is necessary to formulate HVD standards, and to do so, an evaluation tester to evaluate HVD is required.

In developing the HVD evaluator, we changed the evaluation of the coupon material to the evaluation of the disk, so we gathered a particularly excellent optical disk engineer in Pulstec, made a project team, and started development.

In 2006, Pulstec Industrial Co. Ltd. developed HVD-compliant evaluation tester SHOT-2000 (Standard Holographic Optical Tester for HVD system) and greatly contributed to the acquisition of HVD standards.

In this session, details of the SHOT-2000 system will be explained.

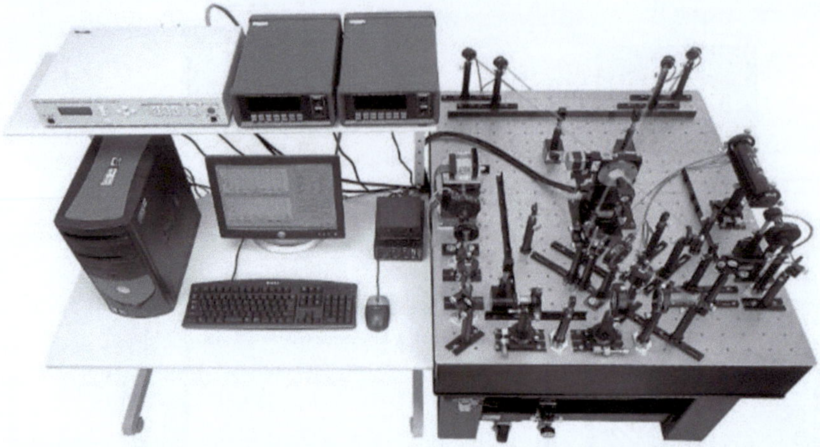

**Figure 2.41** SHOT-500G (Standard Holographic Optical Tester for plane-wave system using green laser). Source: Masaki Igarashi.

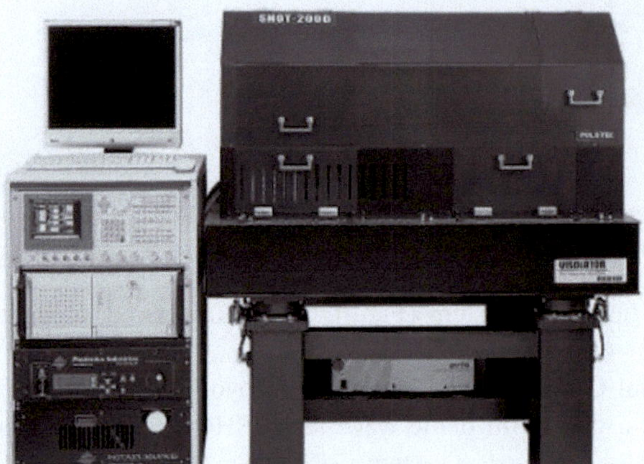

**Figure 2.42** SHOT-2000 (Standard Holographic Optical Tester for HVD system). Source: Masaki Igarashi.

## 2.6.2 SHOT-2000 System Outline

### 2.6.2.1 Device Appearance

see Figure 2.42.

### 2.6.2.2 Actuality of Disc Evaluation

see Figure 2.43.

### 2.6.2.3 Block Diagram

SHOT-2000 is roughly divided into the following five blocks. See Figure 2.44.

***Hologram Recording/Reconstruction Optical System, Disk Spin Stand*** In this block, all optical parts, which can record and reconstruct on hologram material by collinear holographic system, are built.

**Figure 2.43** Setting the Disk on Spindle Unit. Source: Masaki Igarashi.

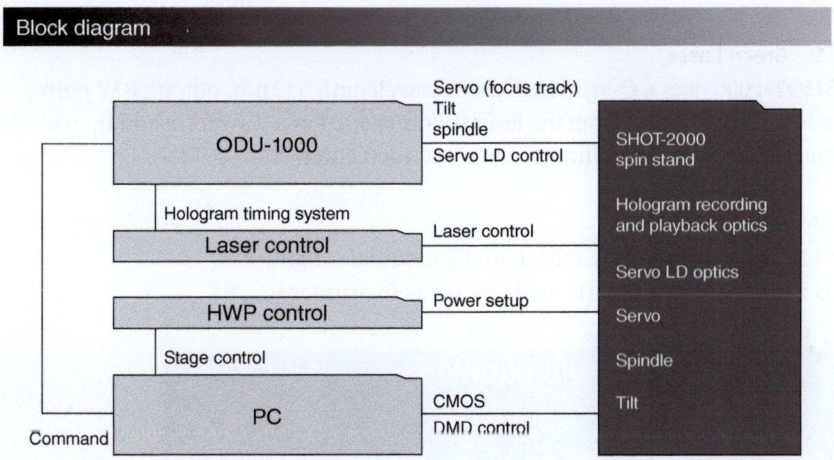


**Figure 2.44** Block diagram. Source: Masaki Igarashi.

In addition to optical parts, mechanical parts such as actuators and feed stages for recording/reconstruction a hologram at the correct position of HVD, spin stand for rotating HVD, LED for pre-cure and post-cure are built.

It also has a unit that reads HVD address information and provides current address information to the "Disc Control Unit ODU-1000 (Optical Disc Unit-1000)": Disc control unit.

***Disc Control Unit ODU-1000 (Optical Disc Unit-1000)*** This unit accurately controls the rotation speed of HVD so that it becomes the target rotation speed. Further, when recording/reconstruction a hologram, this unit always follows the vertical vibration (Fo direction) and radial vibration (Tr direction) generated by rotation of the HVD. It controls so as to focus on the position. (Fo servo, Tr servo control)

***Laser Control Unit for Recording/Reconstruction*** This unit performs current control and temperature control so that the laser for recording/reconstruction a hologram (wavelength 532 nm) will oscillate steadily.

***Motor Control Unit (Spin Stand Position Control)*** This unit controls the feed stage to move the spin stand to the target position during recording/reconstruction a hologram.

***Main Control PC*** This PC operates as a man-machine interface such as setting various parameters during recording/reconstruction a hologram, display of reconstruction image, camera control for image capture, SLM (DMD) control, reconstruction image data management and so on.

### 2.6.3 Optical System (Light Source, SLM, Actuator, Camera, etc.)

An outline of the SHOT-2000 optical system is shown in Figure 2.45.

#### 2.6.3.1 Green Laser

The SHOT-2000 uses a Q-switched laser. (wavelength 532 nm, output: 8 W pulse)

The laser beam output from the laser unit is shaped as a waveform and then made into parallel light by the collimator lens and then enters SLM (DMD).

#### 2.6.3.2 SLM (DMD)

The SLM (DMD) is a device that spatially modulates light.

This SHOT-2000 uses DMD made by Texas Instruments.

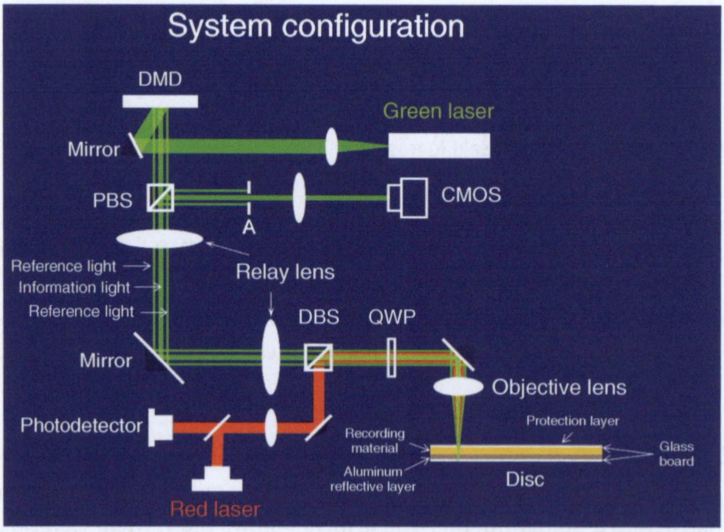


**Figure 2.45** SHOT-2000 system configuration (collinear holography optics). Source: Masaki Igarashi.

A DMD is a display element (MEMS device) in which a large number of movable micromirror surfaces are arranged in a planar manner on an integrated circuit formed by a CMOS process, and the device used also as a projector or the like is.

The specifications of the DMD used in SHOT-2000 are as follows.

Mirror array size: 1024 × 768
Mirror pitch: 13.7 × 13.7 μm
Mirror tilt: ±12°

We select the DMD such that it is compatible with the wavelengths and each mirror tilt shows the minimum possible variation.

#### 2.6.3.3 Objective Lens Unit

The pattern light generated by the DMD enters the objective lens unit via the relay lens.

For Fo and Tr servo control of conventional optical discs, we controlled the objective lens only in the direction of Fo and Tr using the actuator, but since it is necessary to propagate the image in hologram recording/reconstruction, it is necessary to move the objective lens. So in SHOT-2000 a special servo mechanism is adopted. The actuator to control radial direction and a mirror are installed in the object lens unit, even when the objective lens moves, the laser beam always passes through the center of the objective lens. In addition, Fo servo control has a piezo servo mechanism for controlling the whole unit in the Fo direction.

The objective lens numerical aperture (NA) is 0.5.

#### 2.6.3.4 Structure of HVD

The structure of HVD is from the incident light side as shown in Figure 2.46. #1 Cover layer, #2 recording layer (hologram material) ,#3 gap layer, #4 dichroic miller, #5 aluminum refractive film, #6 substrate with pit and it has six layers.

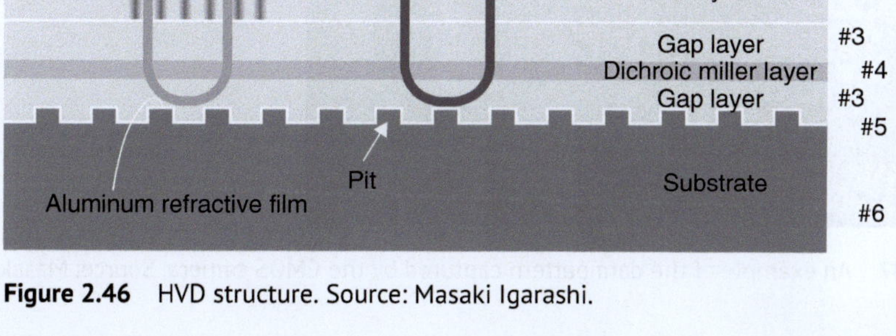


**Figure 2.46** HVD structure. Source: Masaki Igarashi.

#4 Dichroic miller is designed to reflect the green laser beam (wavelength 532 nm) and pass the red laser beam (wavelength 655 nm).

Therefore, it is possible to acquire HVD Fo, Tr servo information and address information from PIT information with a red laser beam and at the same time record and reconstruction hologram with a green laser beam.

#### 2.6.3.5 CMOS Camera

The hologram-reconstructed image is reflected by #5 aluminum refractive film, returns to the objective lens and relay lens, and is incident on PBS × 8 (polarized beam splitter). The reconstructed image beam is bent to the camera side with PBS and only the data pattern is input to the CMOS camera after the reference light pattern is cut at the aperture.

An example of the data pattern captured by the CMOS camera is shown in Figure 2.47.

#### 2.6.3.6 Red Laser for Servo

The output servo red laser is mixed with green laser with DBS.

After that, the red laser beam passes through the objective lens, is reflected by substrate of HVD, returns on the same path, and finally converges on the photodetector which was divided into four. The light condensed on each photodetector is converted into an electric signal, and Fo and Tr servo control is performed by calculating the electric signal. At the same time, current address information is generated from information obtained from PIT of HVD. Using the generated address information, position control of hologram recording and reconstruction is performed.

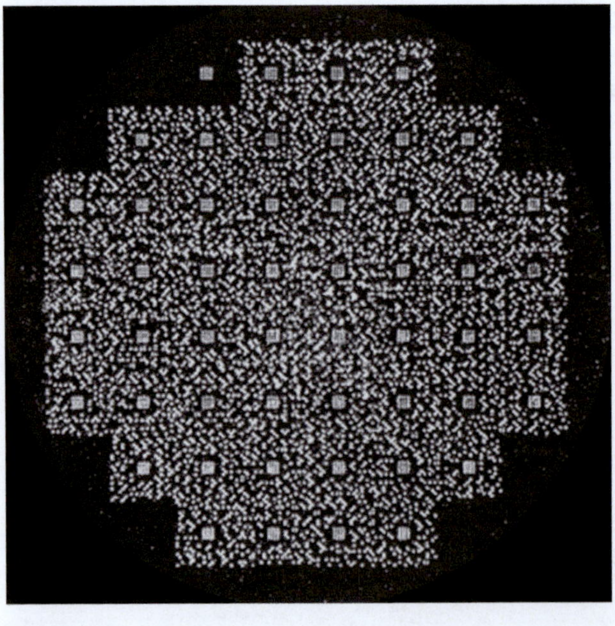

**Figure 2.47** An example of the data pattern captured by the CMOS camera. Source: Masaki Igarashi.

The astigmatism method is used for Fo servo control and the differential phase detection (DPD) method is used for Tr servo control. It is also possible to check whether the Fo and Tr servos are properly controlled by externally monitoring each electrical signal.

### 2.6.4 Control System

The DC spindle motor incorporated in the spin stand of SHOT-2000 is a DC motor whose rotation speed is very stable.

This motor is a DC motor which is used in ODU-1000 of Pulstec Industrial Co. Ltd. which is a global standard evaluation tester of CD, DVD, Blu-ray, and has a very proven motor.

The range of the speed of revolutions that can be evaluated is 90–300 rpm.

In addition, a very fine pulse encoder of 81 000 pulse per revolution is mounted on the motor shaft, and by dividing the disk by 328 000 division per revolution, it is possible to evaluate the disk which does not conform to the HVD format (not having the address PIT).

Disk radial movement control adopts a ball-screw mechanism of lead 1.00 mm per revolution, and position control with 0.01 mm accuracy is possible.

The spin stand also carries a tilt mechanism, and it can also cope with physical fluctuations such as warping and drooping on the outer circumference side of the disc.

### 2.6.5 Control Software (Recording, Reconstruction, Sequence Control, and so on)

#### 2.6.5.1 Parameter Setting Screen

When the application exclusive to SHOT-2000 is activated, after confirming the communication status with each unit and confirming the movement of each stage to the initial position, Figure 2.48 screen is displayed.

On this screen, all sets of parameters necessary for evaluation can be performed. In addition, it is possible to perform arbitrary automatic evaluation operation by setting the operation Sequence on the Sequence setting screen.

#### 2.6.5.2 Evaluation Result Display Screen

When you reconstruct the hologram, the evaluation result display screen pops up.

Figure 2.49 is an example of the evaluation result screen.

The image on the left side of the screen is a raw construction hologram image captured with a CMOS camera.

In the lower part of the image SNR, Error Count, bit error rate (BER) will be displayed. By switching tags, it is possible to display only Error part, SNR distribution by block and so on.

The graph on the right side of the screen is a histogram showing the distribution of the pixel ($\mu_{ON}$) determined as ON and the pixel ($\mu_{OFF}$) determined as OFF.

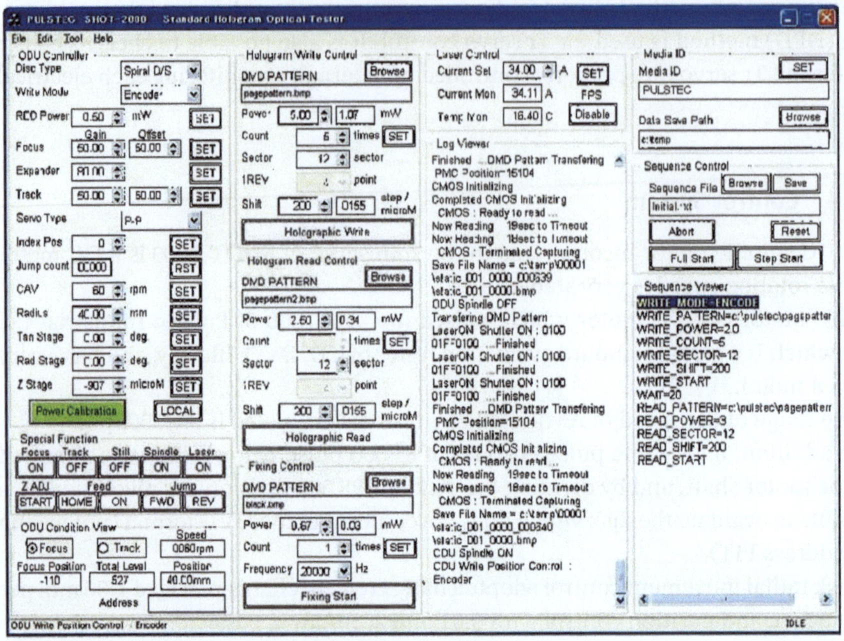


Figure 2.48 Parameter setting screen. Source: Masaki Igarashi.

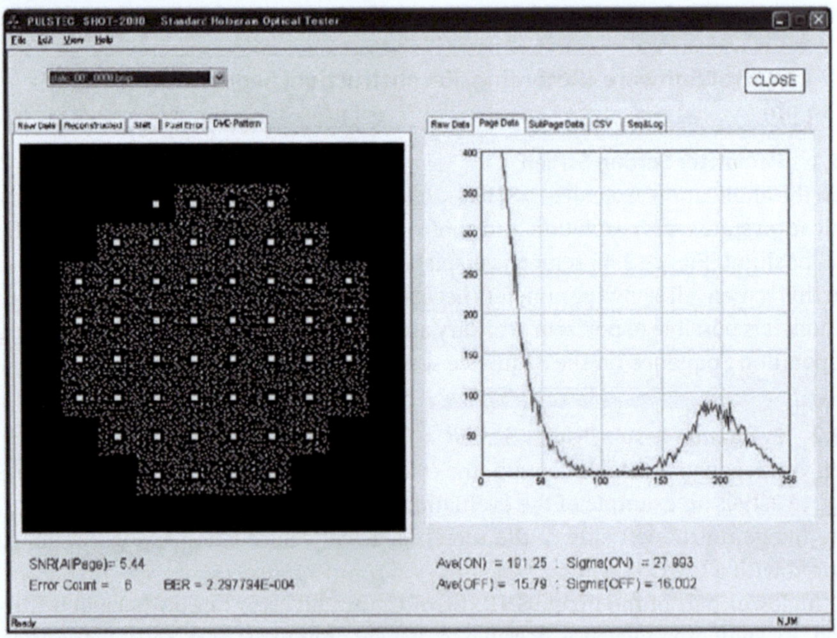


Figure 2.49 Example of the evaluation result screen. Source: Masaki Igarashi.

The scale on the X-axis is 0-255, and the larger number means that the pixel portion captured by the CMOS camera is brighter. When hologram recording/reconstruction is performed properly, the evaluation result is a graph where $\mu_{ON}$ (right side of histogram) and $\mu_{OFF}$ (left side of histogram) are cleanly separated as shown in the histogram of Figure 2.49. If the top of the two histograms is more separated and the variance of each histogram is smaller, the SNR value will be larger.

#### 2.6.5.3 Evaluation Items

There are many items that can be evaluated with SHOT-2000, but this section describes typical evaluation items.

***SNR Evaluation*** SNR shows the ratio of signal to noise, and the larger the SNR value, the more valuable it indicates that the hologram is properly reconstructed.

Below is the formula of SNR used in SHOT-2000.

$$\mathrm{SNR} = \frac{\mu_{\mathrm{on}} - \mu_{\mathrm{off}}}{\sqrt{\sigma^2\ \mathrm{on} + \sigma^2\ \mathrm{off}}} \tag{2.6}$$

$\mu_{\mathrm{on}}$: pixel determined as ON
$\mu_{\mathrm{off}}$: pixel determined as OFF

The average

$$\mu = \frac{1}{n}\sum xi \tag{2.7}$$

$\mu_{\mathrm{on}}$ and $\mu_{\mathrm{off}}$ in Eq. (2.8) use the average $\mu$.
$\sigma_{\mathrm{on}}$: distribution of pixels determined as ON
$\sigma_{\mathrm{off}}$: distribution of pixels determined as OFF

$$\sigma = \sqrt{\frac{1}{n}\sum (xi - \mu)^2} \tag{2.8}$$

***BER Evaluation*** BER shows the ratio of the number of pixels judged as error to the total number of valid pixels.

The smaller the BER value is, the more valuable it indicates that the hologram reproduced is properly reconstructed.

$$\mathrm{BER} = \frac{\text{Error Pixel Number}}{\text{All Effective Pixel Number}} \tag{2.9}$$

### 2.6.6 Current Research and Development Using SHOT-2000

Currently (in 2017), SHOT-2000 has partially modified its function and is used for research and development of optical correlator at the University of Electro-Communications Eriko Watanabe Laboratory.

Further, at the University of Tokyo Shimura laboratory to which I belong, R&D of time sequential signal collinear holographic memory is being carried out with SHOT-1000B coupon material evaluation tester having the same optical system as SHOT-2000.

We are convinced that collinear holographic technology is a very effective technology when looking into the future large capacity data age.

Because we believe that evaluation equipment technology cultivated in SHOT-2000 will be greatly required in the future, it is desirable to have an evaluation tester compatible with future newly developed technologies.

## Glossary

| | |
|---|---|
| SHOT | Standard Holographic Optical Tester |
| HVD | holographic versatile disc $\varphi$ 120 mm |
| ECMA | European Computer Manufacturers' Association |
| Fo (Focus direction) | vertical direction to HVD |
| Tr (Tracking direction) | radial direction to HVD |
| SLM | spatial light modulator |
| DMD | digital micromirror device |
| PBS | polarized beam splitter |
| DBS | dichroic beam splitter |
| DPD | differential phase detection |
| SNR | signal-to-noise ratio |
| BER | bit error rate |
| MEMS | micro electro mechanical systems |

## References


1 Bouwhuis, G., Braat, J., Huijser, A. et al. (1985). *Principles of Optical Disc Systems*. Bristol and Boston, MA: Adam Hilger Ltd.

2 Gabor, D. (1948). A new microscope principle. *Nature* 161: 777.

3 Leith, E.N. and Upatnieks, J. (1962). Reconstructed wavefronts and communication theory. *J. Opt. Soc. Am.* 52: 1123.

4 堀米秀嘉 (1984). 井上明、見山友裕、藤井壽崇、鈴木新一、草鹿履一郎、徳島忠夫、非晶質 TbFe 薄膜を用いた磁気ホログラフィによる高密度記録. 日本応用磁気学会誌 8 (2): 109–112.

5 鶴田匡夫:続 光の鉛筆-光技術者のための応用光学-(新技術コミュニケーションズ社, 1988).

6 Horimai, H., Seo, K., Toyota, K., and Saito, K. (1995). Brand-new method for magneto-optical signal detection with ultra-resolution. *Int. Soc. Opt. Eng.* 2514: 118–128.

7 Coufal, H.J., Psaltis, D., and Sincerbox, G.T. (ed.) (2000). *Holographic Data Storage: Optical Sciences*, vol. 76, Springer-Verlag.

8 Psaltis, D., Levene, M., Pu, A. et al. (1995). Holographic storage using shift multiplexing. *Opt. Lett.* 20 (7): 782–784.

9 Barbastathis, G., Levene, M., and Psaltis, D. (1996). Shift multiplexing with spherical reference waves. *Appl. Opt.* 35 (14): 2403–2417.

**10** Orlov, S.S. and Hesselink, L. (2003). Holographic shift multiplexing in thin volumetric media. *J. Opt. Soc. Am. B* 20 (9): 1912–1921.

**11** Orlov, S.S., Philips, W., Bjornson, E. et al. (2004). High-transfer-rate high-capacity holographic disk data-storage system. *Appl. Opt.* 43 (25): 4902–4914.

**12** Bernal, M.-P., Coufal, H., Grygier, R.K. et al. (1996). A precision tester for studies of holographic optical storage materials and recording physics. *Appl. Opt.* 35: 2360–2374.

**13** Tanaka, T., Takahashi, K., Sako, K. et al. (2007). Littrow-type external-cavity blue laser for holographic data storage. *Appl. Opt.* 46 (17): 3583–3592.

**14** Horimai, H., Tan, X., and Li, J. (2005). Collinear holography. *Appl. Opt.* 44 (13): 2575–2579.

**15** Takasaki, K., Kobayashi, S., Sugiki, M. et al. (2004). Line segment hologram writer using single mode blue laser and one-dimensional spatial light modulator. International Symposium on Optical Memory (ISOM'04), Proceedings, Korea.

**16** Ishioka, K., Tanaka, K., Kojima, N. et al. (2005). Optical collinear holographic recording system using a blue laser and a random phase mask. *International Symposium on Optical Memory/Optics Data Storage (ISOM/ODS'05), Proceedings, USA.*

# 3

# Theories of Collinear Holography

## 3.1 Analysis of a Collinear Holographic Storage System: Introduction of Pixel Spread Function

A simple model to understand the image formation in a collinear holographic storage system [1, 2] is presented in this section, and we will consider the image formation process of the signal pixels of a spatial light modulator (SLM) using various reference patterns (RPs). We will show that the image of a single pixel spreads on the detector array due to Bragg diffraction and the degree of spread changes with media thickness. We will also introduce an image evaluation metric called the pixel spread function (PxSF), which has a function similar to that of the point spread function (PSF) [3] used in a linear imaging system. We will explain that the PxSF is determined only by the optical system and is independent of the data image pattern. Thus, by examining the variation of PxSF as a function of the distance from the optical axis of the system, we can quantify the sharpness of the image. Notably, the purpose of this chapter is not performing the simulation to predict the experimental results quantitatively, but giving a physical insight into the image formation in the collinear holographic system.

A model of the optical system used to analyze the image reconstruction in the collinear holographic storage system is shown in Figure 3.1. In practice, an SLM and a recording medium are placed at the two focal planes of an objective lens. In the model, because a real holographic disk has a mirror on the back surface of the recording medium, the thickness of the recording medium was doubled, and a mirror image of the system was placed after the medium. At the image position of the SLM, a two-dimensional detector array was placed. We assumed that each pixel of the SLM was perfectly imaged at a corresponding pixel of the detector array. A page of data was displayed on an inner area of the SLM, and the pixels in the surrounding area provided a reference wave [1, 2].

We treated each pixel of the SLM as a small square aperture, which was opened or closed according to the ON or OFF state of the pixel. The pixel arrangement was assumed to be a regular rectangular array, and the pixel pitches were the same for both $x$ and $y$ directions. The SLM was illuminated by a coherent plane

*Collinear Holography: Devices, Materials, Data Storage*, First Edition.
Xiaodi Tan, Hideyoshi Horimai, Tsutomu Shimura, and Xiao Lin.

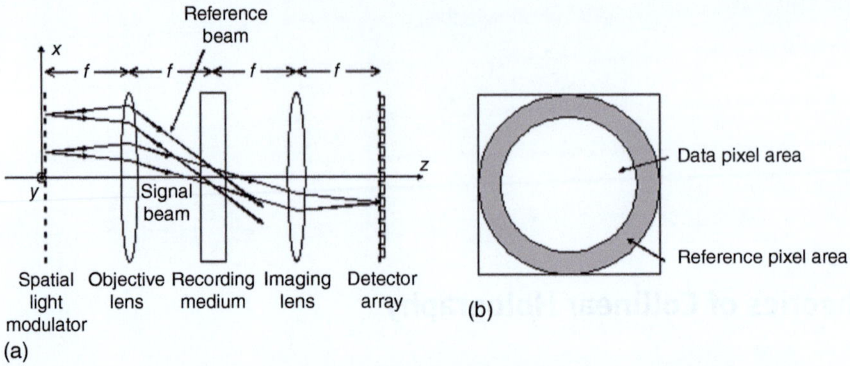


**Figure 3.1** Model for image-formation analysis in the collinear holographic storage system. (a) The optical system, and (b) the arrangement of pixels in the spatial light modulator. Source: Reproduced with permission from Shimura et al. [4], Optical Society of America.

wave. Light transmitted through a pixel propagated to the objective lens. As typical dimensions for the system, when we assumed that, for example, the aperture size of the pixel was 15 μm × 15 μm, the focal length of the lens $f$ was 5 mm, and the wavelength was 0.5 μm. With these dimensions, the Fresnel number was 0.09. This means that the diffracted wave just before the lens was in the Fraunhofer regime, and had a spherical wavefront and an amplitude represented by a two-dimensional sinc function. After the lens, the wavefront propagating from each pixel was a plane wave.

Here, we represent the position of a pixel $(i, j)$ on the SLM as $(x_i, y_j)$, and then the wave vector $\mathbf{k}_{ij}$ of the corresponding plane wave is determined uniquely. When the objective lens maintains the sine condition, the wave vector of the plane wave from the pixel $(i, j)$ in the recording medium is expressed as

$$\mathbf{k}_{ij} = \left( -k_0 \frac{x_i}{f}, -k_0 \frac{y_j}{f}, \sqrt{(nk_0)^2 - \left(k_0 \frac{x_i}{f}\right)^2 - \left(k_0 \frac{y_j}{f}\right)^2} \right) \tag{3.1}$$

where $k_0$ is the angular wave number of the light in vacuum, and $n$ is the refractive index of the recording medium. Thus, the overall interference of all waves from the SLM can be regarded as a superposition of the simple interference of two plane waves, so long as the response of the recording medium is linear. Gratings were written by the interference of all combinations of pairs of plane waves from all of the "ON" pixels in the SLM.

In the read-out process, plane waves from the reference pixels were diffracted by all existing gratings in the recorded medium. Here, let us consider diffraction of a plane wave from a pixel $(k', l')$ by a grating $\mathbf{K}_{klij} = \mathbf{k}_{ij} - \mathbf{k}_{kl}$, which was written by the interference between the waves from a data pixel $(i, j)$ and a reference pixel $(k, l)$. The diffracted field can be calculated based on coupled wave theory [5]. Under the infinite plane wave approximation, Bragg mismatches along the $x$ and $y$ directions are not allowed, i.e. $\Delta k_x = \Delta k_y = 0$, and the wave vector of the diffracted light is the same as $\mathbf{k}_{(k'+i-k),(l'+j-l)}$. We assumed that the diffracted wave was detected only by the pixel $(k' + i - k, l' + j - l)$ of the detector array. Thus, the Bragg mismatch was obtained as:

$$\Delta k_z = (k_{k'l'z} + K_{klij_z}) - \sqrt{(nk_0)^2 - k_{(k'+i-k),(l'+j-l)_x}{}^2 - k_{(k'+i-k),(l'+j-l)_y}{}^2} \tag{3.2}$$

To obtain the light intensity detected by the pixels of the detector array for given data-pixel and reference-pixel patterns, we summed all beams coming from the "ON" reference pixels and diffracted by all existing gratings. Notably, in our calculations, we only considered the gratings formed between a data pixel and a reference pixel to avoid escalating the required calculation time. We ignored the gratings formed between data pixels and those between reference pixels because the contribution of these gratings was considered to be relatively small.

Some results of the calculation are shown in Figure 3.2, together with experimental results. All parameters were the same for both calculation and experiment. The calculation results show good agreement with the experimental results and well represent the differences in the reconstructed image depending on the RPs.

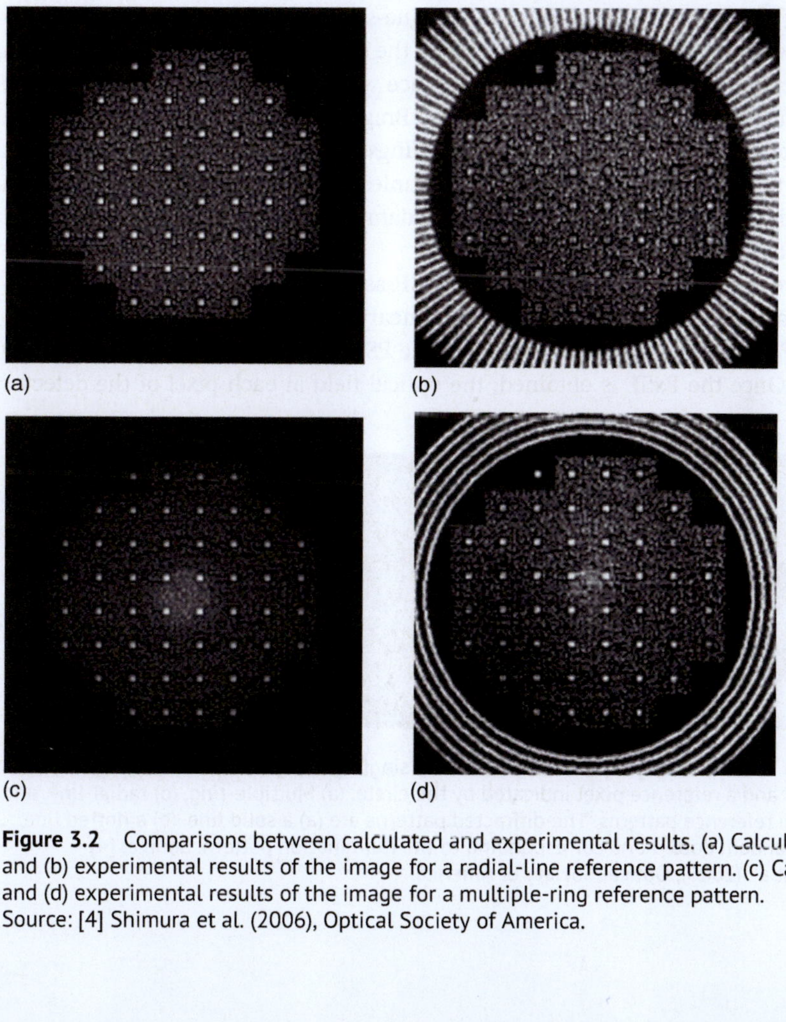


**Figure 3.2** Comparisons between calculated and experimental results. (a) Calculated and (b) experimental results of the image for a radial-line reference pattern. (c) Calculated and (d) experimental results of the image for a multiple-ring reference pattern.
Source: [4] Shimura et al. (2006), Optical Society of America.

Uniformity of the image is better for a radial-line RP (Figure 3.2a,b). The haziness observed at the center of an image with a multiple-ring RP is well reproduced in the calculated results (Figure 3.2c,d).

Here, to clarify the image formation process, let us consider the diffraction by a single grating $\mathbf{K}_{ijkl}$ illuminated by the reference pixels. In an ideal case, the diffracted light should reach only the data point $(k, l)$. However, for certain data points that maintain the condition

$$|\mathbf{k}_{k',l'} \cdot \mathbf{K}_{ijkl}| \approx \frac{|\mathbf{K}_{ijkl}|}{2} \tag{3.3}$$

$\Delta k_z$ becomes nearly zero even for $(k', l') \neq (k, l)$. This is called Bragg degeneracy[6]. Because of this effect, the diffracted light pattern is line-shaped, as shown in Figure 3.3, in which RPs are (a) multiple rings, (b) radial lines, and (c) random. The line-shaped diffracted patterns are perpendicular to the projection of $\mathbf{K}_{ijkl}$ on the $xy$ plane. The width of the lines is determined by the Bragg selectivity. The lines appearing in Figure 3.3b,c are broken according to the shape of the RPs.

An actual reconstructed image of one data pixel $(i_0, j_0)$ can be obtained by summing the complex amplitude of the line-shaped diffracted pattern over all gratings written by the interference between the data pixel and the reference pixels. The directions of the lines vary in accordance with the directions of the gratings, and the data pixel $(i_0, j_0)$ acts as a pivot. The Bragg mismatch of the wave diffracted to the pixel $(i_0, j_0)$ is always zero for all gratings $\mathbf{K}_{kli_0j_0}$, though the diffracted light intensity to other pixels is negligibly small unless the Bragg degeneracy condition is maintained. Thus, the image of the single data pixel has a peak at the target pixel but spreads slightly to neighboring pixels.

Now, we define a pixel spread function, PxSF, as the complex amplitude of the optical field imaged at the detector array, with a linear response of the recording medium. The PxSF has a function similar to that of the PSF normally used in linear imaging systems. Once the PxSF is obtained, the optical field at each pixel of the detector

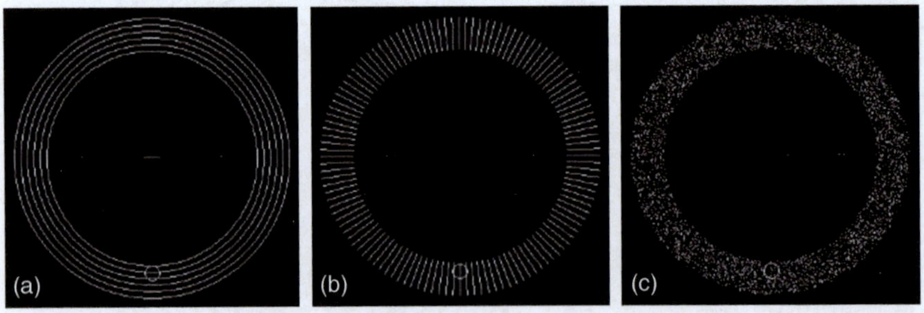


**Figure 3.3** Diffracted light pattern produced by single grating written by a data pixel at the center and a reference pixel indicated by the circle. (a) Multiple-ring, (b) radial-line, and (c) random reference patterns. The diffracted patterns are (a) a solid line, (b) a dotted line, and (c) a random dotted line corresponding to the illumination pattern. Source: [4] Shimura et al. (2006), Optical Society of America.

array can be calculated by the convolution of the object pattern of the signal pixels and the PxSF, as long as the response of the recording medium is linear. Notably, the PxSF of the holographic data-storage system is a complex function that represents the amplitude of the image of a point object due to coherent illumination. The PxSF is not necessarily uniform in the image area, like the PSF, and it describes the variation of the imaging characteristics in the field of view. In general, to calculate the image formation of the whole data page, we have to take into account the diffraction from the gratings formed by pairs of signal pixels. This effect appears as noise on the detector array, but as long as the density of the "ON" data pixels is constant and their distribution is quasi random, it is expected that the noise pattern will have a small dependence of the data pixel pattern. We also have to take account of the existence of the gratings formed by pairs of reference pixels. This effect appears as a fixed pattern noise. Thus, the PxSF is quite effective to estimate the image formation properties, even when we ignore these gratings to substantially reduce the computation time. Notably, the PxSF provides an estimation of the intrapage crosstalk, although the interpage crosstalk could also be calculated with the same principle of the simulation. Studies focused in this direction will be conducted by our group in the near future.

Some examples of the calculated PxSFs are shown in Figure 3.4. Although the PxSF is a complex value, only the intensity is shown in Figure 3.4. The illumination patterns were multiple rings, radial lines, and random patterns. Figure 3.4a shows the results when no phase modulation was applied by the SLM. The spread of the pixel image is the largest for the multiple-ring RP (solid circles), and the intensity at the neighboring pixel is the smallest for the radial-line RP (crosses). For the random RP (open rectangles), the shape of the plot is similar to that for the multiple-ring RP, but the value is more than 10 times smaller.

We now consider the reason for these differences. For the multiple-ring RP, the diffracted light pattern by a single grating extends like a line (Figure 3.3a), and the PxSF is calculated by the summation of these lines whose direction varies according to the gratings. On the other hand, for the radial-line RP (Figure 3.3b), the diffracted

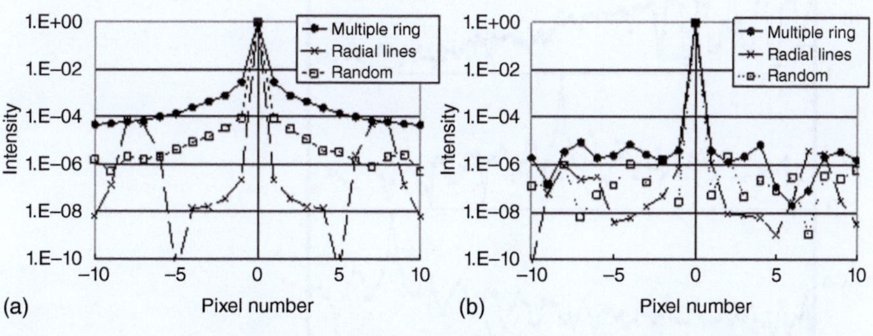


**Figure 3.4** Calculated intensity of pixel spread function (a) without and (b) with random phase modulation. •: multiple-ring, ×: radial-line, and □: random reference patterns. Source: Reproduced with permission from Shimura et al. [4], Optical Society of America.

pattern is a line of dots. In this case, the light intensity at the neighboring pixel to the target pixel is always very close to zero when summing all gratings, and the PxSF becomes quite sharp. The two peaks at both sides seen in Figure 3.4a correspond to the next line of the RP in Figure 3.3b. For the random RP, the neighboring pixel is either bright or dark, depending on the ratio of the numbers of ON and OFF pixels, and the intensity is between those for the multiple-ring and radial-line RPs.

Figure 3.4b shows the results when a random phase is added to each reference pixel. In these cases, all the waves arriving at the target pixel have the same phase and are thus added constructively, because the gratings are read out by the reference pixels used for writing the gratings with the data pixel. The waves diffracted by the gratings and read out by incorrect reference waves have random phase and diminish to zero by averaging during the summation process. That is why the PxSF curves shown in Figure 3.4b all have quite sharp peaks. The improvement for the multiple-ring RP is particularly significant.

The difference of the PxSF at the center, middle, and outer position of the data area is shown in Figure 3.5. The intensity is normalized to the peak value of the PxSF at the center. The peak value of PxSF at middle and outer position is slightly smaller than one. Spread of PxSF becomes larger with the increase of the distance from the center.

The model allows the writing and reading processes in the system to be elucidated. Even a simplified model with some approximations gave results in good agreement with experiment. The model revealed the reason why the radial-line and random RPs gave good experimental results and also predicted that an additional random phase to the reference pixels improves the imaging performance. Though we can discuss the qualitative properties of the images in the collinear system with our model at present stage, we can discuss the relative properties of the system when some of the system parameters are changed.

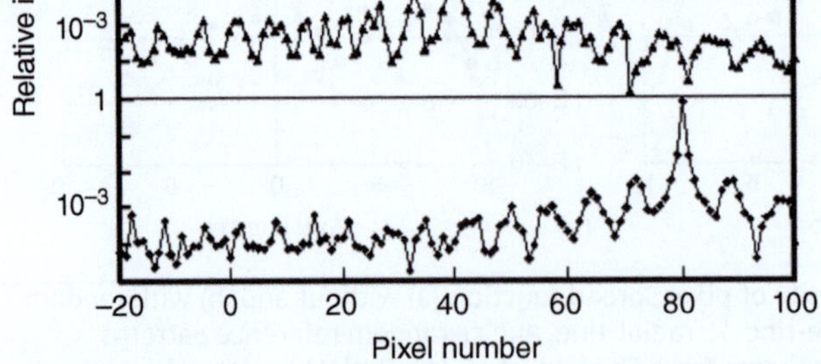


**Figure 3.5** Comparison of PxSF at center, middle, and outer position of the data area. Source: Tsutomu Shimura.

## 3.2 Theoretical Treatment of the Collinear System and Formulate Numerical Evaluation Technique Based on the Scalar Diffraction Theory

In holographic data storage (HDS) [7, 8], the collinear system [2, 9, 10] is one of multiplexing methods using optical autocorrelation and multiplex holograms by superimposing with the shift of the recording medium. In this method, sharper shift selectivity appears due to a sharp autocorrelation function generated by complexly modulated reference light. This section describes theoretical treatment of the collinear system and formulate numerical evaluation technique based on the scalar diffraction theory. Some numerical analyses are also performed.

### 3.2.1 Principle of Optical Autocorrelation Method

Shift selectivity caused by optical autocorrelation can be described as follows. Assuming that the signal, reference and diffracted light are $U_s$, $U_r$ and $U_d$, $U_d$ is given by [8]

$$U_d(\mathbf{r};\delta\mathbf{r}_j) \propto \sum_i \int U_s(\mathbf{r}-\mathbf{r}';\delta\mathbf{r}_j-\delta\mathbf{r}_i)g(\mathbf{r}';\delta\mathbf{r}_i)\mathrm{d}r' \tag{3.4}$$

where $i$ and $j$ are the indexes corresponding to each hologram, $\delta r_i$ and $\delta r_i$ are shift amounts at recording and readout process. The function $g = g_B \cdot g_C$ is defined as

$$g_B(x) = L\mathrm{sinc}\left(\frac{\pi x L}{\lambda f}\right) \tag{3.5}$$

$$g_C(\mathbf{r};\delta\mathbf{r}) = \int U_r(\mathbf{r}'')U_r^*(\mathbf{r}''+\delta\mathbf{r})\ \exp\left(\frac{2\pi i}{\lambda f}\mathrm{r}\cdot\mathrm{r}''\right)\mathrm{d}r'' \tag{3.6}$$

where $\lambda$ is the wavelength, $L$ is the medium thickness, and $f$ is the focal length of the lens. Functions $g_B$ and $g_C$ represent Bragg selectivity and optical autocorrelation, and $x$ is the Bragg direction component of $\mathbf{r}$. The diffracted light $U_d$ is expressed as the convolution of $U_s$ and g. Optical autocorrelation function $g_C$ is determined by the spatial distribution of $U_r$, and it becomes a constant in the case of the plane wave. By choosing a spatially random distribution as $U_r$, $g_C$ becomes a delta function like shape, and very sharp shift selectivity can be obtained. In the collinear system, shift selectivity is mainly determined by the autocorrelation of the Fourier transformed RP because reference light is focused by the Fourier transfer lens.

Figures 3.6a–c shows a sample RP, power spectrum of the reference, and autocorrelation of the power spectrum, respectively. The mesh size $\Delta$ is set as 5 μm, and each pixel of the RP is sampled with a $2\times2$ mesh. Other conditions are shown in Table 3.1. Figure 3.6c shows first null point appears on $x = 0.53$ μm. Thus, by calculation of the autocorrelation of the spectrum, it is possible to easily estimate the shift selectivity. Here, the mesh size on the focal plane (Fourier plane) $\delta$ is expressed as $\delta = f\lambda_e/N\Delta$ with the effective wavelength $\lambda_e = \lambda/n$, refractive index of the medium $n$, and sample size $N$. As can be seen from this relation, to sample sufficiently dense on the Fourier plane, it is necessary to make computational region $N\Delta$ sufficiently large.

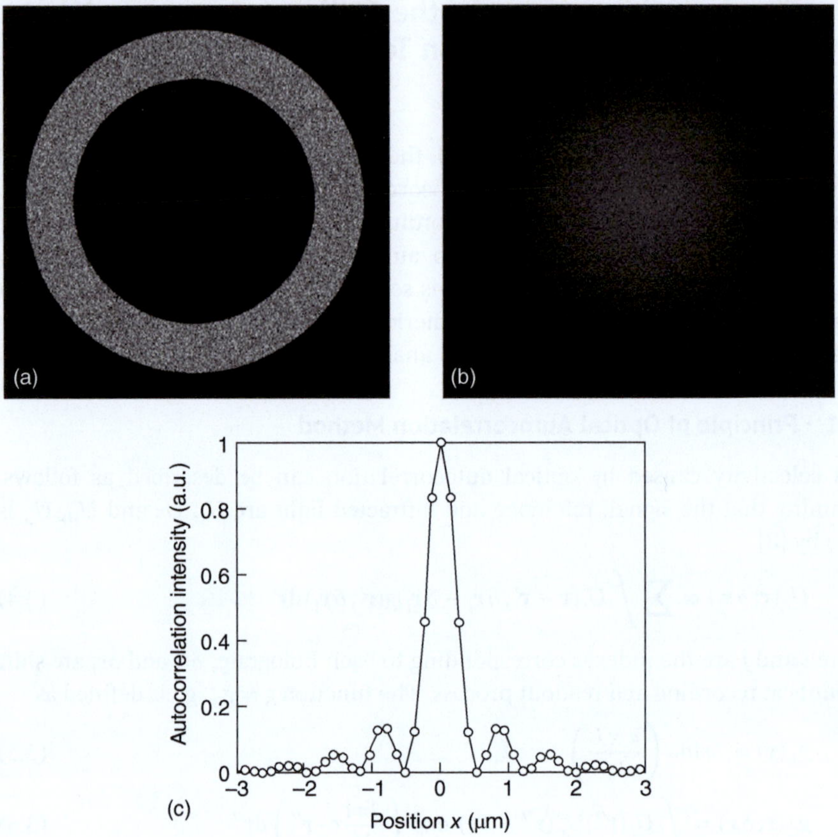


**Figure 3.6** (a) Sample of reference light with random binary pattern. (b) Magnified power spectrum of sample reference light. (c) Autocorrelation intensity of power spectrum. Source: Shuhei Yoshida.

**Table 3.1** Simulation conditions.

| Condition | Symbol | Value |
|---|---|---|
| Wavelength (nm) | $\lambda$ | 405 |
| Refractive index | $n$ | 1.5 |
| Focal length (mm) | $f$ | 20 |
| SLM pixel size (um) | $p$ | 10 |
| Medium thickness (mm) | $L$ | 1.0 |
| Nyquist aperture size | $D$ | $2f\lambda/p$ |
| Number of layers | $2M$ | 60 |
| Sample size | $N_x \times N_y$ | $8192 \times 8192$ |

### 3.2.2 Simulation Scheme for Collinear System

In this section, diffraction from the volume hologram is formulated as a volume integral equation (VIE) based on the scalar diffraction theory. The Born approximation and the angular spectrum method are applied to the VIE, whereby a numerical analysis method applicable to the collinear system is introduced.

#### 3.2.2.1 Formulization of VIE

As is well known, the scalar diffraction theory follows the Helmholtz equation [11].

$$\left[\nabla^2 + k_0^2 \varepsilon_r(\mathbf{r})\right] U(\mathbf{r}) = 0 \tag{3.7}$$

where $\varepsilon_r$ is the relative permittivity, and $k_0^2$ is the wavenumber in the free space. Assuming $\varepsilon_{\text{ave}}$ is the average relative permittivity, permittivity modulation $\Delta\varepsilon$ is defined as $\Delta\varepsilon = \varepsilon_r - \varepsilon_{\text{ave}}$. By dividing the scalar field $U$ into incident light $U_i$ and diffracted light $U_d$ to make $U = U_i + U_d$, we obtain following relation:

$$(\nabla^2 + k^2) U_d(\mathbf{r}) = -k_0^2 \Delta\varepsilon(\mathbf{r}) U(\mathbf{r}) \tag{3.8}$$

where $U_i$ satisfies $(\nabla^2 + k^2)U_i = 0$ by assuming that $U_i$ is a monochromatic wave, and $k^2$ is $k^2 = \varepsilon_{\text{ave}} k_0^2$. Given the propagation kernel of Helmholtz operator as $G$, applying Green's theorem and first-order Born approximation ($U = U_i$) under appropriate boundary conditions, we obtain the following VIE [11].

$$U_d(\mathbf{r}) = \frac{k_0^2}{4\pi} \iint_v \Delta\varepsilon(\mathbf{r}') U_i(\mathbf{r}') G(\mathbf{r} - \mathbf{r}') d^3 r' \tag{3.9}$$

where V is the hologram size. Assuming that the sample size of the hologram on a side is $N$, the computational complexity of Eq. (3.9) is expressed as $o\,(N^3 \log N)$ even when the fast Fourier transform (FFT) is used for three-dimensional convolution. However, since it is sufficient to obtain a two-dimensional wavefront at $z = 0$, we can reduce the computational complexity to $o\,(N^2 \log N)$ by reformulating Eq. (3.9) as the two-dimensional convolution and one-dimensional integral [12–14].

$$\begin{aligned} U_d(x, y; 0) &\propto \int_{-\frac{L}{2}}^{\frac{L}{2}} [\Delta\varepsilon(x', y'; z') U_i(x', y'; z')] * G(x', y'; -z') dz' \\ &\propto \sum_{k=-M}^{M-1} F^{-1}[F[\Delta\varepsilon(x', y'; k\Delta z) U_i(x', y'; k\Delta z)] H(u, v; -k\Delta z)] \end{aligned} \tag{3.10}$$

where $*, F, F^{-1}$, and $H$ represent the two-dimensional convolution operator, Fourier transform, inverse Fourier transform and transfer function, respectively. $\Delta z$ is $\Delta z = L/2M$, and $2M$ is the layer number of the medium. Transfer function $H$ is [15].

$$H(u, v; z) = \exp[-ikz\sqrt{1 - (\lambda u)^2 - (\lambda v)^2}] \tag{3.11}$$

where $u$ and $v$ are spatial frequencies along $x$- and $y$-axes, respectively. The relation $H = F[G]$ holds between propagation kernel $G$ and transfer function $H$.

#### 3.2.2.2 Recording Process

Recording process of the hologram corresponds to obtaining the permittivity modulation $\Delta\varepsilon$. Assuming that the hologram is recorded in an ideal state and giving $\Delta\varepsilon$ as $\Delta\varepsilon = U_s U_r^*$, the three-dimensional distribution of $\Delta\varepsilon$ is given by

$$\Delta\varepsilon(x, y; z) = U_s(x, y; z)U_r^*(x, y; z) \quad U_{s,r}(x, y; z) = F^{-1}[F[U_{s,r}(x, y; 0)]H(u, v; z)] \tag{3.12}$$

where $U_s(x, y; 0) = \text{shift}(F[S], \delta\mathbf{r})$, $U_r(x, y; 0) = \text{shift}(F[R], \delta\mathbf{r})$, $S$ and $R$ are the signal and RP displayed on the SLM. Further, shift and $\delta\mathbf{r}$ represent the shift operator and shift amount. When holograms are multiplex recorded, $\Delta\varepsilon$ is added each while changing $\delta\mathbf{r}$.

### 3.2.3 Numerical Evaluation

The collinear system is evaluated by the numerical analysis in this section. In the simulation, signal and reference light are coaxially arranged on the SLM and phase-modulated by 0 and $\pi$ binary phase mask after being relayed. DC component is suppressed by the phase mask. Further, as the phase is modulated randomly, the signal and reference light interfere with each other in a wider area. The Nyquist aperture cuts unnecessary harmonic components of the signal and reference light, and they are focused by the objective lens and interfere with each other on the medium. In the readout process, only the reference light is displayed on the SLM, and the diffracted light $U_d$ is obtained by above-described method with $U_i = U_r$. Simulation conditions are shown in Table 3.1.

Figure 3.7 shows the evaluation results of the diffraction efficiency. Figure 3.7a,b shows the shift selectivity along $x$-axis (in-plane direction) and $z$-axis (defocus), and (c) shows the wavelength selectivity. The horizontal axis of Figure 3.7a,b shows the shift amount, and Figure 3.7c shows the wavelength gap between recording and readout process. The vertical axis expresses normalized diffraction efficiency. From these results, the shift selectivity along $x$-axis is consistent with the autocorrelation shown in Figure 3.6c and does not depend on the thickness of the medium. On the other hand, it can be seen that the wavelength selectivity becomes sharply as the medium becomes thicker.

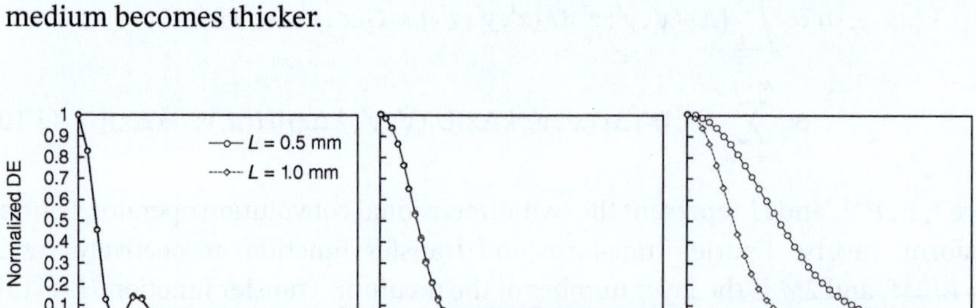


**Figure 3.7** Diffraction efficiency variation for (a) $x$-axis shift, (b) $z$-axis shift, (c) wavelength change. Source: Shuhei Yoshida.

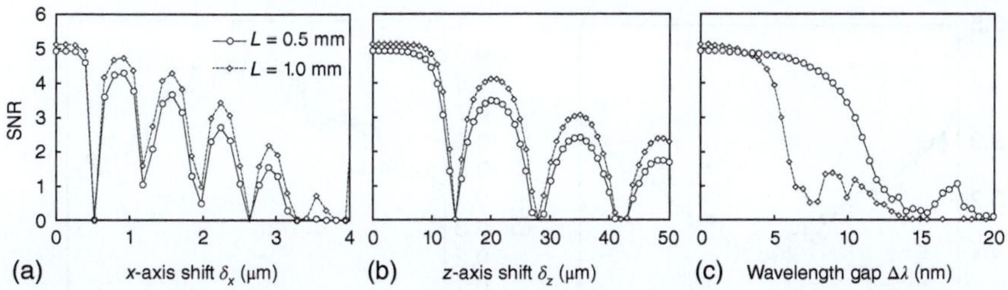


**Figure 3.8** SNR variation for (a) $x$-axis shift, (b) $z$-axis shift, (c) wavelength change. Source: Shuhei Yoshida.

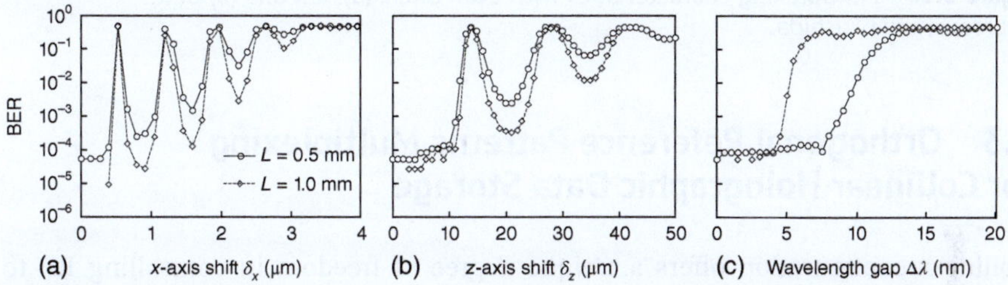


**Figure 3.9** BER variation for (a) $x$-axis shift, (b) $z$-axis shift, (c) wavelength change. Source: Shuhei Yoshida.

Next, signal quality for $x$- and $z$-axes shift and wavelength change are evaluated. These results are shown in Figures 3.8 and 3.9, respectively. The signal quality was evaluated by signal-to-noise ratio (SNR) and bit error rate (BER). For the signal evaluation, page data modulated with a constant weight code of 3/16 white rate and coding efficiency 9/16 was used. The definition of the SNR is

$$\mathrm{SNR} = \frac{\mu_1 - \mu_0}{\sqrt{\sigma_1^2 + \sigma_0^2}} \tag{3.13}$$

where $\mu$ and $\sigma^2$ represent the average and variance of the luminance values of the reconstructed image, and the suffixes 1 and 0 represent "on" and "off." Evaluation results show the SNR and BER periodically fluctuate with respect to the $x$- and $z$-axes shifts. In addition, it can be confirmed that the signal quality is better when a thick medium is used. The period of the fluctuation corresponds to null points of the diffraction efficiency. For $L = 1.0$ mm, it can be seen that error occurs when $\delta_x$, $\delta_z$, and $\Delta\lambda$ exceeds 0.4 μm, 3 μm, and 1.5 nm.

Finally, multiplexing characteristics for $L = 1.0$ mm are evaluated and shown in Figure 3.10. Numbers on the horizontal axis correspond to each multiplexed hologram, and the vertical axis represents SNR and BER. In this simulation, 50 holograms were multiplexed with the shift pitch corresponding to the third null of the diffraction efficiency. From these results, it can be seen that the signal quality is the worst at the center where the largest number of holograms is superimposed. Further, the signal quality is deteriorated almost linearly with respect to the number of overlapping holograms.

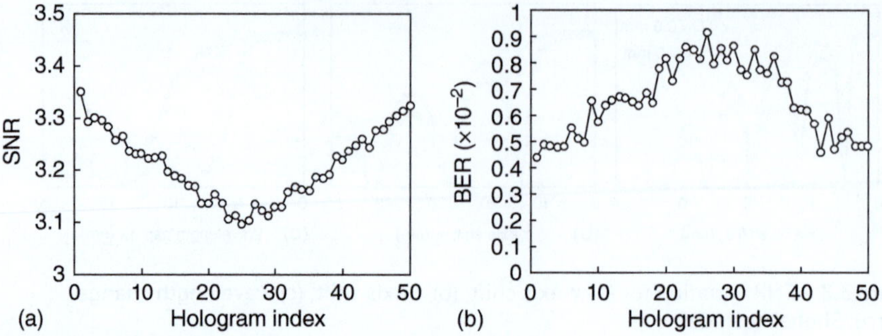


**Figure 3.10** Multiplexing characteristics with 2 μm pitch. (a) SER and (b) BER. Source: Shuhei Yoshida.

## 3.3 Orthogonal Reference Patterns Multiplexing for Collinear Holographic Data Storage

Collinear configuration offers a unique degree of freedom in controlling RP to improve the performance of the system because the RP is also implemented by SLM. By applying orthogonal RPs, holograms can be multiplexed in the same volume of the recording medium to increase data density [16–18]. However, the interpage crosstalk due to Bragg degeneracy and Bragg mismatch readout for this orthogonal reference pattern multiplexing (ORPM) method cannot be eliminated because the RP is along a two-dimensional plane, and the angle between the reference beam and the signal beam is relatively small. The orthogonal condition, the optical characteristics, and crosstalk performance for ORPM method are presented in this chapter. The hybridization of ORMP and shifting multiplexing methods to suppress the crosstalk are also described.

### 3.3.1 Principle of the ORPM Method

The schematic of the ORPM for the collinear holographic data storage system (CHDSS) is shown in Figure 3.11. Fourier geometry is used for the recording of hologram. $N$ different RPs ($RP_1$, $RP_2$, …, $RP_N$) are used for the multiplexing of $N$ data pages in the same area of the holographic medium. $Lens_1$ in the Fourier geometry converts a pixel on the SLM into a plane wave propagating with a unique angle corresponding to that pixel's coordinate [4]. For amplitude SLM, different RPs correspond to a group of plane waves propagating along different directions. Figure 3.12 illustrates the difference between different RPs for amplitude SLM.

For the recording process, as shown in Figure 3.11a, the SLM produces the signal beam from the inner area and the reference beam in the outer area. The nth data page is stored by the reference beam modulated by the $n$th RP. And $N$ data pages can be recorded in the same volume of the medium by this ORPM method. A recorded hologram can be treated as the accumulation of the gratings written by

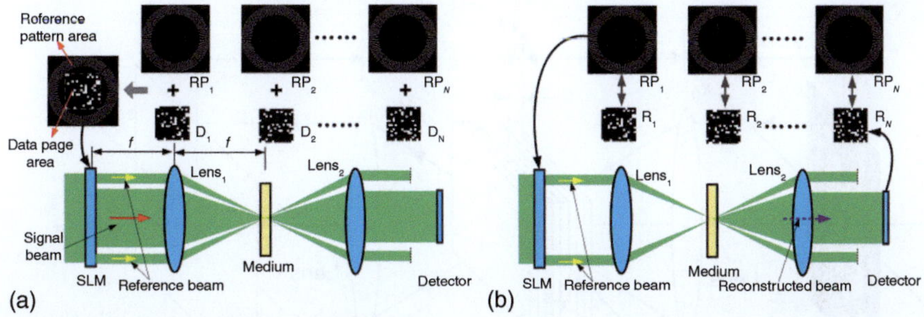


**Figure 3.11** Schematic of the RPM for the transmission type of the CHDS configuration. (a) Recording process; (b) reconstruction process. SLM, spatial light modulator; $RP_n$, the $n$th reference pattern; $D_n$, the $n$th date page; $R_n$, the $n$th reconstructed data pattern; $f$, the focal length of $Lens_1$. Source: Reproduced with permission from Cao et al. [18], Optical Society of America.

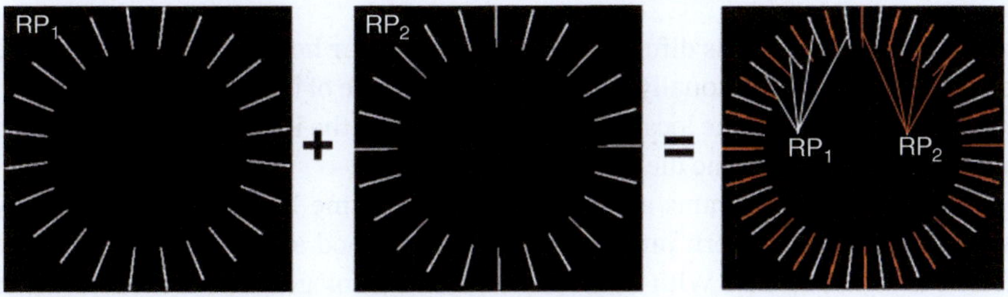


**Figure 3.12** Illustration of the difference between $RP_1$ and $RP_2$. Source: Jianhua Li.

every pair of the plane waves which are from the RP and data page, respectively. During the reconstruction process, as shown in Figure 3.11b, only the RP is uploaded on the SLM. Due to the Bragg angle selectivity for the gratings, each data page can be respectively accessed by applying the corresponding modulation RP.

The access speed for this multiplexing method can be much higher than that of shift multiplexing because data pages can be accessed by changing the RP on the SLM without shifting the recording medium. Data density can also be significantly increased by combining this method with the shift multiplexing while keeping the same shift pitch, given that the medium has enough dynamic range.

#### 3.3.1.1 Theoretical Formulation for the RPM Method

Figure 3.13 shows the optical model of the transmission CHDSS system described in Figure 3.11. The SLM is placed at the plane $(x_0, y_0)$ and the center of the holographic medium is placed at the Fourier transform plane $(x_f, y_f)$. The detector is placed at the plane $(x_d, y_d)$. The holographic medium is divided into many continuous layers along the $z$ direction, and each layer represents a planar hologram. The refractive index of each thin layer can be seen as constant along the $z$ direction when the thickness of each layer is small enough. Based on Born's approximation [19], the diffracted field from the volume hologram can be treated as the coherent

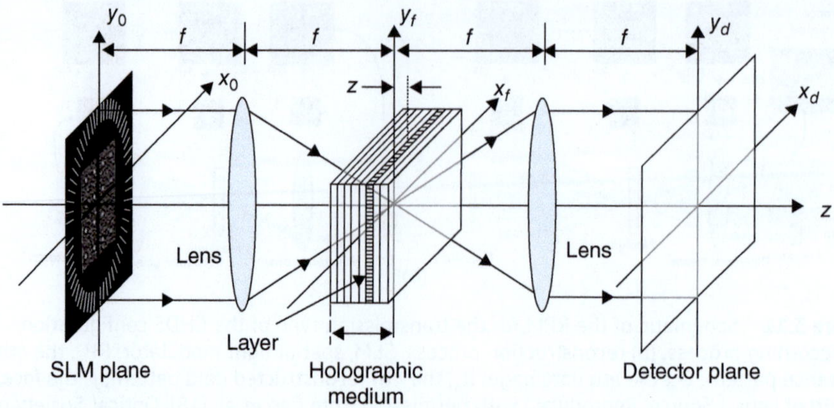


**Figure 3.13** Optical model for the transmission type of CHDSS system. Source: Reproduced with permission from Cao et al. [18], Optical Society of America.

superposition of the fields diffracted from all the planar holograms comprising the volume hologram. Additionally, the complex amplitude of the diffraction from each layer is proportional to the local complex amplitude of the incident optical field and the local modulation of the dielectric constant.

Assume that $N$ holograms are stored inside a volume holographic medium by using the reference pattern multiplexing (RPM) method and these holograms are labeled as $n = 1, 2, \ldots N$. With the presence of these holograms, the modulation of the dielectric constant of the medium for the planar hologram at the distance of $z$ from the Fourier-transform plane of the SLM plane can be written as

$$\Delta\varepsilon(x_f, y_f, z) \propto \sum_{n=1}^{N} R_n{}^*(x_f, y_f, z) S_n(x_f, y_f, z) + \text{c.c.} \tag{3.14}$$

where $R_n$ and $S_n$ are the $n$th complex amplitudes of the reference beams and signal beams, respectively, and c.c. represents the complex conjugate of $R_n^* S_n$. Such a refractive index modulation can be created in a photorefractive polymer or crystal with use of a multiple-exposure schedule [20, 21]. Using the standard Fourier-optics analysis, $R_n$ and $S_n$ can be expressed as

$$R_n(x_f, y_f, z) = \frac{\exp(2jkf)}{j\lambda f} \exp(jkz) \Im\left\{ \mathrm{RP}_n(x_0, y_0) \exp\left(-j\frac{\pi z}{\lambda f^2}\left(x_0^2 + y_0^2\right)\right)\right\}_{f_x=\frac{x_f}{\lambda f}, f_y=\frac{y_f}{\lambda f}} \tag{3.15}$$

$$S_n(x_f, y_f, z) = \frac{\exp(2jkf)}{j\lambda f} \exp(jkz) \Im\left\{ DP_n(x_0, y_0) \exp\left(-j\frac{\pi z}{\lambda f^2}\left(x_0^2 + y_0^2\right)\right)\right\}_{f_x=\frac{x_f}{\lambda f}, f_y=\frac{y_f}{\lambda f}} \tag{3.16}$$

where $f$ is the focal length of the lens, $k = 2\pi/\lambda$ and $\lambda$ is the wavelength, $\Im$ denotes the Fourier transform symbol, $\mathrm{RP}_n$ stands for the $n$th RP uploaded on the SLM for the recording of the $n$th data page $\mathrm{DP}_n$.

During the reconstruction process, when the $m$th RP are used for the reading of this planar hologram, the diffraction signal can be written as

$$R_m(x_f, y_f, z)\Delta\varepsilon(x_f, y_f, z) \propto R_m(x_f, y_f, z)\sum_{n=1}^{N} R_n{}^*(x_f, y_f, z)S_n(x_f, y_f, z) \tag{3.17}$$

where we only maintain the first term of Eq.(3.14) and ignore the complex conjugate term because this term does not significantly contribute to the diffracted field. The optical field on the detector plane for this diffraction field can be expressed as

$$U_m(x_d, y_d, z) \propto \exp(-jkz)\exp\left(j\pi z\frac{x_d^2+y_d^2}{\lambda f^2}\right) \times \Im\left\{R_m(x_f, y_f, z)\sum_{n=1}^{N} R_n{}^*(x_f, y_f, z)S_n(x_f, y_f, z)\right\}_{f_x=\frac{x_d}{\lambda f}, f_y=\frac{y_d}{\lambda f}} \tag{3.18}$$

Then we can obtain the diffraction wave from the whole volume of the hologram with a thickness of $t$ by integral the above expression along the $z$ direction

$$E_m(x_d, y_d) \propto \int_{-t/2}^{t/2} U_m(x_d, y_d, z)\mathrm{d}z \tag{3.19}$$

Using Eqs. (3.15), (3.16), and (3.18), the term in Eq. (3.19) can be evaluated as

$$\begin{aligned} E_m(x_d, y_d) \propto \sum_{n=1}^{N}\int_{-t/2}^{t/2} &\exp\left(j\pi z\frac{x_d^2+y_d^2}{\lambda f^2}\right) \\ &\times\left[\mathrm{RP}_m(-x_d, -y_d)\exp\left(-j\frac{\pi z}{\lambda f^2}(x_d^2+y_d^2)\right)\right] \\ &\star\left[\mathrm{RP}_n(-x_d, -y_d)\exp\left(-j\frac{\pi z}{\lambda f^2}(x_d^2+y_d^2)\right)\right] \\ &\otimes\left[\mathrm{DP}_n(-x_d, -y_d)\exp\left(-j\frac{\pi z}{\lambda f^2}(x_d^2+y_d^2)\right)\right]\mathrm{d}z \end{aligned} \tag{3.20}$$

where ★ and ⊗ stand for the correlation and convolution operation, respectively. As evident from Eq. (3.20), the diffracted field on the detector plane is the summation of the field in each layer of the recording medium, and the quality of the reconstructed data pages is expected to be affected by the correlation between the reading and recording RPs.

#### 3.3.1.2 Orthogonal Condition

For clarity and simplicity, we set the signal to be a point source in the center of the SLM for all the data pages and express the data pages as

$$\mathrm{DP}_n(x_d, y_d) = A_n\delta(x_d, y_d) \tag{3.21}$$

where $A_n$ is either a 0 (OFF) or a 1 (ON) representing the digital value for the signal. Then the reconstructed optical field for this point source signal on the detector plane, which is called PSF [4, 22], readout by using the $m$th RP can be written as

$$\mathrm{PSF}_m(x_d, y_d) \propto \sum_{n=1}^{N} A_n \int_{-t/2}^{t/2} \exp\left(j\pi z \frac{x_d^2 + y_d^2}{\lambda f^2}\right) \times \left[\mathrm{RP}_m(-x_d, -y_d) \exp\left(-j\frac{\pi z}{\lambda f^2}(x_d^2 + y_d^2)\right)\right] \star \left[\mathrm{RP}_n(-x_d, -y_d) \exp\left(-j\frac{\pi z}{\lambda f^2}(x_d^2 + y_d^2)\right)\right] \mathrm{d}z \tag{3.22}$$

By evaluate the integration of $z$, Eq. (3.22) can be rewritten as

$$\mathrm{PSF}_m(x_d, y_d) \propto \sum_{n=1}^{N} A_n \iint \mathrm{RP}_m(x_0, y_0) \mathrm{RP}_n{}^*(x_0 + x_d, y_0 + y_d) \times t \,\mathrm{sinc}\left[\frac{t}{\lambda f^2}\left(x_d{}^2 + y_d{}^2 + x_d x_0 + y_d y_0\right)\right] \mathrm{d}x_0 \mathrm{d}y_0 \tag{3.23}$$

The sinc function in Eq. (3.23) is defined as $\mathrm{sinc}(x) = \sin(\pi x)/(\pi x)$ and is caused by Bragg mismatch. This function relates to the Bragg-matching condition in volume holography. When a specific grating component is Bragg matched (argument of the sinc function is equal to 0), the diffraction efficiency for this grating reaches the maximum; otherwise the amplitude diffraction coefficient decreases as a sinc function.

Because of the infinite plane wave approximation in the derivation, the Bragg mismatches in $x$ and $y$ direction are not allowed. Then when the plane wave from coordinate $(x_0, y_0)$ is used to reading the holograms, the diffracted plane wave going to the point $(x_d, y_d)$ on the detector plane is diffracted by the gratings, which are recorded by the plane wave from coordinate $(x_0 + x_d, y_0 + y_d)$.

Considering the image position of the point source on the output plane, $x_d = y_d = 0$, which corresponds to the Bragg-matched diffraction for all the data pages and

$$\mathrm{PSF}_m(0,0) \propto \sum_{n=1}^{N} A_n \iint \mathrm{RP}_m(x_0, y_0) \mathrm{RP}_n{}^*(x_0, y_0) \,\mathrm{d}x_0 \mathrm{d}y_0 \tag{3.24}$$

To obtain the reconstruction of the desired data page without inter-page crosstalk due to Bragg-matched readout, the RPs for the reference beam have to fulfill the condition

$$\iint \mathrm{RP}_m(x_0, y_0) \mathrm{RP}_n{}^*(x_0, y_0) \,\mathrm{d}x_0 \mathrm{d}y_0 = \mathrm{Constant} \cdot \delta_{m,n} \tag{3.25}$$

where $\delta_{m,n} = 0$ for $m \neq n$ and $\delta_{m,n} = 1$ for $m = n$. The RPs that fulfill the condition can be called orthogonal RPs. For amplitude SLM, orthogonality means the positions of the ON pixels for RPs are totally different from each other, and no overlap position exists.

#### 3.3.1.3 Parallel Optical Image Superimposition

By uploading the linear combination of RPs together during the reconstruction process, for example, using $\mathrm{RP}_i(x_0, y_0) + \mathrm{RP}_j(x_0, y_0) + \mathrm{RP}_k(x_0, y_0)$ instead of $\mathrm{RP}_m(x_0, y_0)$

in Eq. (3.24), we will get

$$\mathrm{PSF}_{ijk}(0,0) = A_i + A_j + A_k \tag{3.26}$$

This relation means the corresponding linear combination of the optical signal is realized at the output of the system. This principle allows one to realize parallel optical image addition for ORPM method by the method similar for phase-code multiplexing [23]. In addition, this principle offers a way to prevent unauthorized readout of desired data page. If the RPs form a complete orthonormal set, any incorrect RP can be expressed as a weighted linear combination of the correct orthogonal RPs. The signal read out by this incorrect RP is then an analog linear combination of the multiplexed data pages, which cannot be used to extract information correctly in a binary system.

#### 3.3.1.4 Orthogonal Radial-Line RPs

Figure 3.14 shows an example of two kinds of RPs for this method ($N = 3$). One is radial line (Figure 3.14a), and the other is multiple rings (Figure 3.14c). For radial line, $RP_2$ and $RP_3$ are generated by rotating $RP_1$ along the center of pattern clockwise by $\varphi/3$ and $2\varphi/3$, respectively, where $\varphi$ represents the angle between the nearest radial lines of $RP_1$. For multiple rings pattern, $RP_2$ and $RP_3$ are formed by changing the radius of the rings of $RP_1$ by $\Lambda/3$ and $2\Lambda/3$, respectively, where $\Lambda$ denotes the radius difference between the adjacent rings. The key parameters such as $\varphi$ and $\Lambda$ for the RPs are illustrated in Figure 3.14. The fill factor of both RPs is 1/6. Here, the fill factor is defined as the ratio of the ON pixel numbers to the total pixel numbers in the pattern area. Figure 3.14b,d is enlarged version of the rectangle in Figure 3.14a,c, respectively. For amplitude RPs, both the fill factor ($\eta$) of the RPs and the crosstalk for the RPM method impose the limitation on the maximum number ($N$) of the multiplexed holograms and

$$N_{\max} \leq 1/\eta \tag{3.27}$$

### 3.3.2 Crosstalk Analysis for ORPM

When the RPs fulfill condition, we can obtain the reconstruction of the desired data page without crosstalk due to Bragg-matched readout. However, due to

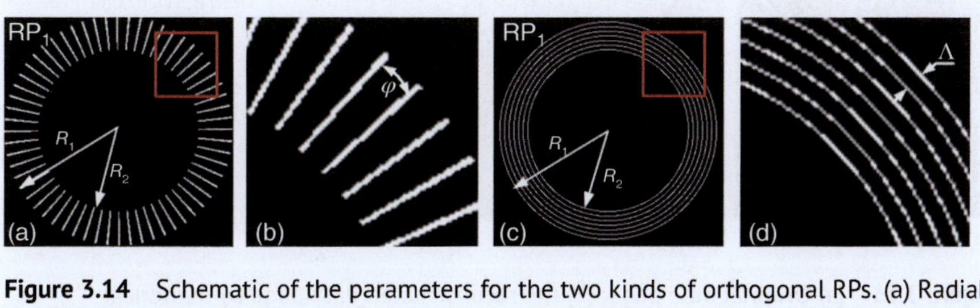


**Figure 3.14** Schematic of the parameters for the two kinds of orthogonal RPs. (a) Radial line RP; (b) enlarged version of the rectangular part in (a); (c) multiple rings RP; (d) enlarged version of the rectangular part in (c). Source: Jianhua Li.

Bragg-mismatch readout and Bragg degeneracy, $PSF_m(x_d, y_d) \neq 0$ when $x_d \neq 0$ or $y_d \neq 0$, which will introduce both intra-page and inter-page crosstalk for this multiplexing method. The intensity of the PSF has been used as an image evaluation metric for the estimate of intra-page crosstalk [4, 22]. Once the PSF is obtained, the optical field on the detector plane can be calculated by the convolution of the original data page and the PSF, provided the CHDSS system is a linear space-invariant system. However, based on Eq. (3.20), the PSF is not a linear space-invariant function due to the exponential factor in the term of $DP_n(-x_d, -y_d) \exp\left(-j\pi\Delta z/\lambda f^2 \left(x_d^2 + y_d^2\right)\right)$. It represents the complex amplitude of the reconstructed optical field of a point source. Therefore, the intensity of the PSF can only be used as an estimation of the intra-page crosstalk. For the calculation of both the intra-page and inter-page crosstalks, image-based simulation is desired, and the results will be more accurate and rigorous.

Based on Eq. (3.20), FFT method can be used for the calculation of the convolution and correlation. The integral can be implemented by dividing the medium into many discrete layers, which are positioned at equidistant points in the $z$-direction with the same thickness, and integration is represented by the summation of the value for each layer. SNR and BER are used to specify the total crosstalk of the reconstructed data pages and are defined as

$$\mathrm{SNR} = \frac{\mu_{\mathrm{on}} - \mu_{\mathrm{off}}}{\sqrt{\sigma_{\mathrm{on}}^2 + \sigma_{\mathrm{off}}^2}} \tag{3.28}$$

$$\mathrm{BER} = \frac{\text{Number of error pixels}}{\text{Total number of pixels}} \tag{3.29}$$

where $\mu_{\mathrm{on}}$ and $\mu_{\mathrm{off}}$ are the average values of the ON pixels and OFF pixels, respectively. $\delta_{\mathrm{on}}$ and $\delta_{\mathrm{off}}$ are the standard deviation for the values of the ON and OFF pixels, respectively. The number of error pixels for BER is calculated by comparing the original data page with the reconstructed data page.

The simulation result based on Eq. (3.20) is shown in Figure 3.15. Only one hologram is recorded. Figure 3.15a is the original data page and is stored by using the RP shown in Figure 3.14a. As shown in Figure 3.15b, the reconstructed data page accurately reflects the original data page. The histogram of the ON and OFF

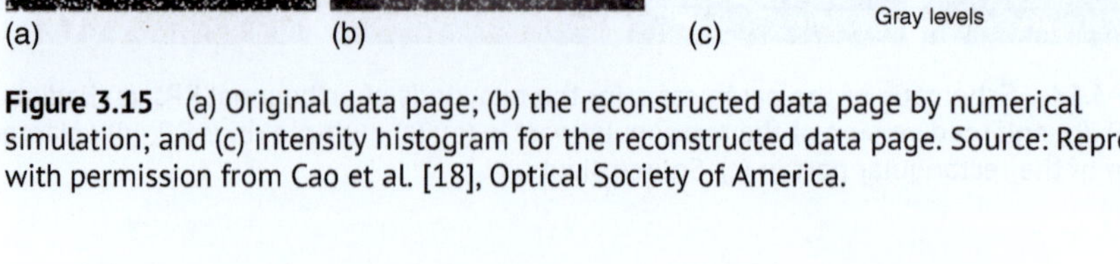


**Figure 3.15** (a) Original data page; (b) the reconstructed data page by numerical simulation; and (c) intensity histogram for the reconstructed data page. Source: Reproduced with permission from Cao et al. [18], Optical Society of America.

**Table 3.2** Parameters for numerical evaluation of the crosstalk.

| Parameter | Symbol | Value |
|---|---|---|
| Focal length for objective lens | $f$ | 4 mm |
| Wavelength | $\lambda$ | 0.532 μm |
| Medium thickness | $t$ | 2 mm |
| Medium refractive index | $n$ | 1.5 |
| Pixel size for SLM | $d$ | 10.8 μm |
| Pixel number for SLM | — | 300 × 300 |
| Outer radius for RP | $R_1$ | 150 pixels |
| Inner radius for RP | $R_2$ | 105 pixels |
| Fill factor for RP and data page | $\eta$ | 1/6 |
| Number of sampling for SLM | — | 300 × 300 |
| Number of layers for storage medium | $N_z$ | 300 |
| Pixel number of the data page | — | 148 × 148 |
| Angle between the adjacent radial lines | $\varphi$ | 6° |
| Radius difference between the adjacent rings | $\Lambda$ | 6 pixels |

pixels is shown in Figure 3.15c. The ON and OFF pixels are well separated, the raw BER by using a global threshold detection method is 0.0069 [19] and the discrimination level for output logical is set at the intersection level, as indicated by the arrow in Figure 3.15c. The parameters for the simulation are presented in Table 3.2. The effective focal length is set as 6 mm instead of 4 mm, and the effective wavelength is 354.7 nm instead of 532 nm, to simplify the calculation of the refraction on the boundary between the air and the holographic medium.

As mentioned previously, the holographic medium is divided into many layers, and the simulation was executed with different number of layers to evaluate the precision of the results. The SNR value for the reconstructed data page shown in Figure 3.15b is 4.15 for 100 layers, 3.29 for 200 layers, 3.28 for 300 layers and 3.28 for 400 layers. Considering the time required and the convergence of the SNR, 300 layers is appropriate for calculation.

The simulation was also executed with the same parameters presented in Table 3.2 except the medium thickness changing from 0 to 2000 μm for different kinds of RPs with the same fill factor. As shown in Figure 3.16, all the SNR increase approximate linearly with the increasing medium thickness. The radial line RP has a higher SNR than that with a multiple rings RP shown in Figure 3.14d. Both RPs with random binary phase (RBP) modulation have a much higher SNR than that without RBP. The RBP modulation has been used to homogenize the spectrum and suppress the DC component on the Fourier plane and therefore effectively utilize the dynamic of the recorded medium [9]. This simulation shows that RBP can also greatly increase the quality of reconstructed data page. It shows good agreement with the analyses in references [4, 22, 24, 25].

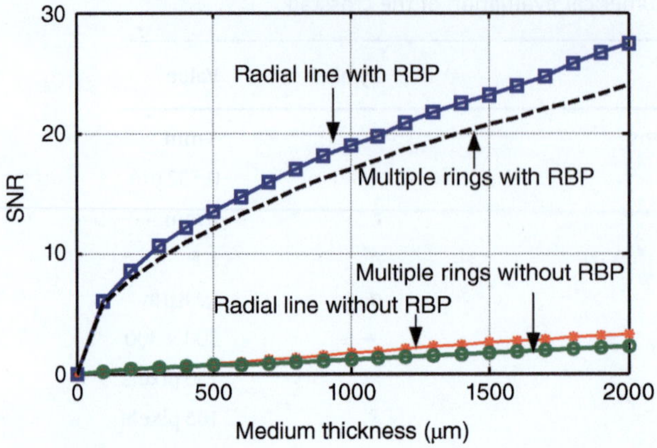


**Figure 3.16** SNR versus medium thickness by using different RPs, with and without RBP. Source: Jianhua Li.

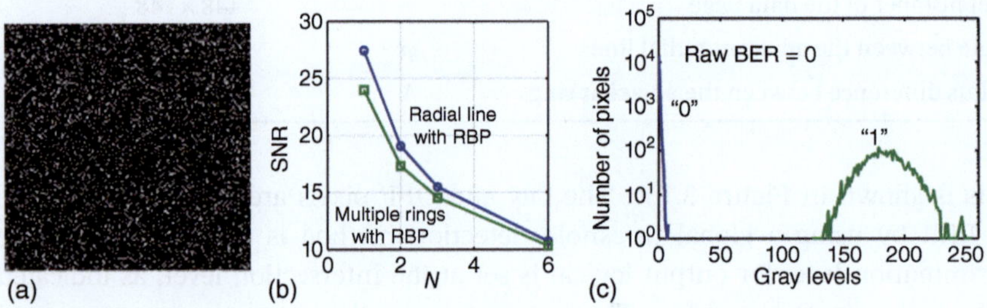


**Figure 3.17** (a) Reconstructed data page by numerical simulation for $N = 6$ with radial line RPs with RBP; (b) simulated SNR with RBP for both kinds of RPs; and (c) intensity histogram for the reconstructed data page of (a), the raw BER is zero. Source: Jianhua Li.

When $N$ volume holograms are multiplexed in the same volume of the medium by using $N$ orthogonal RPs, the reading reference beam interacts with all the holograms in parallel, and the diffracted patterns are superimposed on the detector plane. To analyze the SNR, we assume that all the stored data pages are random patterns with the fill factor of 1/6; other parameters are presented in Table 3.2. During the reconstruction process, one RP is used to read the volume holograms. When RBPs are used for the RMP method, the SNR is much higher than that without RBP, and the raw BER is almost zero for the simulation, as shown in Figure 3.17. According to the above analysis, it can be concluded that ORPM can be realized in the CHDSS system, and the crosstalk can be suppressed by using an random binary phase modulation (RBPM).

### 3.3.3 Orthogonal Reference Patterns Modulated Shifting Multiplexing

The crosstalk of the ORPM can also be suppressed by hybridization of ORMP and shifting multiplexing method. Holograms are multiplexed by not only using orthogonal RPs but also shifting the media with a small distance separately in the recording

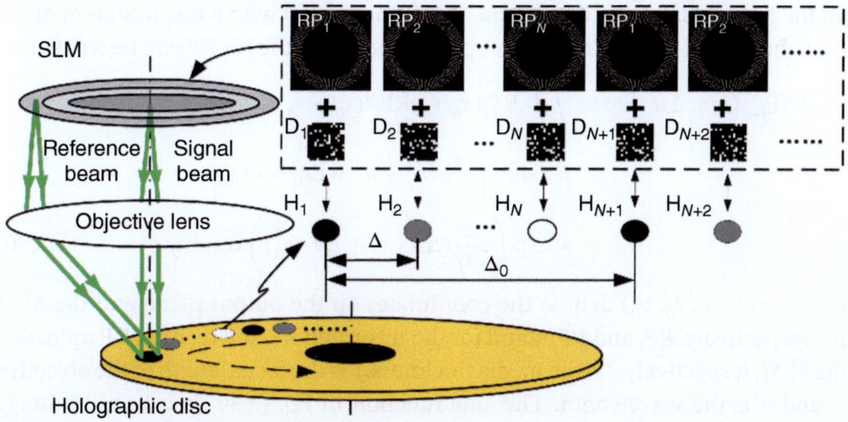


**Figure 3.18** Schematic of the orthogonal RPs modulated shift multiplexing method by using $N$ RPs. SLM, spatial light modulator; $RP_i$, the $i$th RP; $D_i$, the $i$th date page; $H_i$, the $i$th hologram; $\Delta_0$, traditional shift pitch; $\Delta$, new shift pitch. Source: Reproduced with permission from Li et al. [17], Optical Society of America.

process. The shift pitch for this hybrid shift multiplexing method can also be substantially reduced by utilizing shift selectivity in conjunction with Bragg selectivity.

The schematic of the orthogonal RPs modulated shift multiplexing is shown in Figure 3.18. $N$ orthogonal RPs are used for this method. The objective lens in the Fourier geometry converts a pixel on the SLM into a plane wave with a unique angle corresponding to pixel's coordinate. A recorded hologram can be treated as the accumulation of the gratings written by every pair of the plane waves from the ON pixels in the SLM. For the recording process, as shown in the dashed line rectangle, both the RP and data page are simultaneously uploaded on the SLM to encode the reference and signal beams. Both beams will interfere with each other and form a hologram inside the holographic disc.

On recording, the first data page ($D_1$) is recorded as the first hologram ($H_1$) by using the first RP ($RP_1$). Then the disc is shifted by a distance $\Delta$ and $RP_2$ is used instead of $RP_1$ for the recording of $D_2$. $D_3$ is also recorded in this way, changing the RP and shifting the disc at the same time. If $N$ RPs are employed for this method, the $(N + 1)$th data page is then recorded again by using $RP_1$. By repeating these two operations, all the data pages can be multiplexed. On reconstruction, the desired data page can be readout by using the corresponding RP. For this method, the shift pitch can be set to $1/N$ of that of the traditional shift multiplexing method with a single RP since holograms can even be multiplexed in the same volume of the media by using orthogonal RPs. However, the shift selectivity of the hologram imposes a limit on the maximum of $N$ because the inter-page crosstalk will increase with the decrease of the shift pitch. For amplitude SLM, the upper limit of $N$ is also determined by the filling rate ($\eta$) of the RP and $N_{max} \leq 1/\eta$.

To analyze the shift selectivity and inter-page crosstalk for this hybrid method, we set the signal to be a point source in the center of the SLM and consider the intensity of the signal on the output plane reconstructed by using an RP. Based on Eq. (3.22), when the hologram is recorded by the $i$th RP and shifted with a displacement of $\Delta u$ and $\Delta v$, the optical field on the output plane readout by the $j$th RP can be written as

$$\mathrm{PSF}_{j\to i}(\varepsilon,\eta,\Delta u,\Delta v)=\iint \mathrm{RP}_j(x_0,y_0)\mathrm{RP}_i^*(x_0+\varepsilon,y_0+\eta)$$
$$\times\, t\,\mathrm{sinc}\left[\frac{t}{\lambda f^2}(\varepsilon^2+\eta^2+\varepsilon x_0+\eta y_0)\right]$$
$$\times\exp\left(i\frac{2\pi}{\lambda f}(\Delta u\cdot x_0+\Delta v\cdot y_0)\right)\mathrm{d}x_0\mathrm{d}y_0 \tag{3.30}$$

where $(\varepsilon, \eta)$ and $(x_0, y_0)$ denote the coordinates on the output plane and the SLM plane, respectively. $RP_i$ and $RP_j$ stand for the recording RP and reading RP uploaded on the SLM, respectively. $t$ is the media thickness, $f$ is the focus length of the objective lens, and $\lambda$ is the wavelength. The sinc function in Eq. (3.30) is caused by Bragg mismatch and the exponential function is introduced by the disc shift.

If the disk is kept at the same place for reading and only considering the image position of the point source on the output plane, $\varepsilon = \eta = \Delta u = \Delta v = 0$ and $\mathrm{PSF}_{j\to i}(0, 0, 0, 0)$ is proportional to the cross-correlation between $\mathrm{RP}_i$ and $\mathrm{RP}_j$. When the $\mathrm{RP}_i$ is orthogonal with $\mathrm{RP}_j$ $(i \neq j)$, $\mathrm{PSF}_{j\to i}(0,0,0,0)$, which means multiple holograms can be multiplexed without shifting the disc by using orthogonal RPs. However, due to Bragg mismatch and Bragg degeneracy, $\mathrm{PSF}_{j\to i}(\varepsilon, \eta, 0, 0) \neq 0$ when $\varepsilon \neq 0$ or $\eta \neq 0$ which will introduce inter-page crosstalk for this scheme.

Shift selectivity can be defined as the variation of the integrated intensity of the reconstructed signal on the output plane when the disc shifts. It can be expressed as

$$I_{j\to i}(\Delta u,\Delta v)=\iint |\mathrm{PSF}_{j\to i}(\varepsilon,\eta,\Delta u,\Delta v)|^2\,\mathrm{d}\varepsilon\mathrm{d}\eta \tag{3.31}$$

Figure 3.19 shows the simulation of Eq. (3.31) for $i = j$ and $i \neq j$. In the simulation, $\lambda$ is 532 nm, $f$ is 4 mm, and $t$ is 1 mm. The pixel size of the SLM is 10.8 μm and the total pixel number used is 300 × 300. The outer radius $R_1$ of the RPs is 150 pixels, the inner

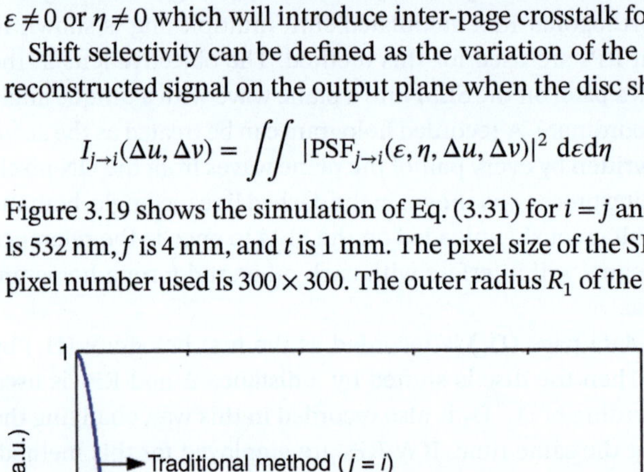


**Figure 3.19** Shift selectivity for the traditional and orthogonal RPs modulated shift multiplexing method. Source: Reproduced with permission from Li et al. [17], Optical Society of America.

radius $R_2$ is 105 pixels, and $\varphi$ is 3°. Twenty-five (5 × 5) pixels are used for the integral in Eq. (3.31). It shows that when a different RP is used ($i \neq j$) to read the hologram, the diffraction intensity has been substantially attenuated without shifting of disc, and this intensity (inter-page crosstalk) will decrease in fluctuation with the shift distance. It also shows the new shift pitch Δ can be almost as small as 1/3 of the traditional shift pitch $\Delta_0$ with similar crosstalk with these parameters, which are based on the following experimental setup.

### 3.3.4 Experimental Results

The optical setup for the demonstration of ORPM and orthogonal RPs modulated shifting multiplexing is shown in Figure 3.20. A diode-pumped solid-state laser (DPSSL) at the wavelength of 532 nm is employed as the light source. After passing through the beam expander, the laser beam is collimated into a plane wave. The intensity of the plane wave is adjusted to 97 mW/cm$^2$ by the neutral density filter (NDF). A digital micromirror device (DMD, Texas Instrument, DLP 0.55 XGA) is used as the SLM. The relay lens accommodating with iris$_1$ filtering out the higher orders of the DMD. The objective lens (Nikon, numerical aperture (NA) = 0.55) Fourier transforms the DMD pattern into a phenanthrenequinone:polymethyl methacrylate (PQ:PMMA) polymer media, which is mounted on a precise micro-translation stage (PI, M-112). The resolution of the stage is 70 nm. On reconstruction, iris$_2$ is used to prevent the reference beam from entering the complementary metal-oxide semiconductor (CMOS) image sensor (Mikrotron, MC1310), which is employed to capture the reconstructed data pages.

The required shift pitch for the traditional shift multiplexing method was measured. A data page was stored into the media by using RP$_1$. Then the media was shifted and the intensity of the reconstructed data page was monitored. As shown in Figure 3.21, the normalized diffraction intensity decreases in fluctuation with the shift distance. When the media was shifted 4.5 μm away from the original position, the diffraction intensity reached the minimum and a new data page can be recorded in this position for the traditional shift multiplexing method. This value agrees well with the theoretical value in Figure 3.19.

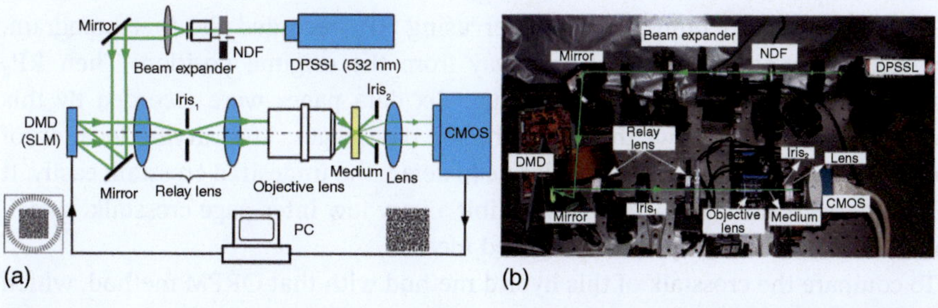


**Figure 3.20** (a) Schematic diagram and (b) photo of experimental setup of the CHDSS system for the RPM method and parallel optical image addition. DPSSL, diode-pumped solid-state laser; NDF, neutral density filter; CMOS, complementary metal-oxide semiconductor; DMD, digital micromirror device; PC, personal computer. Source: Jianhua Li.

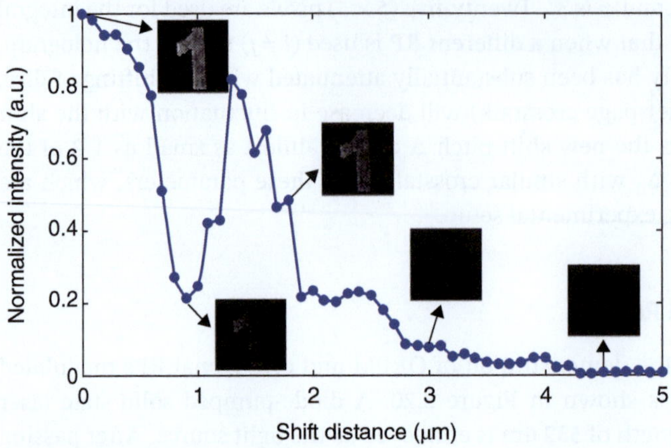


**Figure 3.21** Measured intensity of the reconstructed data page versus the shift distance. Source: Based on Yoshida et al. [13].

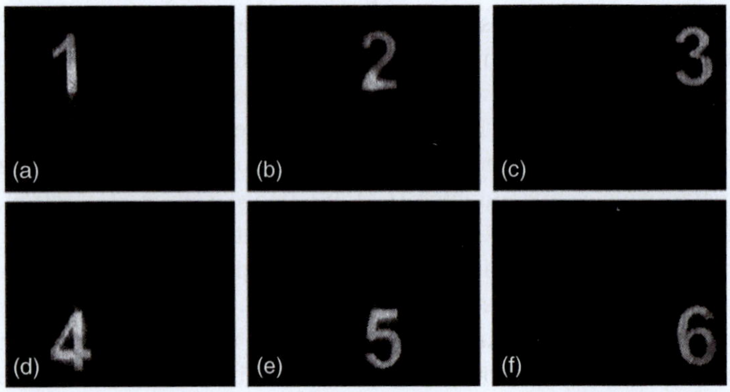


**Figure 3.22** Reconstructed data pages for the proposed method with the shift pitch of 1.5 μm by using three RPs. (a) and (d) Reconstructed by $RP_1$; (b) and (e) reconstructed by $RP_2$; (c) and (f) reconstructed by $RP_3$. Source: [17] Li et al. (2012), Optical Society of America.

Three amplitude RPs as shown in Figure 3.14a were used for the orthogonal RPs modulated shift multiplexing. After using $RP_1$ recorded the first hologram, the media was shifted only 1.5 μm away from the original position. Then $RP_2$ was used to record the second hologram. Six data pages were recorded by this method. Figure 3.22 shows the reconstructed data pages. The varying location of the numbers in the data pages enables us to detect the undesired crosstalk easily. It shows the reconstructed data pages exhibit a very low inter-page crosstalk, which has verified the feasibility of the proposed method.

To compare the crosstalk of this hybrid method with that ORPM method, which multiplexes holograms in the same volume of the media, three data pages were multiplexed in a single spot by using the same three amplitude RPs used for Figure 3.22. As shown in Figure 3.23, the reconstructed data pages exhibit a relatively higher

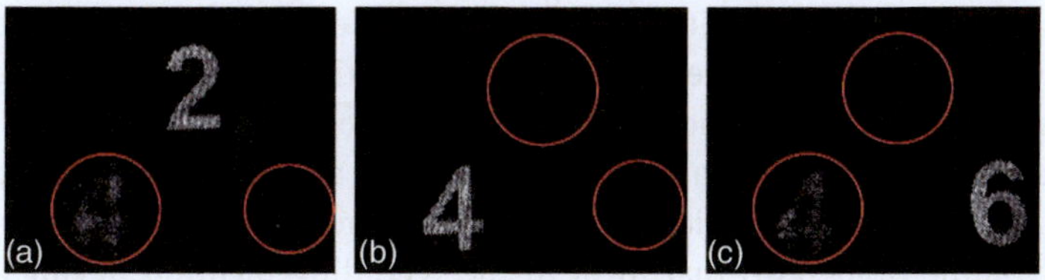


**Figure 3.23** Reconstructed data pages after multiplexing three holograms in the same volume by using three RPs. (a)–(c) Reconstructed by $RP_1$, $RP_2$, and $RP_3$, respectively. The inter-page crosstalk is indicated by the circles. Source: [17] Li et al. (2012), Optical Society of America.

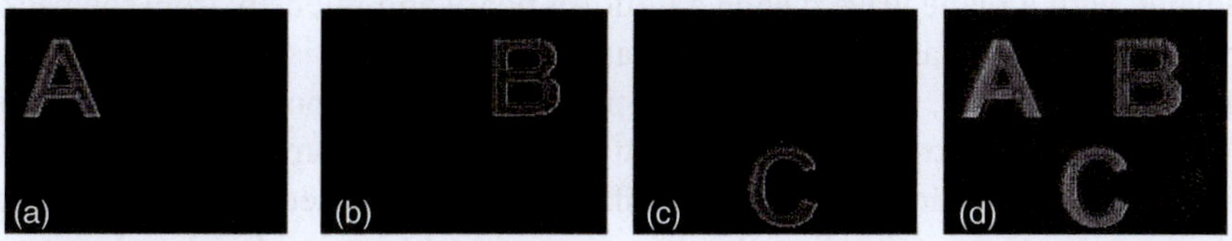


**Figure 3.24** Reconstructed images by the ORPM method. (a)–(c) Reconstructed by radial line $RP_1$, $RP_2$, and $RP_3$, respectively; (d) reconstructed by uploading the three RPs simultaneously on the DMD. Source: Jianhua Li.

inter-page crosstalk than that in Figure 3.22, which shows good agreement with the theoretical analysis.

Three data pages were multiplexed in the same volume of the medium by the ORMP method with the radial RPs shown in Figure 3.24. The reconstructed image by the ORPM method is shown in Figure 3.24a–c. The addition of the recorded three images is detected when the three RPs are uploaded on DMD simultaneously shown in Figure 3.24d. This property can be used to realize parallel optical image addition or encrypt the stored data pages. It is also possible to realize the logical operations OR for image addition by the ORMP method.

## 3.4 Simulation of Interference Fringes Formed by Collinear Method

At the initial stage of research and development, it was not possible to quantitatively explain what kind of interference fringes are formed by the collinear method. However, considering geometrical optics, if the reference beam and the information beam diverge after they travel on the same axis and are collected by the objective lens, they should cross each other and interfere at least near the focal point. Therefore, it was considered that the crossing state of light rays near the focal point could not be visualized by simulation.

Although various methods have been tried, we finally focused on the fact that the reference light and the information light travel coaxially as a two-dimensional image in the collinear method, so that they are observed as diffraction patterns as they move away from the DMD. That is, if a two-dimensional diffraction pattern of a two-dimensional image at a position separated by a desired distance can be numerically calculated, the obtained result is plane information, but the amplitude and phase pattern at an arbitrary position in the recording medium can be obtained. I thought it would be required. If a large number of plane amplitudes and phase patterns are obtained at different positions in the recording medium, then the interference fringes (volume hologram) that form in the recording medium having a certain thickness can be finally visualized. The visualization approach and actual simulation results are shown below.

Assume a mask image having the same pattern of minute openings as the page data, with the pixels displayed by the DMD being square minute openings. This is irradiated with a plane wave having a uniform phase and intensity from the back, and the wavefront obtained when the distance from the mask is changed is numerically calculated. First, this mask image is placed on the entrance pupil plane of the objective lens, and a two-dimensional pattern of phase and amplitude is obtained by numerical calculation when Fourier diffraction is performed to the main plane of the lens. Next, assuming that this two-dimensional (amplitude, phase) pattern is diffracted while being collected as a spherical wave by the objective lens, a two-dimensional pattern of amplitude and phase at an arbitrary plane position not reaching the focal point is obtained. Next, for example, a numerical calculation process is performed in which this distance is obtained again at a plane position that has traveled in the optical axis direction by 1 μm, and a two-dimensional pattern of amplitude and phase is obtained again and further at a plane position that has shifted by 1 μm in the optical axis direction. In this operation, the interference fringes collected by the objective lens are sampled by a two-dimensional image group having a phase and an amplitude sliced in a plane every 1 μm and can be obtained by Eq. (3.32).

$$s(\vec{r}) = C \times \exp(ikz) \iint s(x_0, y_0) \exp\left[-i\frac{2\pi}{\lambda f}(xx_0 + yy_0)\right] \times \exp\left(-i\frac{2\pi}{\lambda f}\left(x_0^2 + y_0^2\right)\right) dx_0 dy_0 \tag{3.32}$$

The sliced two-dimensional image group obtained by the simulation method covers almost all information of the three-dimensional interference fringe formed when collected by the objective lens and is used in the medical field. Like a CT scan, it has the advantage that the interference fringes of a volume hologram can be intuitively grasped through any cross-section. In addition, since the obtained image is scalable, the interval and size of the interference fringes that are actually formed vary depending on the wavelength, pixel size, and focal length of the lens. Figure 3.25 schematically shows data actually obtained through this processing. Here, an intensity pattern is generated and displayed from the amplitude data. To save space, only five images were selected from the huge simulated images.

When observing the image generated by the simulation, a similar image is repeated so that it returns to the original from a certain point, and the image of the center position of the return is displayed on the focus position of the objective lens (displayed in the DMD) (Fourier transform image of page data). Figure 3.25 shows

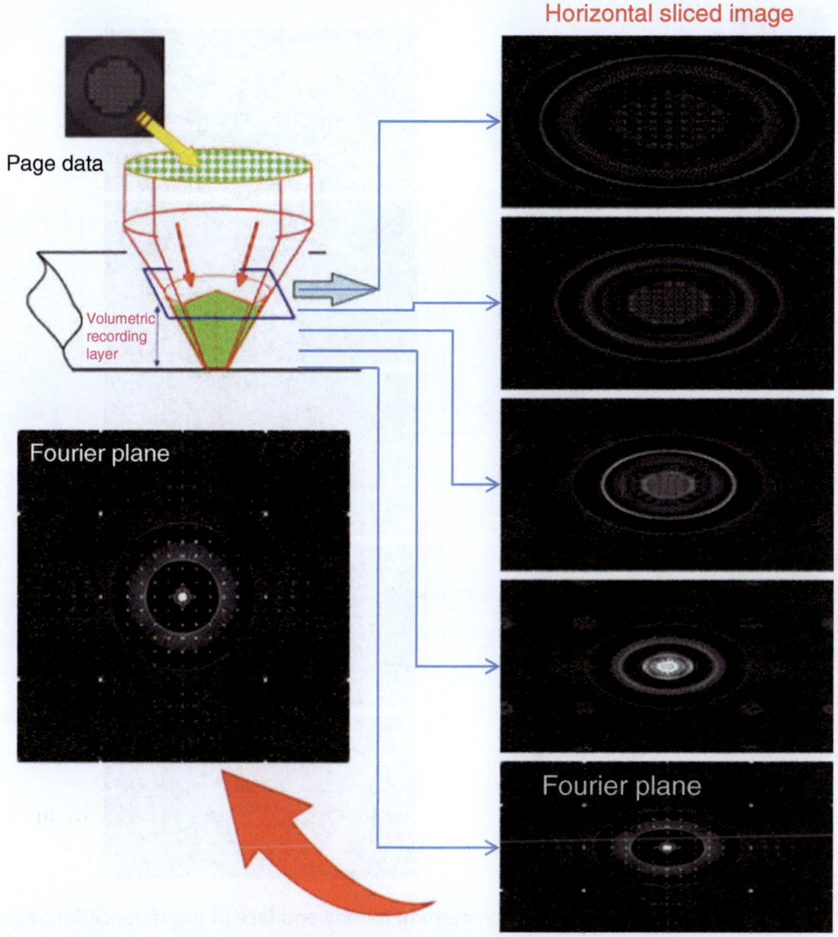


**Figure 3.25** Cross-section of interference fringe near the focal plane of objective lens by simulation. Source: Hideyoshi Horimai.

a horizontal sectional slice image from the focal position to a thickness of about 100 μm when the wavelength is 532 nm, the objective lens NA = 0.55, and the focal length $f = 4$ mm.

In the photopolymer recording material [14] that has greatly evolved in hologram recording, a refractive index difference is generated according to the intensity pattern of the interference fringes, and the hologram is recorded as phase interference fringes. Therefore, assuming that the intensity pattern is linearly converted into the phase pattern as it is, a phase interference fringe pattern in which a slice image group of the intensity pattern as shown in Figure 3.25 is formed in the thickness direction of the recording medium by the collinear method and is considered to be almost equivalent.

Therefore, a visualization model image that is cut in the thickness direction at an arbitrary cross-section in the recording medium can be easily generated from the

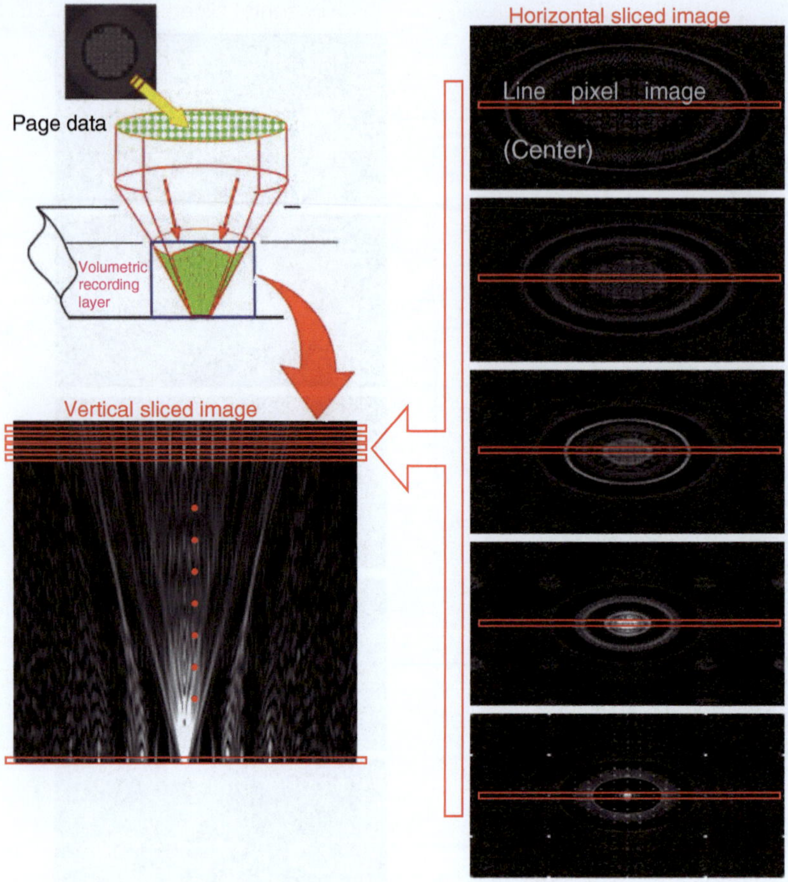


**Figure 3.26** Visualized image generation method for any end face of interference fringes in collinear method. Source: Hideyoshi Horimai.

group of horizontal slice images combined in the thickness direction. The method is schematically shown in Figure 3.26. Figure 3.26 is an example in which a vertical cross-section at the center of the optical axis is visualized. A line image of one to several pixels in the central portion corresponding to the cross-section to be observed is cut out from the plane slice image, and this is combined in accordance with the interval at which the slice image is sampled (in this case, 1 μm). We succeeded in re-synthesizing and displaying the cut image.

By developing the above-mentioned technology, an image having only a reference light pattern and an image having only an information light pattern were also simulated. Figure 3.28a is a vertical slice visualization image of only the information light and the reference light (Figure 3.28b) in the page pattern shown in Figure 3.27. Conversely, Figure 3.29 shows a so-called interference fringe pattern in which reference light and information light are displayed simultaneously. From this simulation result, what kind of phase pattern is formed in the recording medium by the reference and information lights using the collinear method alone, and

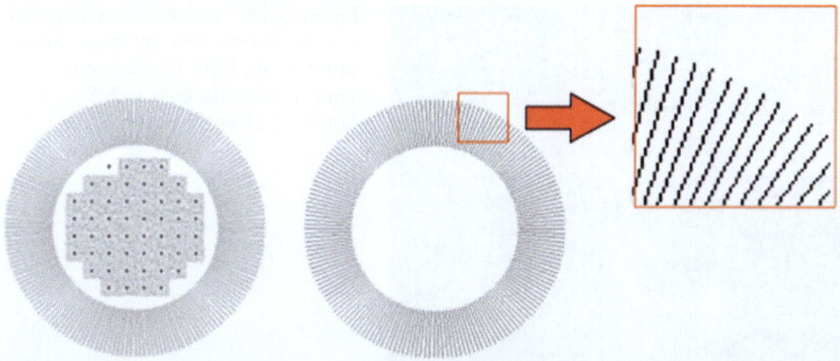

**Figure 3.27** Page pattern used for visualizing the interference fringes (radial reference light). Source: Hideyoshi Horimai.

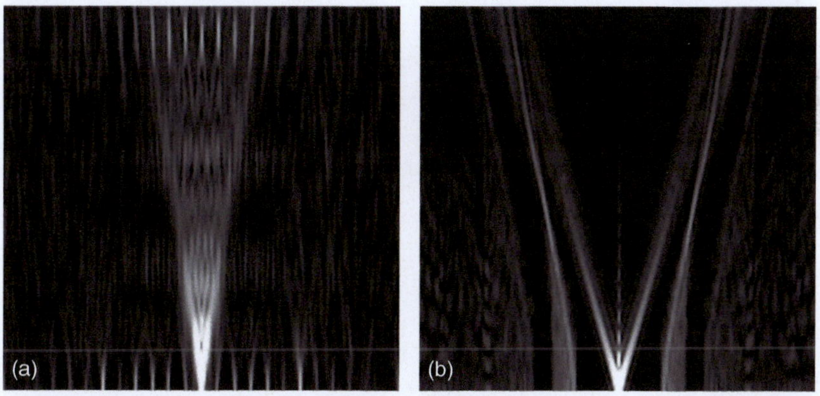


**Figure 3.28** Visualization image of interference fringes: (a) Information light only and (b) reference light only. Source: Hideyoshi Horimai.

what hologram is formed when it causes phase interference were evaluated for the first time.

During reproduction, the interference fringe pattern shown in Figure 3.28b is projected onto the recorded collinear interference fringe pattern shown in Figure 3.29, and the convolution reversing the interference fringe pattern is shown in Figure 3.28a. It can be understood that a reconstructed image corresponding to Fourier transform is obtained.

In contrast, Figures 3.30 and 3.31 are cross-sections of the interference fringe pattern when the reference light is a random pattern. Comparing Figures 3.29 and 3.31, it was found that although the information light patterns are the same, completely different three-dimensional interference fringes are formed due to the difference in the reference light pattern.

From the above observation results, it is clear that when the reference light pattern and the information light pattern are combined, the cross-sectional patterns of different interference fringes are obtained. It was confirmed by visualization that

**Figure 3.29** Visualization image of the interference fringes: page pattern of reference light + information light (radial reference light). Source: Hideyoshi Horimai.

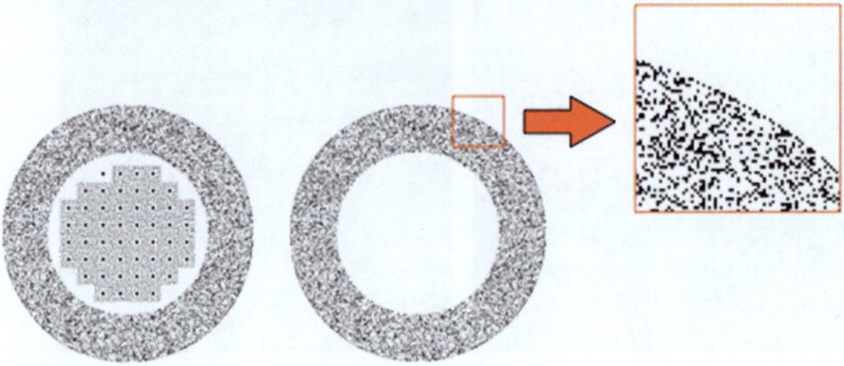

**Figure 3.30** Page pattern used for visualizing the interference fringes (random reference light). Source: Hideyoshi Horimai.

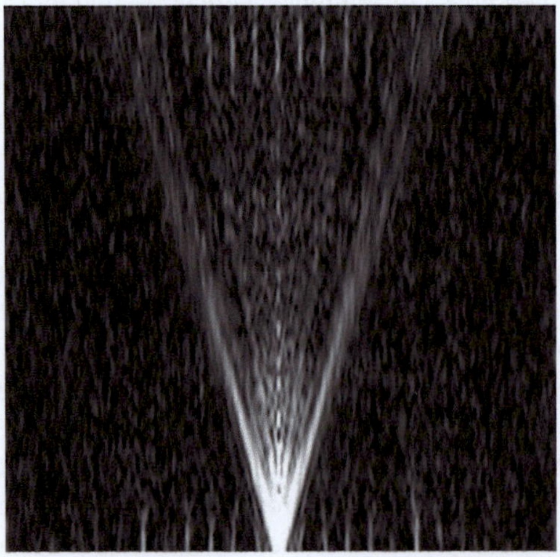

**Figure 3.31** Visualization image of the interference fringes: page pattern of reference light + information light (random reference light). Source: Hideyoshi Horimai.

the collinear system proceeding on the same axis also formed a volume hologram by mixing each other in the vicinity of the focal plane of the objective lens.

Simulating the collinear holography visualization enables the analysis of the three-dimensional structure of interference fringes by repeating the Fourier diffraction of a two-dimensional image propagating on the same axis. For example, by using pattern notation that adds phase to the two-dimensional image information, we believe that it can be used for optimizing the design of random phase shifters in the collinear method, and examining the coding in phase multilevel recording and reproduction.

## References


1 Horimai, H. and Tan, X.D. (2005). Advanced collinear holography. *Opt. Rev.* 12: 90–92.

2 Horimai, H., Tan, X.D., and Li, J. (2005). Collinear holography. *Appl. Opt.* 44: 2575–2579.

3 Williams, C.S. and Becklund, O.A. (2002). *Introduction to the Optical Transfer Function*. Bellingham, WA: SPIE Press.

4 Shimura, T., Ichimura, S., Fujimura, R. et al. (2006). Analysis of a collinear holographic storage system: introduction of pixel spread function. *Opt. Lett.* 31: 1208–1210.

5 Yariv, A. and Yeh, P. (1984). *Optical Waves in Crystals*. New York: Wiley.

6 Orlov, S.S. and Hesselink, L. (2003). *J. Opt. Soc. Am. B: Opt.Phy.* 20: 1912.

7 Coufal, H.J., Psaltis, D., and Sincerbox, G.T. (ed.) (2000). *Holographic Data Storage*. Springer.

8 Curtis, K., Dhar, L., Hill, A. et al. (ed.) (2010). *Holographic Data Storage: From Theory to Practical Systems*. Wiley.

9 Tanaka, K., Hara, M., Tokuyama, K. et al. (2007). Improved performance in coaxial holographic data recording. *Opt. Express* 15 (24): 16196–16209.

10 Fukumoto, A. (2008). Coaxial holographic data recording. *Opt. Photonics News* 19 (11): 28–33.

11 Born, M. and Wolf, E. (1999). *Principles of Optics*, 7e. Cambridge University Press.

12 Gombkötő, B., Koppa, P., Sütő, A., and Lőrincz, E. (2007). Computer simulation of reflective volume grating holographic data storage. *J. Opt. Soc. Am. A* 24 (7): 2075–2081.

13 Yoshida, S., Kurata, H., Ozawa, S. et al. (2013). High-density holographic data storage using three-dimensional shift multiplexing with spherical reference wave. *Jpn. J. Appl. Phys.* 52 (9): 09LD07.

14 Yoshida, S., Matsubara, T., Kurata, H. et al. (2013). Multi-dimensional shift multiplexing technique with spherical reference waves. *IEICE Trans. Electron.* E96-C (12): 1520–1524.

15 Goodman, J.W. (2004). *Introduction to Fourier Optics*, 3e. Freeman.

**16** Toishi, M., Hara, M., Tanaka, K. et al. (2007). Novel encryption method using multi reference patterns in coaxial holographic data storage. *Jpn. J. Appl. Phys.* 46: 3775–3781.

**17** Li, J.H., Cao, L.C., Gu, H.R. et al. (2012). Orthogonal-reference-pattern-modulated shift multiplexing for collinear holographic data storage. *Opt. Lett.* 37: 936–938.

**18** Cao, L.C., Liu, J.Q., Li, J.H. et al. (2014). Orthogonal reference pattern multiplexing for collinear holographic data storage. *Appl. Opt.* 53: 1–8.

**19** Coufal, H.J., Psaltis, D., and Sincerbox, G.T.E. (ed.) (2000). *Holographic Data Storage*, Springer Series in Optical Science, 31–35. Berlin: Springer-Verlag.

**20** Hsu, K.Y., Lin, S.H., Hsiao, Y.N., and Whang, W.T. (2003). Experimental characterization of phenanthrenequinone-doped poly(methyl methacrylate) photopolymer for volume holographic storage. *Opt. Eng.* 42: 1390–1396.

**21** Mok, F.H., Tackitt, M.C., and Stoll, H.M. (1991). Storage of 500 high-resolution holograms in a $LiNbO_3$ crystal. *Opt. Lett.* 16: 605–607.

**22** Sun, C.C., Yu, Y.W., Hsieh, S.C. et al. (2007). Point spread function of a collinear holographic storage system. *Opt. Express* 15: 18111–18118.

**23** Denz, C., Dellwig, T., Lembcke, J., and Tschudi, T. (1996). Parallel optical image addition and subtraction in a dynamic photorefractive memory by phase-code multiplexing. *Opt. Lett.* 21: 278–280.

**24** Yu, Y.W., Chen, C.Y., and Sun, C.C. (2010). Increase of signal-to-noise ratio of a collinear holographic storage system with reference modulated by a ring lens array. *Opt. Lett.* 35: 1130–1132.

**25** Song, W., Tao, S., and Wang, D. (2010). Investigation on influence of wave-front property of reference beams on the quality of images reconstructed from holograms. *Proc. SPIE* 7730: 77301V.

# 4

# Key Devices and Components

## 4.1 Lasers

### 4.1.1 Introduction

Lasers are key components of the holographic data storage system (HDSS) [1]. The full name of laser, i.e. light amplification by stimulated emission of radiation, fully expresses the main process of laser manufacturing and principle. When the electrons in an atom absorb energy, they are excited from the low-energy levels to the high-energy ones, and subsequently return to the low-energy levels by releasing energy in the form of photons. The optical properties of the induced (excited) photon beam (laser) are highly consistent. This makes the laser monochromatic, bright, and directional more than ordinary light sources. In addition to HDSS, lasers are also widely used in laser marking, laser welding, laser cutting, optical fiber communication, laser ranging, radars, weapons, laser correction, laser beauty, laser scanning, laser mosquito killer, nondestructive testing technology, and various other applications.

In HDSS, many different kinds of lasers can be used as there is no strict limit on the required laser characteristics. However, the interference pattern should be clearly formed by the information and reference beams, which are divided by the laser source. Due to the different working materials, excitation methods, and application requirements, the operational modes and working states of the lasers are also different, and accordingly, they can be broadly divided in the following main types:

**Continuous lasers.** they operate via the excitation of the working substance, and the corresponding laser output remains continuous over a long period of time; examples include solid-state lasers (excited by a continuous light source), gas lasers, and semiconductor lasers (excited by continuous electronic excitations). These lasers are overheated because of their continuous operation, and thus, most of them require appropriate cooling measures.

**Single-pulse lasers.** in these lasers, the working material is excited by single pulses, and the corresponding laser emissions are also single pulses. General solid-state lasers, liquid lasers, and some special gas lasers operate via the absorption and emission of single pulses. Moreover, the thermal effect of these devices can be ignored; thus, special cooling measures are not required.

*Collinear Holography: Devices, Materials, Data Storage*, First Edition.
Xiaodi Tan, Hideyoshi Horimai, Tsutomu Shimura, and Xiao Lin.

**Repetitive pulse lasers.** these lasers emit a series of repetitive laser pulses, indicating that they can be excited continuously or by using repetitive pulses. However, the laser oscillation process is modulated to obtain repetitive pulses as the output. Generally, effective cooling measures are also required for these device.

**Q-switched lasers.** these are pulse lasers based on certain laser switching technologies to obtain high output powers. In these lasers, no laser oscillation occurs (the switch is in the off state) after the particle number of the working substance is inverted. When the number of accumulated particles increases beyond a certain threshold value, the switch is instantaneously turned on (for example, $10^{-10}$ seconds) resulting in very strong laser oscillations and a high-power pulsed laser output. Q-switched lasers can be divided into four types: electro-optic, acousto-optic, saturable absorption, and mechanical mirror Q-switched lasers according to the different media and working modes of the Q-switched elements. Among these, the electro-optic and acousto-optic Q-switched ones are widely used. Electro-optic Q-switching is realized by using the linear electro-optic effect of some crystals and exhibits the advantages of a short switching time, high efficiency, precise control of the Q-switching time, stable system operation, high repetition frequency, narrow output pulse (10–20 ns), high peak power (more than tens of megawatts), and so on. Acousto-optic Q-switching is used to control the beam-deflection-induced losses in the resonator because of diffraction effects, which arise when the laser beam passes through the ultrasonic field in the acousto-optic medium. The acousto-optic Q-switched lasers exhibit a stable performance, high repetition rates (1–20 kHz), and low modulation voltages (generally, <200 V) and are suitable for obtaining medium and small output powers with high pulse repetition rates. Electro-optic or acousto-optic Q-switching is an active Q-switched method, which uses all kinds of physical effects of the light traversing the electro-optic or acousto-optic medium under the action of an electric or acoustic field to control the reflection loss in the cavity and realize a sudden change in the Q value. In contrast, the saturable absorption Q-switching technology is a passive Q-switching method, which uses the absorption characteristics of the saturable absorber itself. This is a non-linear absorption medium, and its absorption coefficient gradually decreases with the increasing light intensity until it is saturated under the action of a relatively strong laser, showing the transparent characteristics of the light. In this case, the Q value is adjusted by controlling the absorption loss of the cavity; these lasers are simple, convenient, and practical.

**Mode-locked lasers.** these are special types of lasers that operate according to the mode-locked technology. Their working characteristics are determined by the phase relationship between the different longitudinal modes in the resonant cavity. Therefore, a series of ultrashort pulses (pulse width: $10^{-10}$ seconds) with equal time intervals can be obtained using these lasers. If the special fast optical switching technology is further adopted, then the aforementioned pulse sequences can also be obtained, and a single ultrashort laser pulse can be selected.

**Single-mode and frequency-stabilized lasers.** single-mode lasers operate in a single transverse or longitudinal mode after adopting a certain mode-limiting technology, whereas the frequency-stabilized lasers are special types of laser devices that use certain automatic control measures to stabilize the output wavelength or frequency within a certain accurate range. In some cases, both the single-mode operation and frequency stabilization characteristics can be combined to obtain a laser device with automatic control capability.

In HDSSs, usually, solid yttrium aluminum garnet frequency-doubled lasers, external-cavity lasers, and laser diodes are used. However, the conventional HDSS, so-called two-axis holography, still have essential issues concerning practicality. First, the reference beam for the readout process is spatially separated with an off-axis optical configuration, which results in a complex and large optical system preventing miniaturization. Second, to increase the storage density, normally, angle multiplexing [2], phase-coded multiplexing [3], wavelength multiplexing [4], and other methods [5, 6] are employed. Spatial fluctuation of the storage media should then be strictly controlled, and a large and heavy damping system is indispensable. This, in turn, prevents the removability and interchangeability of the media as well as the miniaturization of the system. Third, the holographic storage technology is completely different from the other existing storage systems, like bit-by-bit storage methods, and the holographic media is incompatible with the conventional storage media. Collinear technology [7], which is a new technology for HDSS, can solve the aforementioned problems and can aid in the production of small and practical HDSSs more easily than the conventional two-axis holography technique.

### 4.1.2 Solid Lasers for Collinear Technology

Collinear technology as a new writing and reading method for a holographic storage system is very promising and differs from the conventional two-axis holography. The unique feature of this technology is that a two-dimensional (2D) data page is recorded as volumetric holograms that are generated by coaxially aligned information and reference beams. These are simultaneously displayed by the same spatial light modulator (SLM), and interfere with each other in a holographic versatile disc (HVD) after passing through a single objective lens.

Figure 4.1 shows the schematic optical configuration of collinear holography, and Figure 4.2 depicts a pair of the 2D data page patterns. A green (or blue) linear polarization state collimated laser light beam is used for both writing and reading the holograms. In the writing process, a combined image of the information (central part) and reference patterns (outer ring part) is displayed on the digital micromirror device (DMD), which is used as the SLM, as shown in Figure 4.2a. The laser beams modulated by the DMD, with *p*-polarization, pass through the polarization beam splitter (PBS) and are then incident upon a quarter-wave plate (QWP). The laser beams, which are converted from a *p*-polarized state to a circularly polarized state by the QWP, are focused onto the holographic recording disc by an objective lens, where they interfere with each other, and the interference pattern is recorded as a

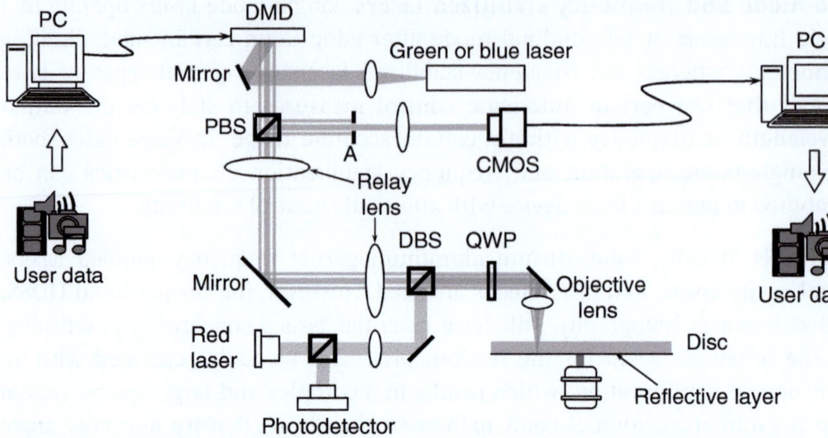


**Figure 4.1** Optical configuration of the HVD system as a new reading and writing technology for an HDSS. The green (or blue) and red lasers represent laser beams with wavelengths of green (or blue) and red lights, respectively. Source: Reproduced with permission from Horimai et al. [8]/SPIE.

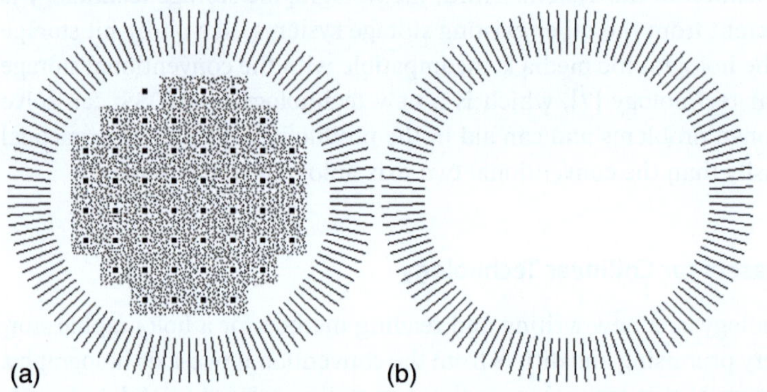


**Figure 4.2** Pair of 2D digital data page patterns: (a) write pattern and (b) read pattern displayed on the SLM. Source: Reproduced with permission from Liu et al. [9], Optical Society of America.

volume hologram in the disc. In the reading process, only the outer ring pattern, as shown in Figure 4.2b, is displayed on the SLM and is used for creating a reference beam. The *p*-polarized laser beam passes through the PBS and is converted to a circularly polarized beam by the QWP; this circularly polarized beam is then focused by the same objective lens as that used in the recording process. The circularly polarized reconstructed information beam, which is diffracted from the hologram and reflected by a reflective layer in the disc, is sent back to the objective lens and passed through the QWP again. The reconstructed beam, which is converted from a circularly polarized state to an *s*-polarized state, is reflected by the PBS and detected by a high-speed complementary metal-oxide semiconductor (CMOS) sensor. A red laser is used for optically controlling the servo to adjust the focal point of the objective lens correctly onto the disc correctly and to locate the hologram address in the disc.

The servo and address information are preformatted in the disc as embossed pits. To eliminate the diffraction noise in the recording media, caused by the embossed pits, a dichroic mirror interlayer is used to reflect the green (or blue) laser beams and to allow the red laser beams to pass through. The green (or blue) and red laser beams are combined on the same axis and separated by a dichroic mirror layer in the disc.

A Q-switched repetitive pulse laser is the first choice for a collinear CHDSS because of its high power and continuous controllability. Additionally, the HVD can rapidly read from and write on a disk rotating at a high speed.

A plethora of studies have been published on external cavity lasers, in which a laser diode generates the single-mode light with the aid of a grating. Such lasers can be divided into two types according to the configuration of their optical components. Littman-type lasers use a grating as well as a mirror to exclude beams of other wavelengths, and as a result, have remarkably long coherence lengths [10]. In contrast, the Littrow-type lasers only use a grating, and thus, their coherence length is shorter than that of a Littman-type laser, but their output power is higher because of their simple structures. An HDSS requires high power, but the coherence length of the light source is relatively short, especially in a CHDSS. Therefore, a Littrow-type laser is most suitable for HDSSs [11]. Figure 4.3 shows the schematic of the laser, illustrating how the multimode light is converted to a single-mode light. The lens collimates the multimode light emitted from a laser diode, and then the grating divides the light into bundled zero-order and several diffracted first-order beams. One first-order beam can be diffracted into the laser diode chip by adjusting the incidence angle of the grating. The first-order diffracted beam is injected into the chip and can produce an effect that the feedback beam can cause the light wave with this wavelength to dominate in the chip and emit a single-mode beam from the laser diode. The zero-order beam is used as the light source in an HDSS. In

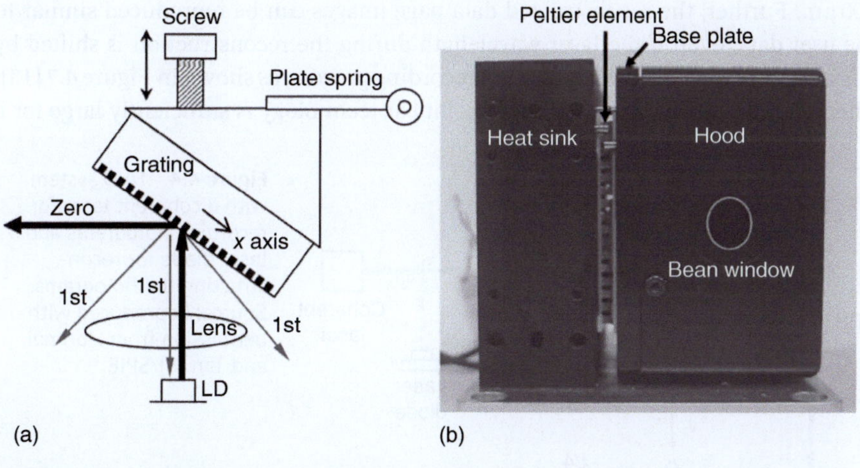


**Figure 4.3** (a) Schematic of the laser; the grating angle is adjusted with a screw and a plate spring. (b) Front view of the laser; the heat sink, Peltier element, and base plate are aligned. Source: (a) Reproduced with permission from Tanaka et al. [11], Optical Society of America. (b) Younse [12]. Reproduced with permission of Optical Society of America.

our CHDSS, a Littrow-type external-cavity blue laser (size: 53 mm × 55 mm × 35 mm (53 mm × 55 mm × 60 mm including heat sink); Sony Corporation) was used to emit a single-mode light up to 80 mW.

### 4.1.3 Laser Diode for HVD-ROM

To develop a small HVD-ROM driver system for practical use, a unique optical configuration design with a small laser light source are required. Although laser diodes are compact and low cost, they are not suitable for a conventional two-axis holographic system because of bad coherence combined with multiple and unstable wavelengths. Because collinear technology requires a light source with a short coherence length, and its wavelength margin is also large [13], a laser diode is the best choice of light source. The construction of the optical setup and the structure of the HVD employed in this technology are presented in Figure 4.4. We demonstrate that a low coherence laser diode is used to retrieve the data from the HVD-ROM system. The experiments and theoretical analysis of the wavelength margin and coherence length of our HVD system are discussed as well.

To develop the HVD-ROM driver system using a laser diode, an optical system, as shown in Figure 4.4, is set up. A coherent laser (wavelength, $\lambda = 407.9$ nm) is used for writing the holograms and a low coherence laser diode ($\lambda \sim 408$ nm; Nichiya Corporation) is used for reconstructing the holograms. The spectra of the two lasers are shown in Figure 4.5, and the reconstructed data page images detected by a high-speed CMOS sensor are shown in Figure 4.6. They are decoded into the user data at a low bit error rate (BER). We can see that the BERs are similar, but their spectra are completely different.

We found that the wavelength margin of collinear holography is over three times larger than that of the conventional two-axis holography at the same thickness of 500 μm. Further, the reconstructed data page images can be reproduced similar to the user data, even if the laser wavelength during the reconstruction is shifted by $\sim\Delta\lambda = 3$ nm from that used during the recording process, as shown in Figure 4.7 [13]. Therefore, the wavelength margin of collinear technology is sufficiently large for a

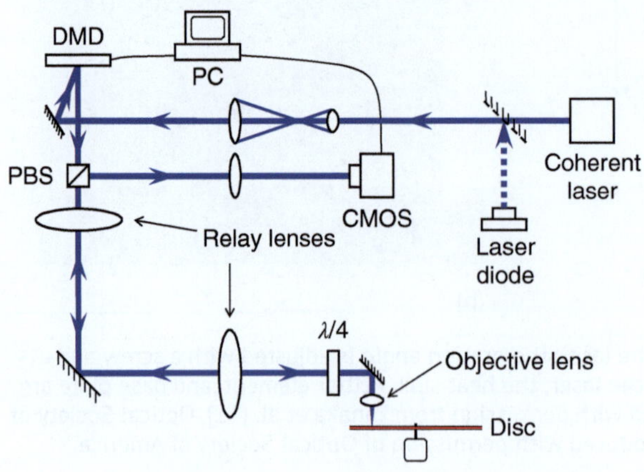


**Figure 4.4** HVD system with a coherent laser for recording holograms and a laser diode for reconstructing the holograms. Source: Reproduced with permission from Horimai and Tan [8]/SPIE.

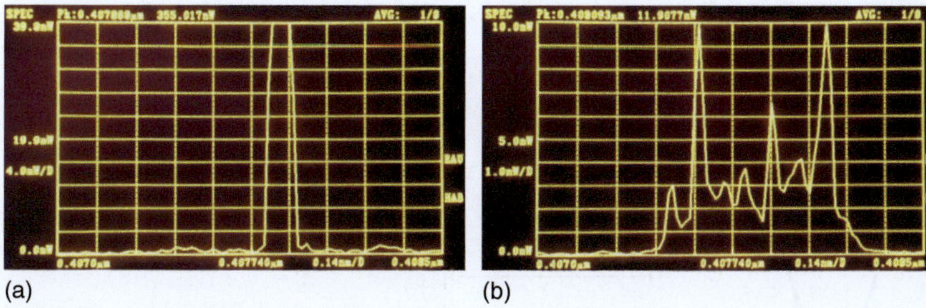


**Figure 4.5** Spectra of (a) a coherent laser and (b) a laser diode. Source: Reproduced with permission from Horimai and Tan [8]/SPIE.

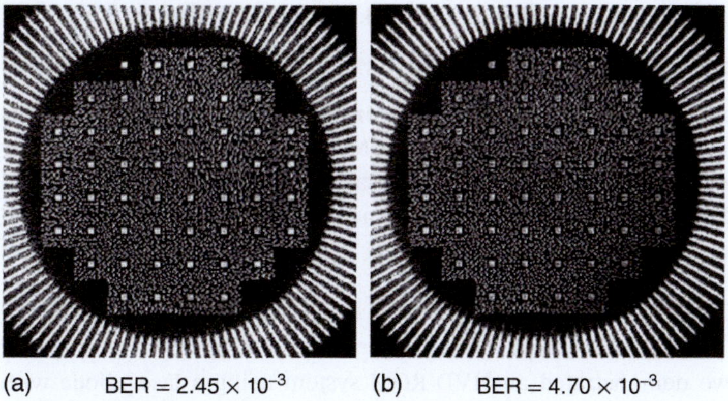


**Figure 4.6** Data page pattern images reconstructed by (a) a coherent laser and (b) a laser diode. Source: Based on Dudley et al. [14].

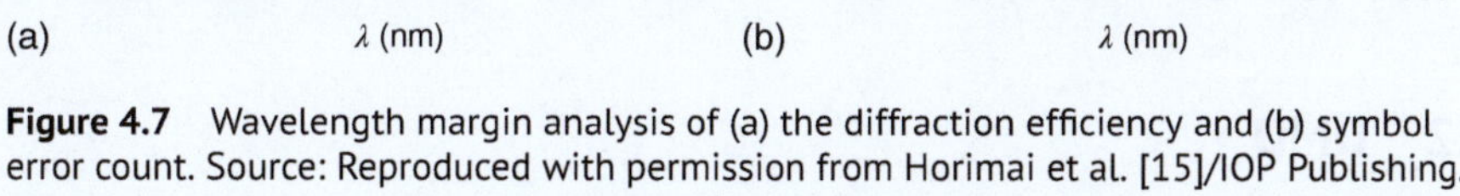


**Figure 4.7** Wavelength margin analysis of (a) the diffraction efficiency and (b) symbol error count. Source: Reproduced with permission from Horimai et al. [15]/IOP Publishing.

laser diode (even if it emits multiple and unstable wavelengths) to be used as the light source of an HVD system. However, in a conventional two-axis holographic system, the reconstructed images cannot be received if the wavelength is shifted by over 1 nm, at which point the laser wavelength in the reconstructing process differs from that of the recording process.

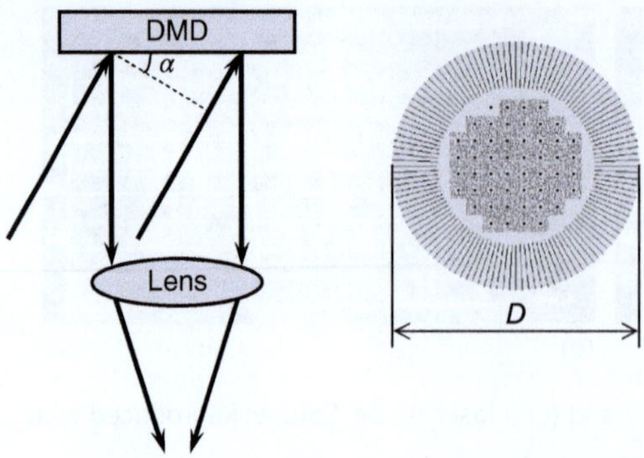


**Figure 4.8** Necessary coherence length of collinear technologies. Source: Reproduced with permission from Horimai and Tan [8]/SPIE.

The necessary coherence length of collinear technologies can be calculated from their unique configurations, as shown in Figure 4.8, using the following equation:

$$L = D \sin \alpha \tag{4.1}$$

where $L$ is the coherence length, $D$ is the dimension of the data page pattern, and $\alpha$ is the angle of the input and output beams on the DMD as shown in Figure 4.1. In our HVD system, $D = 4.4$ mm, $\alpha = 24°$, and the necessary coherence length, $L < 1.8$ mm, was very short. This indicates that collinear technology requires a light source with a short coherence length.

### 4.1.4 Conclusions

In conclusion, we demonstrated an HVD-ROM system using a laser diode with collinear technology. In this system, a 2D data page pattern can be recorded as volumetric holograms generated by information and reference beams that are bundled on the same axis and irradiated on the HVD through a single objective lens. The reconstructing process demonstrates that the proposed HVD system and disc structure are indeed very effective for a low-coherence laser diode emitting unstable wavelengths. Thus, a laser diode is the best choice for an HVD-ROM system as it enables us to design and develop small, practical, and low-cost HVD-ROM systems with more ease.

In addition, theoretical studies are necessary to elucidate why a laser diode can be used in collinear technology. This technology with its unique configuration will enable us to expand its application in other optical information storage systems.

## 4.2 DMD

In the era of big data, CHDSSs have emerged as very promising storage systems owing to their large storage capacities and high transfer rates. DMD as an SLM is the key device of CHDSSs because of its high speed, high precision, and broadband working range.

Usually there are two devices that can be used as the SLM. One is a liquid crystal display (LCD) and the other is DMD. The DMD is a semiconductor-based "light switch" array of thousands of individually addressable, tiltable, mirror pixels [16]. It is the only volume production device that is both a microelectronic mechanical system (MEMS) and an SLM. These tiny mirrors are highly reflective and are used to modulate light, indicating that DMDs can be used both as optical MEMSs and SLMs, and more specifically, as reflective SLMs. LCD is an electronic visual display device that uses the light-modulating properties of liquid crystals and is also a kind of widely used SLM. However, it lacks the properties of high speed, high precision, and broadband capability, which are characteristic of DMDs [12, 15, 17–19].

A DMD is considered as an ideal device for disk drives with holographic media owing to its superior switching speed, high contrast ratio, and overall maturity. In addition, a flexible controller allows system developers to define their own unique data formats instead of varying the color hue and intensity [14, 20–22]. One important parameter of the DMDs is the micromirror tilt angle, which is often fixed after the fabrication of the devices.

Here, we first evaluate the relationship between the diffraction pattern and the micromirror tilt angle of a DMD to obtain the optimal tilt angle, which will provide valuable production guidelines for the selection and fabrication of DMDs to be used in CHDSSs. The results were verified through both theoretical deductions and experiments. Second, we proposed a novel chessboard data page sync mark point based on the optimal tilt angle of the micromirror to obtain the optimal BER in the reading process, while reducing the spatial aperture.

Figure 4.9 shows the schematic optical configuration of collinear holography. Green (or blue) linear-polarized collimated laser beams are used for both writing and reading the holograms. In the write process, a combined image of the

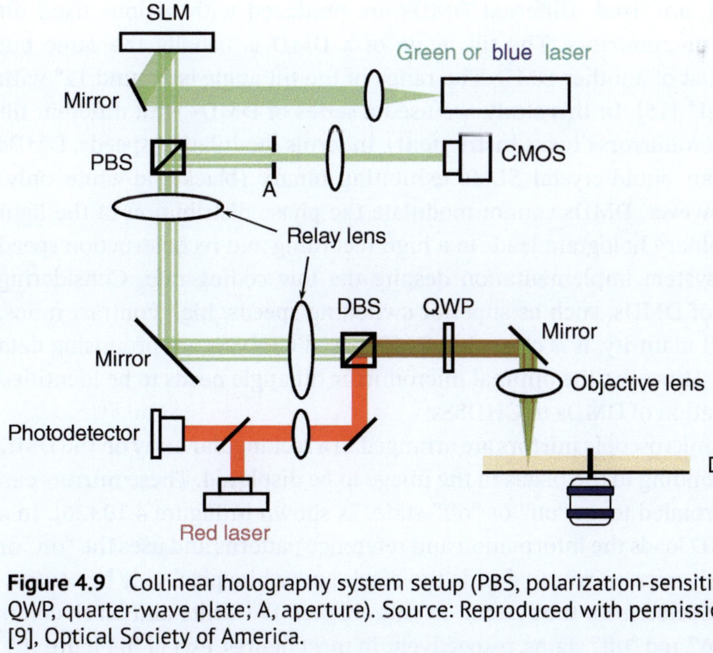


**Figure 4.9** Collinear holography system setup (PBS, polarization-sensitive beam splitter; QWP, quarter-wave plate; A, aperture). Source: Reproduced with permission from Liu et al. [9], Optical Society of America.

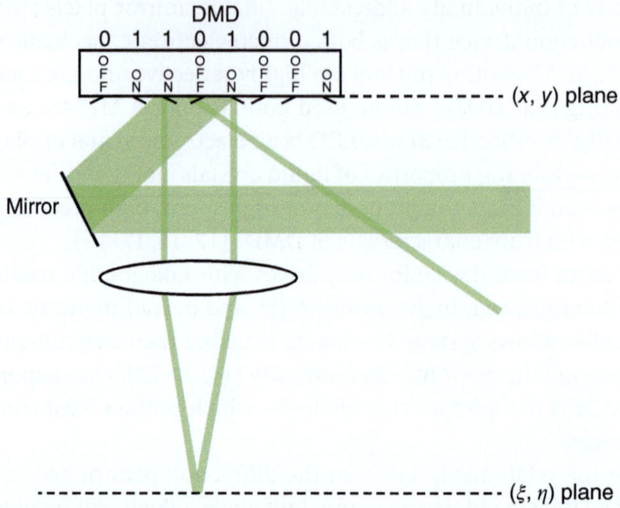


**Figure 4.10** Simplified diagram of the DMD working principle in a CHDSS. Source: Reproduced with permission from Liu et al. [9], Optical Society of America.

information (central part) and reference patterns (outer ring part) is displayed on the SLM, as shown in Figure 4.10.

DMDs, also known as SLMs, have been produced in a wide variety of configurations depending on their applications, such as joint-transform correlator systems [23] and optical neural networks [24, 25]. While fabricating a DMD for different applications, the associated design parameters, such as the size of each pitch and switching speed, are fixed. Different DMDs are produced with various fixed tilt angles of their micromirrors. The tilt angle of a DMD is usually the same but different from that of another DMD. The range of the tilt angle is around 12° with a deviation of ±1° [15]. In this study, we used a series of DMDs with different tilt angles of the micromirrors (Texas Instrument). In terms modulation speeds, DMDs are superior than liquid-crystal SLMs exhibiting binary (black-and-white only) modulation. However, DMDs cannot modulate the phase distribution of the light beam. Using a binary hologram leads to a high recording and reconstruction speed and simplifies system implementation despite the low coding rate. Considering the advantages of DMDs, such as superior switching speeds, high contrast ratios, and their overall maturity, it is clear that DMD-based CHDSSs are promising data storage systems. However, the optimal micromirror tilt angle needs to be identified before the utilization of DMDs in CHDSSs.

Thousands of microscopic mirrors are arranged in a rectangular array on the DMD surface, corresponding to the pixels in the image to be displayed. These mirrors can be individually rotated to an "on" or "off" state, as shown in Figure 4.10 [26]. In a CHDSS, the DMD loads the information and reference patterns and uses the "on" or "off" state of each micromirror to decide the black-and-white pixel only by controlling the light reflected from the system. In this chapter, the white and black colors represent the "on" and "off" states, respectively, in most figures, except in Figure. 4.9.

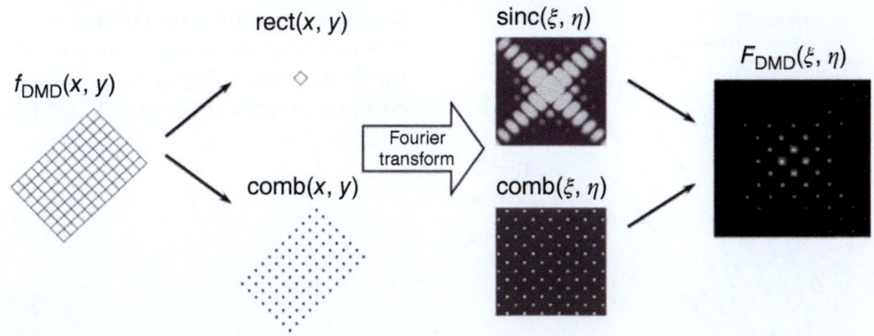


**Figure 4.11** Illustration of the decomposition of a DMD into the convolution of rect and comb functions and its corresponding diffraction pattern at the Fourier plane. Source: Based on Sampsell [26].

Because thousands of micromirrors can only rotate around the diagonal axis of the mirror surfaces, the DMD needs to be rotated by 45° in the clockwise or anticlockwise direction to ensure that the reflected beam is parallel to the optical axis of the system.

Moreover, with thousands of micromirrors on its surface, a DMD can be treated as a grating in the CHDSS. Thus, we treat the DMD as a 2D square array, which can rotate with the diagonal axis. Further, we assume the coordinate to be the $(x, y)$ plane. Thus, the micromirror array can be considered as a repeated distribution of squares, and mathematically, we can separate the micromirror array into a convolution of rect$(x, y)$ and comb$(x, y)$, as illustrated in Figure 4.11.

In the CHDSS, the medium is placed at the real focal plane of the objective lens, namely, the Fourier transform plane of the DMD plane. We assume that the coordinate is the $(\xi, \eta)$ plane. With the decomposition of the DMD, its diffraction pattern can be calculated by multiplying sinc$(\xi, \eta)$ and comb$(\xi, \eta)$, which are the Fourier transforms of rect$(x, y)$ and comb$(x, y)$, respectively, as shown in Eqs. (4.2) and (4.3):

$$f_{\text{DMD}}(x, y) = \text{rect}(x, y) \otimes \text{comb}(x, y) \tag{4.2}$$

$$F_{\text{DMD}}(\xi, \eta) = \text{sinc}(\xi, \eta) \times \text{comb}(\xi, \eta) \tag{4.3}$$

Because the two parts are independent, we can discuss them separately. First, let us consider the "comb" part of the two equations. We can treat a DMD as a grating to analyze its diffraction pattern. As shown in Figure 4.12, the spatial period of the micromirror on a DMD is $a$, incident angle is $\theta_0$, diffraction angle is $\theta$, and wavelength is $\lambda$. Thus, we can write the grating equation as:

$$a(\sin\theta - \sin\theta_0) = m\lambda \tag{4.4}$$

In the CHDSS, we used the vertical direction of the DMD, which means $\theta = 0$, and obtained:

$$m = -a\sin\theta_0/\lambda \tag{4.5}$$

We can see that $m$ is determined only by the incident angle $\theta_0$ and is independent of the tilt angle of the micromirror because both $a$ and $\lambda$ are constants. The diffraction orders to be used in a CHDSS, perpendicular to the DMD, are shown in Figure 4.13.

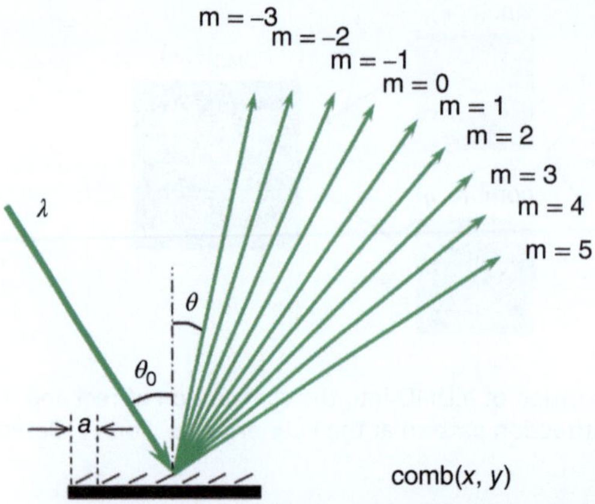


**Figure 4.12** Diffraction orders at different angles for the comb part of the DMD. Source: Reproduced with permission from Liu et al. [9], Optical Society of America.

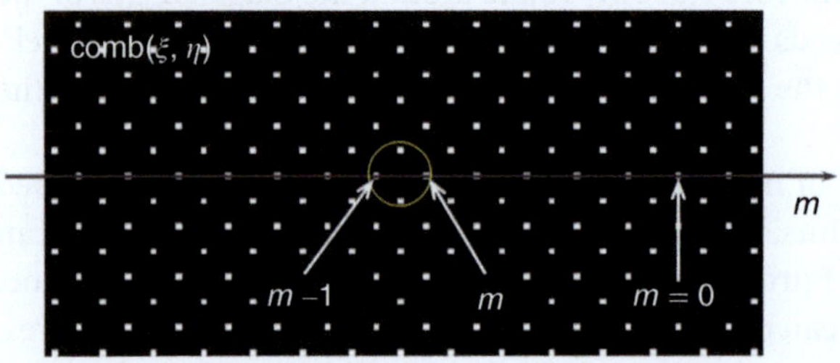


**Figure 4.13** Diffraction orders to be used in the CHDSS, in a direction perpendicular to that of the DMD. Source: Reproduced with permission from Liu et al. [9], Optical Society of America.

To collect as much energy as possible, we want to use as many diffraction orders as possible with a fixed aperture size. In Figure 4.13, the yellow circle represents the energy-collecting range, and the indicated four energy points are the most efficient way to form the interference fringe. Thus, instead of the $m$th fringe, the center of the $(m-0.5)$th fringe lies perpendicular to the DMD plane.

Under this condition we can get the following Eqs. (4.6) and (4.7):

$$m - 0.5 = -a \sin \theta_0 / \lambda \tag{4.6}$$

$$\sin \theta_0 = (0.5 - m)\lambda / a \tag{4.7}$$

Then, we consider the "rect" part of Eq. (4.2), as shown in Figure 4.14. The tilt angle of the micromirror determines the direction of the Fourier transform of the rect function part. The white circle represents the region in the CHDSS, perpendicular to DMD surface. In Figure 4.14, the diffraction pattern is shifted for different DMDs with different tilt angles, indicating that the position shift of sinc$(\xi, \eta)$ depends on the tilt angle of the micromirror.

The diffraction pattern is acquired in the vertical direction of the DMD in the CHDSS. We calculated the relationship between $\theta_0$ and the micromirror tilt angle $\psi$ according to the geometric relationship, and the corresponding result is described by the following equation:

$$\theta_0 = 2\psi \tag{4.8}$$

Substituting Eq. (4.8) into Eq. (4.7), we obtain:

$$\sin 2\psi = (0.5 - m)\lambda / a \tag{4.9}$$

We can use Eq. (4.9) to obtain the optimal micromirror tilt angle of a DMD.

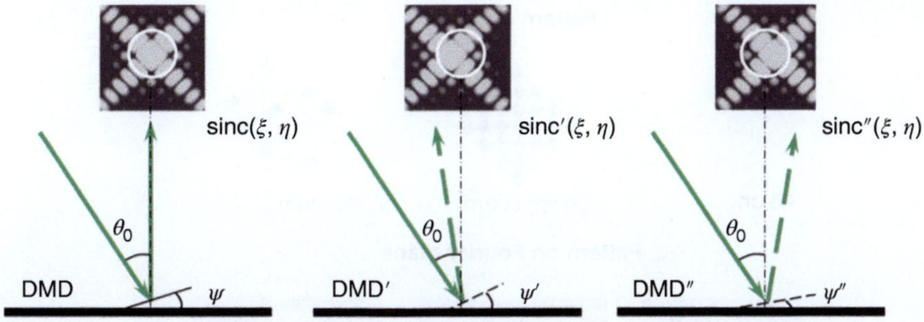


**Figure 4.14** Effect of tilt angle on the diffraction pattern of the rect function. Source: Based on Horimai and Tan [8].

**Table 4.1** Optimal tilt angle at different diffraction orders.

| $M$ | $\theta_0$ | $\Psi$ |
|---|---|---|
| −1 | 4.74° | 2.97° |
| −2 | 7.91° | 3.96° |
| −3 | 11.11° | 5.56° |
| −4 | 14.34° | 7.17° |
| −5 | 17.62° | 8.81° |
| −6 | 20.94° | 10.48° |
| −7 | 24.37° | 12.18° |
| −8 | 27.88° | 13.94° |
| −9 | 31.52° | 15.76° |

Source: Reproduced with permission from Liu et al. [9], Optical Society of America.

Thus, we studied the relationship between the micromirror tilt angle and DMD diffraction pattern as expressed by Eq. (4.9). Further, we used 9.67 μm as the pitch size $a$ of the DMD, and a green laser with $\lambda = 532$ nm. We substituted these values into Eq. (4.9) and obtained the corresponding tilt angles as listed in Table 4.1.

Based on the overall system arrangement as well as the rotation range of the micromirror (~12°) [15], we found that the diffraction order of −7 is suitable for this condition. At this order, the incident angle is 24.37°, and the optimal tilt angle is 12.18°.

We also experimentally confirmed the optimal tilt angle of 12.18°. The experimental light path diagram is shown in Figure 4.10. To simulate different page data information, we applied different encoding patterns on the DMD and observed the diffraction pattern on the Fourier plane. The corresponding results are shown in Figure 4.15, indicating that the rect-function diffraction pattern of the DMD shifts with the tilt angle, irrespective of the applied encoding pattern. At a tilt angle of 12.2°, which is very close to the optimal angle, the overall diffraction pattern lies at the center of the marked board on the Fourier plane. Due to limited precision of angle control during the experiment, we could control the micromirror angle in steps of 0.2°.

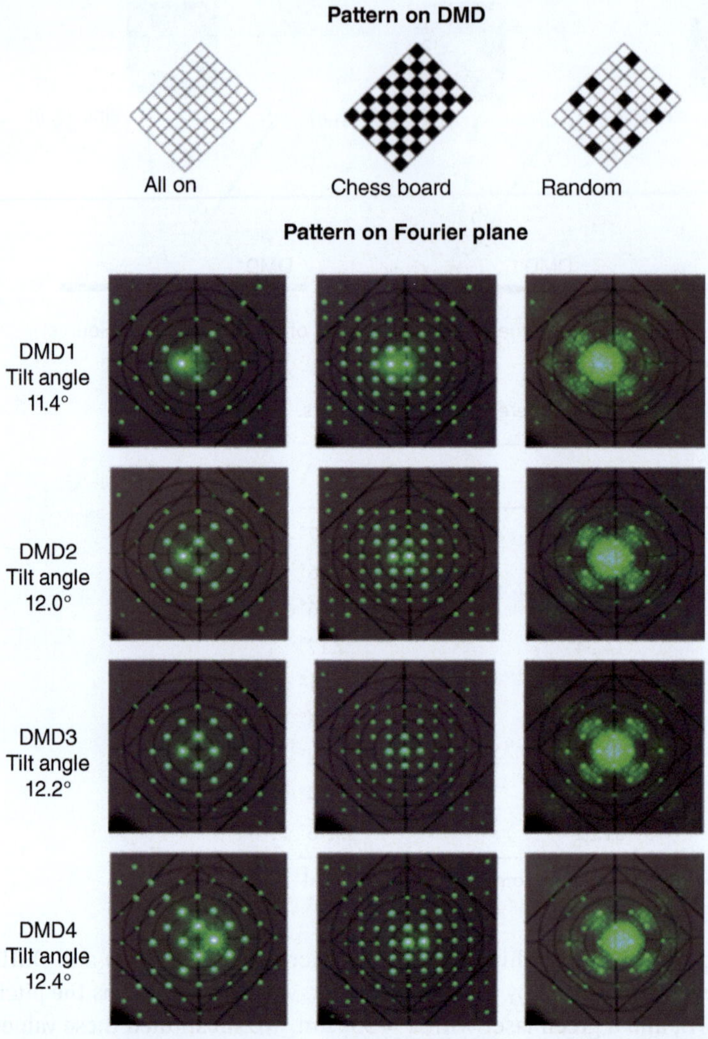


**Figure 4.15** Diffraction patterns of the DMD at different tilt angles of the micromirror for all on, chessboard, and random pattern encoding of the DMD. Source: Reproduced with permission from Liu et al. [9], Optical Society of America.

## 4.3 CMOS

In a CHDSS, a CMOS device is used as the image sensor consisting of 2D array of thousands of discrete pixels. The working principle of a CMOS is illustrated in Figure 4.16, indicating the charge-to-voltage conversion in each pixel. This difference in readout techniques has significant implications for sensor architecture, capabilities, and limitations. The main contribution of the CMOS active pixel sensors is the combination inside the pixel of the detector, the charge-to-voltage conversion, and transistor facilitating buffering and addressing capabilities [27, 28].

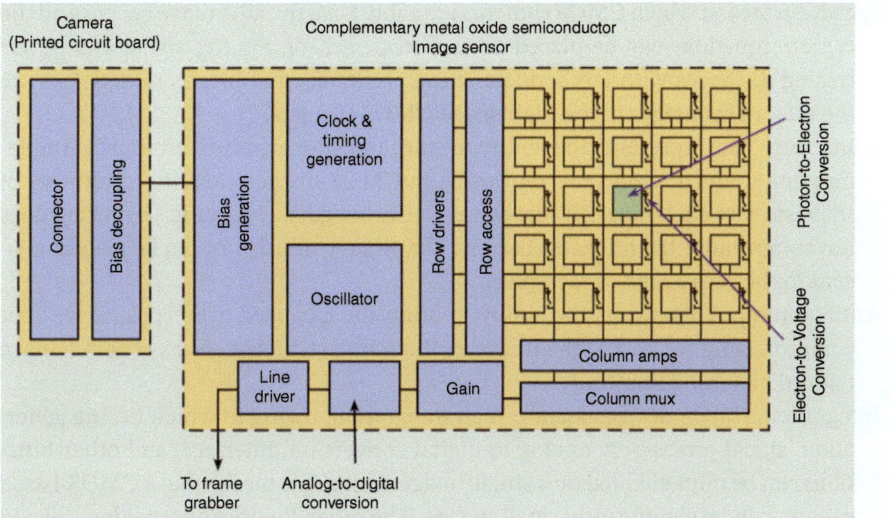


**Figure 4.16** Working principle of a CMOS sensor. Source: [27]/Laurin Publishing Co. Inc.

The following parameters of the CMOS are usually focused on by us.

Light sensitivity. In a CMOS device, a large number of sensors are cluttered on the chip. Thus, the light sensitivity reduces as there are a number of circuits through which the light needs to penetrate, and consequently, a low-quality image is obtained. However, due to the presence of small circuits on the chip, the image is boosted at every stage. Hence, even though the original image may be of low quality, the output image has a better quality.

Power. A CMOS camera converts the data into digital form and sends all the data simultaneously, thus consuming only a negligible power.

Rolling shutter. A CMOS camera has a rolling shutter, wherein all the parts of a frame are not captured at a time. Each part of the frame is captured separately, and then, all these parts are displayed at once. This may induce a time lag in the frames, and the images may wobble or suffer from skewness. However, the high-end CMOS cameras have more efficient sensors.

Image quality. CMOS sensors, because of their layouts, cannot fit more pixels on their surface. Thus, we obtain low-resolution images, which affect the image quality.

Reliability. Both the image chip types are equally reliable in most consumer and industrial applications. In ultra-rugged environments, CMOS imagers have an advantage because all the circuit functions can be exhibited by a single integrated circuit chip. This minimizes leads and solder joints, which are the leading causes of circuit failures in extremely harsh environments.

Windowing. One unique capability of CMOS technology is the ability to read out a portion of the image sensor. This allows elevated frame or line rates for small regions of interest and is an advantageous in some applications, such as high-temporal-precision object tracking in a sub-region of an image.

Speed. An area in which CMOS cameras arguably have the advantage because all the camera functions can be placed on the image sensor. Shorter signal and power tracing distances with less inductance, capacitance, and propagation delays are the major advantages of the high-speed CMOS imagers.

Shuttering. This represents the ability to start and stop exposure arbitrarily. Implementing uniform electronic shuttering in CMOS imagers requires a number of transistors in each pixel. In line-scan CMOS imagers, electronic shuttering does not compromise the fill factor because the shutter transistors can be placed adjacent to the active area of each pixel.

Antiblooming. This means the ability to drain the localized overexposure without compromising the rest of the image in the sensor. CMOS devices generally have natural blooming immunity.

Integration. CMOS devices feature high integration, owing to which timing generation, signal processing, analog-to-digital conversion, interface, and other functions can be implemented on a single imager chip. This means that a CMOS-based camera can be significantly small in size. The other significant considerations in system integration are adaptability, flexibility, and speed of change. Most CMOS image sensors are designed for large consumer or near-consumer applications. They are highly integrated and tailored for few applications.

In the readout process, the reconstructed hologram is imaged onto the output detector array, where the digital data are extracted from the detected signal. Noise can cause errors during the detection process in various ways. The electronic detection process at the camera contributes the same amount of noise, irrespective of the brightness of the hologram, and as the number of holograms or the readout rate increases, the amount of power diffracted into each reconstructed hologram decreases. The detector array does not line up with the pixel array in the reconstructed hologram because of errors in camera registration, rotation, focus, tilt, and image magnification. The detector receives undesired light because of the light scattered from the storage material, crosstalk from other stored holograms, or crosstalk between the neighboring pixels of the same hologram. Although the contribution from the crosstalk increases with the strength of the holograms, scattering depends only on readout power and the components' optical quality. There are brightness variations across the detected image. This can be a problem if a single threshold is used across the image to separate the pixels into bright and dark and to assign binary values. These fluctuations can be caused by the SLM, optical imager, or original laser beams.

Usually, the pixel pitch of the CMOS imager is smaller than that of the SLM, and thus, a simple 4-$f$ system cannot be used. Matching between the CMOS and the SLM must be considered carefully to avoid a high BER. In our amplitude modulation system, we used a DMD as the SLM with a pixel pitch of 13.68 μm; the pixel pitch of the CMOS sensor was 4.5 μm. Therefore, we used the $3 \times 3$ pixel array on the CMOS to capture one pixel on the DMD. Moreover, the 4-$f$ system required a slight adjustment, which zoomed out the pixel on the DMD from 13.68 to 13.5 μm. Then, we captured the correct value and reduced the unwanted noise to some extent.

## 4.4 Servo Actuator

Collinear™ technology as a new reading and writing method for a holographic storage system is very promising and differs from the conventional two-axis holography. The unique feature of this technology is that the 2D page data are recorded as volume holograms generated by coaxially aligned information and reference beams, which are displayed simultaneously by the same SLM and interfere with each other in the recording medium (disc) through a single objective lens.

In the HVD system, the green (or blue) and red laser beams are combined at the same axis and are transmitted through a single objective lens. Consequently, the HVD is designed with a special disc structure. Figure 4.17 shows the schematic of the six layers of the HVD structure. A photopolymer is used as the holographic material in the recording layer. The HVD has a reflective layer with preformatted address information. Based on this layer, the focusing and tracking servo technologies can be used in the recording and reconstructing processes, enabling precise maintenance of the distance and relative position of the objective lens and the disc, and the holograms can be accurately recorded and reconstructed in the HVD, even if there is axial deflection or radial runout. Furthermore, a vibration isolator is no longer necessary [8]. Figure 4.18 shows the schematic optical configuration of the actuator used in collinear technology, and Figure 4.19 shows the actuator design. Figure 4.20 presents the electrical signals of the address and servo information read from the preformatted embossed pits using a red laser. The focus servo signal is shown in Figure 4.21.

Figure 4.22 shows a photograph of the actuator actually incorporated in the ES-1,which is a system developed for the purpose of demonstration of dynamic recording and reproduction of holograms and acceleration of research and development of hologram media, and its peripheral part (in the actual machine and the

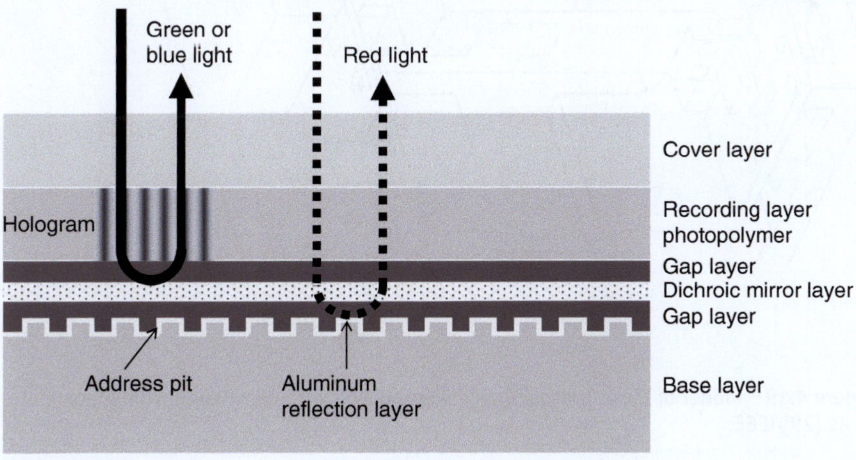


**Figure 4.17** Schematic showing a disc with a dichroic reflective structure. Source: Reproduced with permission from Matsumoto et al. [29]/IEEE.

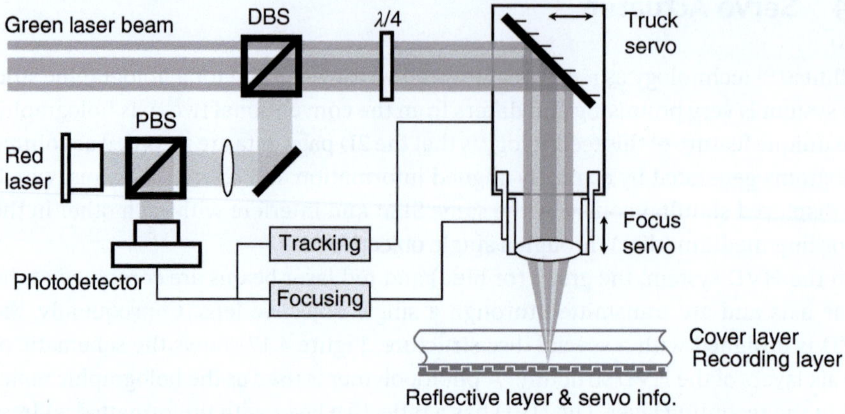


**Figure 4.18** Optical configuration of the actuator used in the HVD system. Source: Reproduced with permission from Matsumoto et al. [29]/IEEE.

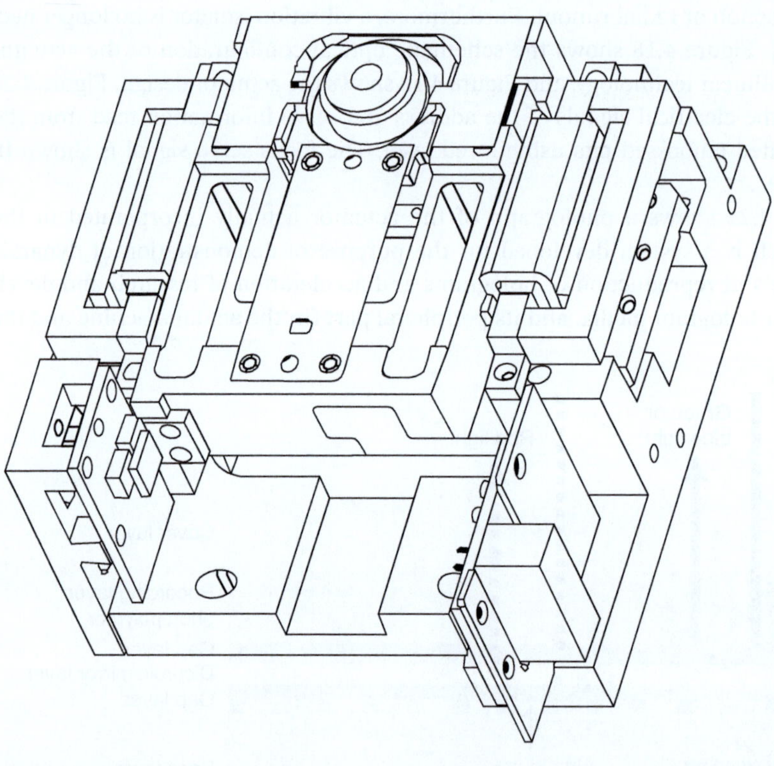

**Figure 4.19** Model of the actuator. Source: Reproduced with permission from Matsumoto et al. [29]/IEEE.

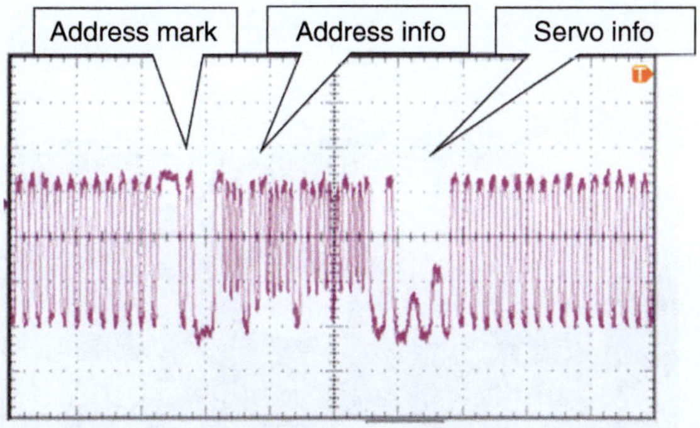


**Figure 4.20** Electrical signal of the preformatted embossed pit. Source: Reproduced with permission from Matsumoto et al. [29]/IEEE.

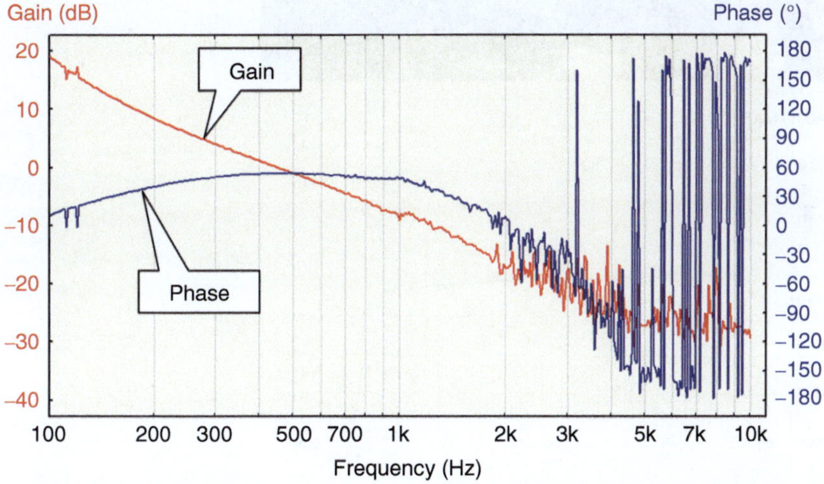


**Figure 4.21** Electrical signal of the focus servo. Source: Reproduced with permission from Matsumoto et al. [29]/IEEE.

positions of the disk; the actuators are upside down) and the measurement results of the focusing and tracking open-loop characteristics. It was developed as a design target that can handle rotation speeds up to 1000 rpm to realize collinear on-the-fly hologram recording/playback following the continuous rotation and eccentricity of the HVD.

For reference, some photos of the devices developed specifically for ES-1 are shown in Figure 4.23. Figure 4.23a shows a dual-wavelength objective lens [1], which is built into the biaxial actuator shown in Figure 4.23b to perform servo focusing and tracking. The transfer rate in the collinear method does not depend on the number of revolutions of the disk, and the upper limit is determined by the frame rate for displaying/reading the page data. Therefore, a DMD with a display speed of 10 000 fps or higher is used for the recording, and a high-sensitivity CMOS sensor that can shoot at an ultra-high frame rate of 5000 fps is used for the playback. A dedicated circuit board suitable for data storage is used. Further, to display the page data on the DMD at a high speed and to decode the page data read

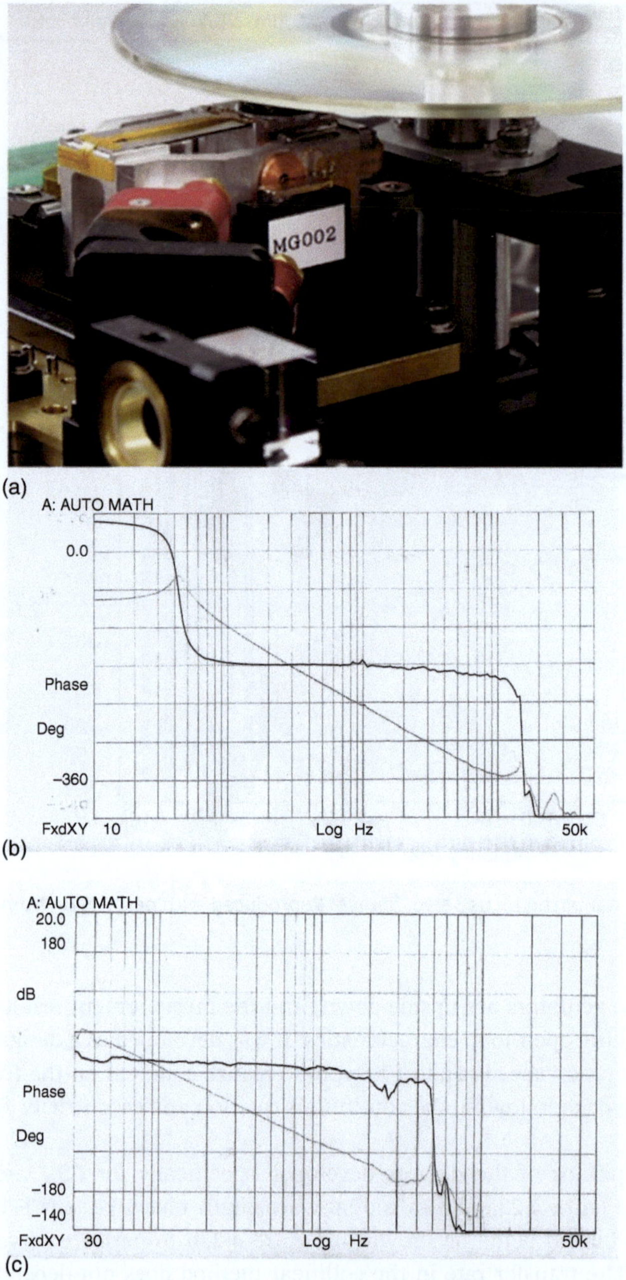


**Figure 4.22** (a) ES-1 actuator (b) its focusing open-loop characteristics and (c) tracking open-loop characteristics. Source: Hideyoshi Horimai.

**Figure 4.23** Various components and signal processing boards developed exclusively for ES-1. Source: Hideyoshi Horimai.

by the CMOS sensor at a high speed, the speed of the data page signal processing is also boosted. Figure 4.23c shows the developed ultra-high-speed CMOS sensor and detection circuit, and Figure 4.23d shows an overview photograph of the developed high-speed data page signal processing board (details are omitted in this paper).

By adopting the opto-mechanical servo technology with the biaxial actuator, dual-wavelength optical system, and wavelength-selective reflection film structure, it is finally possible to move to any position on the rotating HVD based on the physical-format information, which is preformatted on the HVD base substrate. With accurate positioning, the 2D page data can be effectively recorded and reproduced.

The physical-format information is continuously engraved in the circumferential direction of the disc in the form of minute preformatted pits, similar to the optical disc. This preformat is formed by alternately repeating two types of header and tracking data. The header consists of the track number in the radial direction of the disk, sector number in the circumferential direction, and error correction codes. The header information along with the next header information are filled with the tracking data of a single-clock frequency pit. By reading the preformatted pits with a red laser ($\lambda = 650\,\text{nm}$), which is not photosensitive to the holographic recording material, the position information for recording and reproducing the hologram is obtained along with the servo signals for focusing and tracking.

The address and tracking data are engraved according to the reference clock, and hence, the hologram recording/reproducing position can be uniquely determined in the circumferential and radial directions of the disc with a high positioning accuracy ±25 nm or less in the plane.

## References


1 Coufal, H.J., Psaltis, D., and Sincerbox, G.T. (ed.) (2000). *Holographic Data Storage*, Springer Series in Optical Sciences. Berlin: Springer-Verlag.

2 Mok, F.H. (1993). Angle-multiplexed storage of 5000 holograms in lithlium niobate. *Opt. Lett.* 18: 915–917.

3 Denz, C., Pauliat, G., Roosen, G., and Tschudi, T. (1991). Volume hologram multiplexing using a deterministic phase encoding method. *Opt. Commun.* 85: 171–176.

4 Lande, D., Heanue, J.F., Bashaw, M.C., and Hesselink, L. (1996). Digital wavelength-multiplexed holographic data storage system. *Opt. Lett.* 21: 1780–1782.

5 Tan, X., Matoba, O., Shimura, T., and Kuroda, K. (2001). Improvement in holographic storage capacity by use of double-random phase encryption. *Appl. Opt.* 40: 4721–4727.

6 Kang, Y.H., Kim, K.H., and Lee, B. (1997). Volume hologram scheme using optical fiber for spatial multiplexing. *Opt. Lett.* 22: 739–741.

7 Horimai, H., Tan, X., and Li, J. (2005). Collinear holography. *Appl. Opt.* 44: 2575–2579.

8 Horimai, H. and Tan, X. (2005). Holographic versatile disc system. *SPIE Symposium on Optics & Photonics 2005, Organic Holographic Materials and Applications III* (San Diego, California, USA, 2005) (ed. K. Meerholz), *Proceedings of SPIE*, Volume 5939, pp. 1–9.

9 Liu, J., Horimai, H., Lin, X., Liu, J., Huang, Y. and Tan, X. (2017). Optimal micro-mirror tilt angle and sync mark design for digital micro-mirror device based collinear holographic data storage system. *Appl. Opt.* 56 (16): 4779–4784.

10 Liang, M., Stehr, R.L., and Krause, A.W. (1997). Confocal pattern period in multiple-aperture confocal imaging systems with coherent illumination. *Opt. Lett.* 22 (11): 751–753.

11 Tanaka, T., Takahashi, K., Sako, K. et al. (2007). Littrow-type external-cavity blue laser for holographic data storage. *Appl. Opt.* 46: 3583–3592.

12 Younse, J.M. (1995). Projection display systems based on the digital micromirror device (DMD). *Micromachining and Microfabrication International Society for Optics and Photonics,* Volume 2641, pp. 64–75.

13 Littman, M.G. and Metcalf, H.J. (1978). Spectrally narrow pulsed dye laser without beam expander. *Appl. Opt.* 17: 2224–2227.

**14** Dudley, D., Duncan, W., and Slaughter, J. (2003). Emerging digital micromirror device (DMD) applications. *Proceedings of SPIE – The International Society for Optical Engineering*, Volume 4985.

**15** Horimai, H., Tan, X., Li, J., and Suzuki, K. (2005). Wavelength margin analysis in advanced collinear holography. *Jpn. J. Appl. Phys.* 44: 3493–3494.

**16** Hanley, Q.S., Verveer, P.J., Gemkow, M.J. et al. (1999). An optical sectioning programmable array microscope implemented with a digital micromirror device. *J. Microsc.* 196 (3): 317–331.

**17** Wagner, E.P., Smith, B.W., Madden, S. et al. (1995). Construction and evaluation of a visible spectrometer using digital micromirror spatial light modulation. *Appl. Spectrosc.* 49 (49): 1715–1719.

**18** Nesbitt, R.S., Smith, S.L., Molnar, R.A., and Benton, S.A. (1999). *Holographic Recording Using a Digital Micromirror 'Device. Electronic Imaging'99*. International Society for Optics and Photonics.

**19** Douglass, M.R. (1998). Lifetime estimates and unique failure mechanisms of the Digital Micromirror Device (DMD). *Reliability Physics Symposium Proceedings*, 1998 IEEE International 1998, pp. 9–16.

**20** Florence, J.M. (1989). Joint-transform correlator systems using deformable-mirror spatial light modulators. *Opt. Lett.* 14 (7): 341–343.

**21** Tanaka, K., Hara, M., Tokuyama, K. et al. (2007). Improved performance in coaxial holographic data recording. *Opt. Express* 15 (24): 16196–16209.

**22** Yoder, L., Duncan, W., Koontz, E.M. et al. (2001). DLPTM technology: applications in optical networking. *Proceedings of SPIE – The International Society for Optical Engineering*, Volume 4457, pp. 54–61.

**23** Lin, L.-Y., Goldstein, E.L., and Tkach, R.W. (2000). On the expandability of free-space micromachined optical cross connects. *J. Lightw. Technol.* 18 (4): 482–489.

**24** Collins, D.R., Sampsell, J.B., Hornbeck, L.J. et al. (1989). Deformable mirror device spatial light modulators and their applicability to optical neural networks. *Appl. Opt.* 28 (22): 4900–4907.

**25** Gregory, D.A., Juday, R.D., Sampsell, J. et al. (1988). Optical characteristics of a deformable-mirror spatial light modulator. *Opt. Lett.* 13 (13): 10–12.

**26** Sampsell, J.B. (1994). Digital micromirror device and its application to projection displays. *J. Vac. Sci. Technol., B* 12 (6): 3242–3246.

**27** Litwiller, D. (2001). CCD vs. CMOS: facts and fiction. *Photonics Spectra* 35 (1): 154–158.

**28** Magnan, P. (2003). Detection of visible photons in CCD and CMOS: a comparative view. *Nucl. Instrum. Methods Phys. Res., Sect. A* 504 (1–3): 199–212.

**29** Matsumoto, M., Matsuda, H., Sasada, T., Sakai, Y., and Horimai, H. (2006). HVD Servo Actuator. Optical Data Storage Topical Meeting. IEEE.

# 5
# Materials

## 5.1 Holographic Media for Commercial Uses

### 5.1.1 Introduction

Holographic data storage (HDS) has many advantages, including the highest density of removable media, faster transfer rates (much faster than those of the other optical transmission devices), very-low-cost media, resistance to piracy because of volume recording, and design flexibility allowing disc cards and copying ROM, both of which are recordable and rewritable [1–4]. At the beginning of the twenty-first century, a large number of important materials and engineering research work were initiated in the field of HDS. However, advanced optical storage applications that required high-performance and low-cost devices still encountered difficulties related to the procurement of suitable system components and storage media [5]. The lack of suitable storage materials has long been an obstacle to the development of holographic storage technology. Photopolymers are being touted as new materials for holographic storage because of their compelling properties, such as high diffraction efficiencies, high resolutions, high signal-to-noise ratios (SNRs), low costs, simple processes, and high holographic fidelity, which are suitable for making large-area disc storage materials. These features are ideal for applications such as the development of high-capacity and highly dynamic memory. Companies, such as InPhase and Aprilis, systematically developed photopolymer recording materials, which greatly contributed to the commercial development of photopolymers as media for HDS. In this section, we briefly describe the holographic media and recording systems developed by these two companies.

### 5.1.2 Media Developed by InPhase and Aprilis

#### 5.1.2.1 HDS Recording System

Holographic storage facilitates the use of the entire volume of its medium, and not just the surface, to store information. For writing the data in a holographic storage system, a beam splitter divides the input light from a single laser into two parts: a signal beam that carries the data and a reference beam. The hologram is formed where these two beams intersect in the volume of the recording medium. A spatial

*Collinear Holography: Devices, Materials, Data Storage*, First Edition.
Xiaodi Tan, Hideyoshi Horimai, Tsutomu Shimura, and Xiao Lin.

light modulator (SLM) encodes the data onto the signal beam by translating the bit stream of 0 s and 1 s into an optical checkerboard pattern of light and dark pixels. The data are arranged in an array or page of around a million bits, and the exact number of bits is determined by the pixel count of the SLM [6]. When the reference and signal beams intersect in the storage medium, they record the hologram in the light-sensitive storage medium. In photopolymer materials, a photoinduced polymerization is responsible for the index modulation that creates the permanent hologram. By varying the reference beam angle or wavelength or by shifting the position of the medium, many different holograms can be recorded in the same volume of the material, thereby vastly increasing the data density. Superimposing many holograms in the same volume is called multiplexing, which enables the realization of very high densities [7–9]. To read the data, a reference beam with characteristics identical to those used to store the data diffracts off the index modulation (hologram), reconstructing the data page with the stored information. The image of the reconstructed data page is captured by a detector, and the data are read in parallel. This parallel recording and readout of the data provides holography with its potential for very fast transfer rates.

#### 5.1.2.2 InPhase Media

InPhase Technologies, based in Longmont, Colorado, is a company that develops holographic storage devices and media. It was founded in the late 2000s by the core technical team from Bell Labs, who invented and developed the breakthrough holographic storage technology. InPhase Technologies held the record for the "highest commercial data storage" by achieving 1 Gb/in$^2$ of storage media at that time. The major challenge that hinders the implementation of holographic storage is the development of a suitable storage medium. Tapestry material is a new material invented by the InPhase team at Bell Labs. Typical photopolymers use a single chemical process for bonding the molecules together to form the medium and perform the recording. InPhase has developed a polymer system with materials that use two distinct chemistries that are independent but compatible. One chemistry allows the formation of the medium and the control of the mechanical, manufacturing, and data/archive life parameters. The second chemistry plays a role in the recording process. These two chemistries do not interact or interfere with each other during the recording, and therefore, the photopolymers have a high dynamic range with an extreme dimensional stability. The chemistries are also compatible, achieving minimized scattering, good optical quality, and thick, robust media [10]. Embodiments of the polymer matrix can comprise components, including isocyanates, such as NCO groups, and polyol, such as OH groups, which can provide compatibility with photoactive monomers, allowing quick and easy manufacturing of optical articles. Further, the embodiments of the photoactive polymerizable materials in index contrast applications, such as HDS, holographic optical element development, waveguide fabrication, and lithography, can be improved by maintaining an equivalent ratio. The hydroxyl functionality of the polyol component with respect to the isocyanate functionality of the NCO component is from 0.5 to 1.5. A photoinitiator is required during the recording of the

Tapestry material, which comprises an acylphosphine oxide, an oxime ester, or a combination thereof. The properties of the photoactive polymerizable material can be enhanced by improving the type of organic photoinitiator therein [11].

InPhase has developed high-performance media for recording with a blue laser. The blue-sensitized holographic media provide a higher dynamic range and sensitivity compared to the previously disclosed blue-sensitized holographic media. These media can be used in a variety of applications, such as data storage or for optical waveguides, where items provide greater optical limitations and faster manufacturing time [11, 12]. The media developed by InPhase are available for use with recording systems using blue and green wavelength lasers. The characteristics of the media [6] are summarized in Table 5.1.

Experimental results showed that after eight weeks of aging at 60 °C, the media retained dynamic ranges higher than 80% of their respective initial values [13]. In addition, the sensitivity of the HDS media did not degrade. Testing results showed that archival aging leads to the formation of two types of media, both of which exhibit excellent archival stability, retaining more than 95% of initial dynamic range after aging for 2000 hours (86 days) at 80 °C. The InPhase Tapestry materials are the breakthrough commercially-viable storage materials. Overall, the material offers a high dynamic range, high photosensitivity, dimensional stability, optical clarity, manufacturability, nondestructive readout, tunable thickness, and environmental and thermal stability.

In addition to developing a new class of materials, InPhase developed automated methods to manufacture the media. Their system uses the ZeroWave method of InPhase that produces optically flat media without requiring optically flat substrates [2]. InPhase Technologies also developed manufacturing processes such as the ZeroWave manufacturing processes, to enable cost-effective fabrication of high-quality optical media. To record large pages of data containing 1 million bits with a good SNR, the medium must have a very good optical quality as measured by its transmitted wavefront quality. The wavefront quality of a typical media coupon manufactured by the ZeroWave process is about $(\lambda/8)/\text{cm}^2$ at 633 nm [7, 14]. An optical component of this quality would typically be very expensive to manufacture, but with ZeroWave, it can be made for the same or lower cost than that of a DVD.

**Table 5.1** Characteristics of the media developed by InPhase.

| Holographic properties | Green media | Blue media |
|---|---|---|
| Wavelength of use (nm) | 510–540 | 400–420 |
| Dynamic range (M/# at 200 nm) | 5 | 6 |
| Millimeter thickness | >1 mm | >1 mm |
| Dimensional change during recording | <0.1% | <0.1% |
| Optical flatness | <1/2/cm | <1/2/cm |
| Shelf-life of media | Tested at 60 °C | Tested at 60 °C |
| Archival life of stored data | Tested at 80 °C | Tested at 80 °C |

ZeroWave enables high-performance high-density storage while making the media price competitive for mass consumption. The currently offered Tapestry materials of InPhase are recordable (write once, read many times) and work with either blue, green, or red laser sources. Along with ZeroWave, these materials allow the immediate commercialization of recordable and ROM products.

InPhase designed a system with a predictable and sustainable roadmap, enabling families of products, to make holographic storage "truly" commercially viable. To do this, InPhase adopted a technology platform to optimize the performance, used the available components, simplified the interchange, and supported backward compatibility for a series of product generations. The technology choice in any holographic system is in essence the selection of a multiplexing paradigm. To read the data, InPhase makes the reference beam deflect off the hologram; thus, the stored information can be reconstructed. This hologram is then projected onto a detector that reads the data in parallel. It is then processed using the appropriate channel and error correction codes developed at InPhase and Bell Labs to recover the original data. This parallel readout of the data facilitates the faster transfer rates characteristic of holography [10]. A small unit that records and plays a video was developed to demonstrate the feasibility of the technology. A video featuring the data storing operation and high-quality video streaming of the device is shown. The performance of the device is limited by the control and decoding performed by the computer. InPhase also developed servo marks, which are similar to the DVD/CD marks but are modified to allow holographic recording. The combined use of polytopic and angle multiplexing results in the development of products with performances initially starting at 200 GB and evolving to more than 1 TB. Further, densities in excess of 80 Gb/in$^2$ have been demonstrated at 532 nm [15–17]. The development of Tapestry storage media has enabled the designing and construction of a series of first-generation holographic storage demonstration platforms, as shown in Figure 5.1.

This system uses angle multiplexing and records multiple superimposed holograms in the volume of 0.8- to 1-mm-thick photopolymer media. The holograms

**Figure 5.1** Angle multiplexed demonstration platform. Source: Wilson et al. [17].

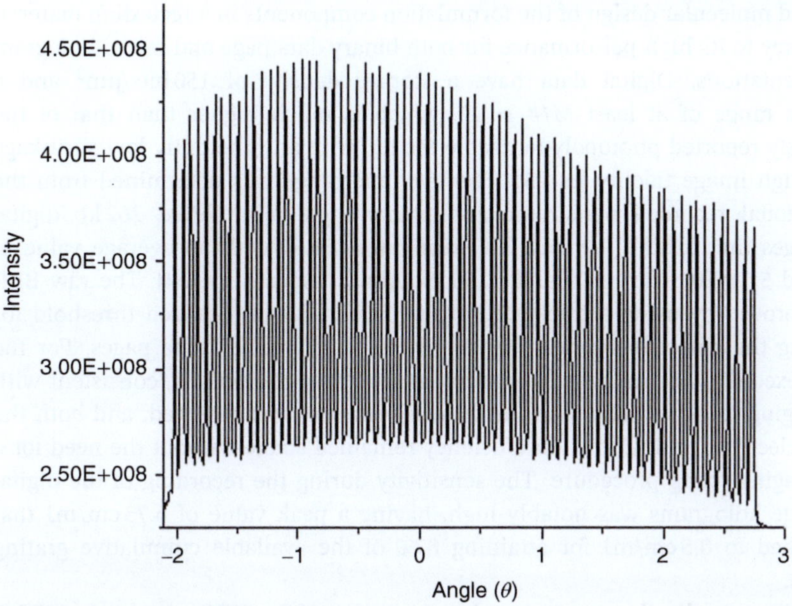


**Figure 5.2** Angle scan of a set of ~90 holograms. Source: Wilson et al. [17].

are stored at a spacing of 0.065°. The optical mechanical assembly contains the modulator, Fourier transform optics, and media mounts. This system recorded 1 Mb data page images with an extremely high fidelity [18]. Single holograms easily exhibit SNRs in the range of 9–10 dB. The superior optical quality of the media makes this fidelity possible. With the current data overhead, more than 5 MB of user data can be stored in a single location. The diffracted intensity of a set of recorded data pages is shown in Figure 5.2. The data page holograms as functions of angle are reported in [5, 6].

#### 5.1.2.3 Aprilis Media

Aprilis, whose business focus was the development of media and optical drive technology for HDS, started in June 1999 and acquired the rights to Polaroid holographic technology. This company achieved a recording speed of 10 Gb/s using holographic recording technology. Since then, the company has produced holographic storage media using cationic ring-opening polymerization (CROP) photopolymers for optical data storage with densities higher than 100 bits/μm$^2$, and has obtained 21 patents covering the basic holographic drive service [19–22].

The HMD-050-G-C-400 photopolymerizable holographic recording material for card and disc media, based on CROP, was optimized for a film thickness of 400 μm [23]. Conventional holographic photopolymers utilize the free-radical linear-chain chemistry to exhibit a significant volume shrinkage, which limits the usable portion of the achievable dynamic range even for moderate inclined angles and media thicknesses. The advantages of the CROP method combined with advanced molecular design of the formulation components in a recording material are the key to its high performance for both binary data page and micro-hologram implementations. Digital data have a storage density of 150 bits/$\mu m^2$ and a dynamic range of at least $M/\# = 22$, which is much higher than that of the previously reported photopolymerizable media, with an inherently low shrinkage and a high image fidelity [5, 23]. The dynamic range was determined from the co-locational recorded peristropic and planar-angle multiplexed 262 kb digital data pages that exhibit low raw bit error rates (BER) with an average value of $10^{-3}$ and $5.5 \times 10^{-3}$ for ~100 and ~ 150 bits/$\mu m^2$, respectively [24]. The raw BER was improved by using a thicker recording material and localized threshold for assigning the detected intensity values of the reconstructed data pages. For the multiplexed digital data page holograms, a good Bragg selectivity, consistent with the imaging thickness and $\text{sinc}^2$ functional behavior, was observed, and both the Bragg selectivity and diffraction efficiency remained stable without the need for a post-imaging fixing procedure. The sensitivity during the recording of the digital data page holograms was notably high, having a peak value of 6.75 cm/mJ that diminished to 0.5 cm/mJ for attaining 87% of the available cumulative grating strength.

In addition to the above-mentioned optimization of the HDS property by CROP, further improvements can be made by designing a media sensitive to shorter laser wavelengths while increasing the dynamic range and recording sensitivity at short wavelengths. The DCE Aprilis DHD CROP holographic recording material, which was sensitive to the short violet wavelengths of the visible spectrum near 407 nm, improved the storage density by recording/reading using a short-wavelength laser. The DHD CROP recording medium exhibited a favorable dispersion compared to a medium sensitive to 532 nm; this increased the refractive index modulation of the decreasing wavelengths in the visible spectrum. This recording method is advantageous for reducing the interpage crosstalk and improving the SNR in binary data page holograms recorded at high storage densities. Additionally, the method is effective for increasing the accessible angular bandwidth, thereby increasing the achievable areal density in the CROP material, which can be significantly thinner than its free radically polymerized counterpart, while still achieving high storage densities [23–25].

### 5.1.3 Conclusion

Storage media based on photopolymers exhibit large storage densities, small volumes, long lifetimes, and low costs. InPhase and Aprilis introduced a number of commercially viable photopolymer products for HDS. The high-performance recording media developed by InPhase demonstrated digital storage capabilities and ease of manufacturing, enabling the commercial realization of holographic storage. The CROP recording media of Aprilis exhibited a substantially high recording sensitivity than that achieved with their free-radical polymerization media counterpart, with inherently low shrinkages, large dynamic ranges, and high image fidelities. These reported developments indicate the potential of photopolymers as the next-generation optical data storage media.

## 5.2 Holographic Photopolymer Materials

### 5.2.1 Introduction

Dry photopolymer materials are being actively studied for practical applications, such as HDS, three-dimensional (3D) displays, wearable displays with diffractive optical elements, wavelength filters, hybrid guided-wave optoelectronics, photoembossing, self-trapping of light for beam manipulation, and so on [15, 26–29]. Their versatility, ease of use, and self-processing ability give them many advantages over traditional recording materials, such as silver halide and dichromate gelatin [30, 31], for holographic applications. The necessity for development and optimization of such dry photopolymer materials with higher capabilities and stabilities has been recognized, leading to a series of investigations on these materials for holography-based applications.

This section is organized as follows. In Section 5.2.2, we explain the photochemical and holographic grating formation processes in two types of free-radical-mediated photopolymer systems: one with binder-based all-organic components and the other with a nanocomposite structure, namely, photopolymerizable nanoparticle–polymer composites (NPCs) [32], which include photopolymers with a high concentration of inorganic or organic nanoparticles. In Section 5.2.3, we describe the basic holographic setup for the optical characterization of holographic volume gratings recorded in these materials. In Section 5.2.4, we begin by briefly reviewing the various photopolymer materials reported to date, many of which are currently being developed for HDS media. We also describe two experiments for improving the spatial frequency response of their recorded holograms via the addition of chain-transfer agents (CTAs) in two types of photopolymer systems: an all-organic acrylamide/polyvinylalcohol (AA/PVA) system and a photopolymerizable NPC system. The corresponding conclusions are presented in Section 5.2.5.

### 5.2.2 Photopolymerization and Grating Formation Processes

#### 5.2.2.1 Photochemical Reactions

In free-radical-mediated photopolymerization systems, the main mechanism involves five fundamental processes: initiation (process I), propagation (process II), termination (process III), inhibition (process IV), and chain transfer (process V), if any, each of which may also involve several physicochemical reactions. We highlight the major chemical reactions in each process below [31, 33–36]:

***Process I. Initiation*** During an inhomogeneous light illumination, the absorption of the photoinitiator leads to the production of initiator radicals, $R^{\cdot}$, which can react with the monomers to produce chain initiators, $M^{\cdot}$ [34]:

$$I \xrightarrow{h\nu} R^{\cdot} \tag{5.1}$$

$$R^{\cdot} + M \xrightarrow{k_i} M^{\cdot} \tag{5.2}$$

where $I$ is the initiator, $h\nu$ indicates the energy absorbed from a photon, $k_i$ is the chain initiation kinetic constant, and $M$ represents a monomer molecule.

***Process II. Propagation*** The chain initiator, $M_1^{\cdot}$, attaches itself to another monomer molecule, M, via addition to the C=C bond, yielding a growing polymer radical with an active tip. Through propagation, the polymer chain grows as [33]:

$$M_n^{\cdot} + M \xrightarrow{k_p} M_{n+1}^{\cdot} \tag{5.3}$$

where $k_p$ is the rate constant of propagation and $M_n^{\cdot}$ and $M_{n+1}^{\cdot}$ are the growing macroradical chains of the $n$ and $(n+1)$ repeating monomeric units $(n \geq 1)$, respectively.

***Process III. Termination*** Termination can occur in three ways, and two of these, disproportionation and combination, involve two growing macroradical interactions, i.e. the bimolecular termination mechanism, which is expressed as:

$$M_n^{\cdot} + M_m^{\cdot} \xrightarrow{k_{tc}} M_{n+m} \tag{5.4}$$

$$M_n^{\cdot} + M_m^{\cdot} \xrightarrow{k_{td}} M_n + M_m \tag{5.5}$$

where $k_{tc}$ and $k_{td}$ are the rate constants of the combination and disproportionation terminations, respectively; $M_n$, $M_m$, and $M_{n+m}$ represent the terminated chains, which have no radical tips, i.e. dead polymers possessing $n$, $m$, and $(n+m)$ chains, respectively.

A third possible termination mechanism involves primary radical termination, represented by

$$M_n^{\bullet} + R^{\bullet} \xrightarrow{k_{tp}} M_nR \tag{5.6}$$

where $k_{tp}$ is the rate constant of the primary radical termination. In this step, a growing macroradical chain reacts with a primary radical (initiator radical), leading once again to the production of inactive or dead polymer chains.

***Process IV. Inhibition*** Inhibitors are chemicals, which react with the initiating and propagating radical species by rapidly removing or scavenging these radicals. Polymerization is completely halted until they are all consumed [33].

$$R^{\bullet} + Z \xrightarrow{k_{Z,R^{\bullet}}} (R + Z^{\bullet} \text{ and/or } RZ^{\bullet}) \tag{5.7}$$

where $Z$ is the inhibitor species, e.g. oxygen; Z· is the concentration of singlet oxygen [37]; and $k_{Z,R^{\bullet}}$ is the rate constant of inhibition of the primary radicals. Inhibition leads to a dead band at the beginning of the exposure, i.e. no initial grating formation occurs at the onset of holographic exposure. The effect of the inhibitors on photopolymerization is especially significant when lower exposure intensities are used, for example, when either a large illumination area must be exposed or short pulses must be used [34].

***Process V. Chain transfer*** In many polymerization systems, the average polymer weight is observed to be lower than that predicted by the chain-transfer reaction [33, 38, 39]. Generally, the chain-transfer process causes premature termination of a growing macroradical chain and arises because of the presence of CTAs [33].

Due to this reaction, a new radical is produced, namely, a reinitiator, which reacts with a monomer molecule to initiate a new growing macroradical chain. The chain-transfer reactions can be written as:

$$M_n^{\cdot} + (RI - X) \xrightarrow{k_{tr,S}} (M_n - X) + RI^{\cdot} \tag{5.8}$$

$$RI^{\cdot} + M \xrightarrow{k_{ri}} RI - M^{\cdot} \tag{5.9}$$

where RI-X is a CTA, -X is the atom or species transferred, and RI· is the reinitiator possessing a radical tip. $k_{tr,S}$ and $k_{ri}$ are the CTA transfer and reinitiation rate constants, respectively. Due to the premature termination reaction with the CTA, RI-X, the propagating polymer chains stop growing earlier than they would have if the CTA was not present. Therefore, $l_{av}$ can be reduced by introducing CTAs into a photopolymer material.

#### 5.2.2.2 Holographic Grating Formation in a Binder-Based Photopolymer

In the previous section, we presented the main photochemical processes, which occur in general free-radical photopolymerization materials. However, the processes may vary in complexity and significance in different photopolymer systems. Furthermore, Figure 5.3 illustrates a simple 1D grating formation process in such a binder-based photopolymer material, showing that polymerization occurs most strongly in the bright regions, representing a high exposure, because of the sinusoidal illumination intensity pattern shown in Figure 5.3(a). A monomer, indicated by white points in Figure 5.3(b), is polymerized in these regions, i.e. a monomer concentration gradient is created. In the weakly illuminated regions, the excess monomer diffuses into the brighter regions to eliminate the resulting concentration gradient, as illustrated in Figure 5.3(c). As a result, a sinusoidal polymer concentration distribution is formed. Assuming all the monomers are converted to polymers by the end of the recording process, the spatial variation in the refractive index of the material is related to the formed polymer concentration distribution in the background binder material. Thus, a permanent refractive index modulation is formed.

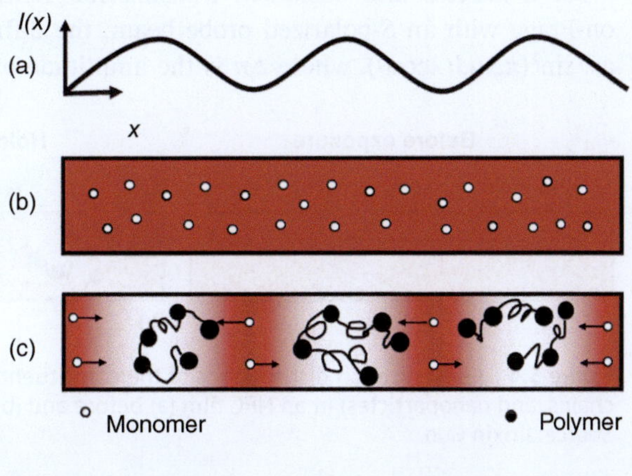


**Figure 5.3** Grating formation in a binder-based photopolymer, where (a) represents the sinusoidal illumination intensity distribution at the plate, (b) represents the photopolymer layer before the recording, and (c) represents the photopolymer layer during the recording. Source: Jinxin Guo.

#### 5.2.2.3 Holographic Grating Formation in an NPC

Figure 5.4 illustrates the grating formation process in a photopolymerizable NPC. It can be seen that the monomer molecules are photopolymerized more than those in the dark regions, owing to a chemical potential difference of the monomer molecules in the bright and dark regions. In contrast, photoinsensitive nanoparticles, which are not consumed, undergo counter-diffusions from the bright to the dark regions as their chemical potential increases in the bright regions owing to the consumption of the monomer. This polymerization-driven mutual diffusion process essentially continues until the photopolymerization reaction is completed. In this way, a periodic assembly of nanoparticles under holographic exposure is formed, and as a result, a refractive index grating is created owing to the compositional and density differences between the bright and the dark regions [40, 41]. Notably, the mutual diffusion of the monomer and nanoparticles may increase with the inclusion of CTAs, which increase their mobility within the developing network [33], indicating a reduction in the viscosity due to a shortening $l_{av.}$ Thus, it is expected that the saturated refractive-index modulation amplitude $\Delta n_{sat}$ increases and the spatial frequency response is improved, as elucidated in Section 5.2.3.

### 5.2.3 Holographic Characterization Methods

To characterize holographic planewave volume gratings recorded in holographic photopolymers, one usually uses Kogelnik's two-wave coupled-wave theory [42]. For gratings that are outside the Bragg diffraction regime and are spatially nonuniform, one has to apply the presented rigorous coupled-wave approach [43] that takes the second derivatives and boundary diffraction into account to perform a more accurate analysis. Another rigorous model [44], based on the electromagnetic theory, has also been developed to explore what happens in a volume (thick) hologram. Compared with the simple two-wave coupled-wave theory, these rigorous models require complex computational techniques and are computationally intensive, although they can deal with more complicated grating structures. For this reason, Kogelnik's two-wave coupled-wave theory is both simpler and more practical as it provides an analytic form of the diffraction efficiency.

For a lossless and unslanted transmission refractive index grating, replayed on-Bragg with an *S*-polarized probe beam, the diffraction efficiency $\eta$ is defined as: $\sin^2(\pi\Delta nd/\lambda\cos\theta)$, where $\Delta n$ is the amplitude of the first harmonic refractive

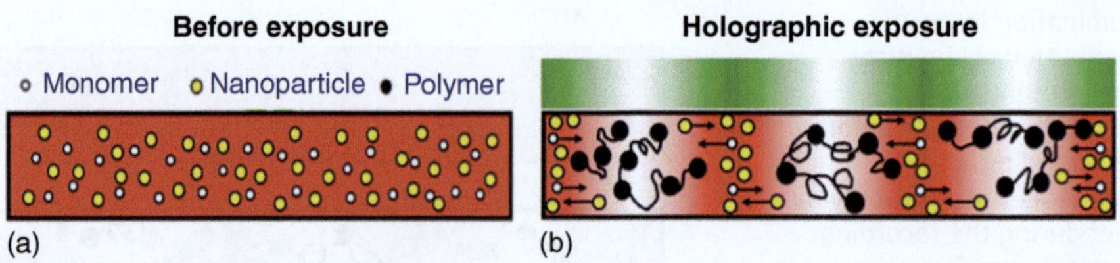


**Figure 5.4** Concentration distributions of the constituents (monomer molecules, polymer chains, and nanoparticles) in an NPC film (a) before and (b) after holographic exposure. Source: Jinxin Guo.

index modulation signal being recorded, $d$ is the grating thickness, $\theta$ is the Bragg angle, and $\lambda$ is the wavelength of the incident probe beam in vacuum. For deriving the above expression, all the boundary reflections have been neglected. This equation can be rearranged to determine the value of $\Delta n$ from the measurement of $\eta$ such that $\Delta n = \lambda \sin^{-1}(\eta)\cos\theta/\pi d$. Note that $d$ can be determined from the Bragg-angle detuning data of $\eta$ as shown in Figure 5.5 [45], which indicates a typical experimental setup for holographic recording (by two beams shown in red) and readout (by a photoinsensitive beam shown in blue) of the buildup dynamics of a planewave transmission grating during exposure. The Bragg-angle detuning curve can be obtained by rotating a film sample around the Bragg angle and by fitting the detuning data of $\eta$ with Kogelnik's analytical formula [44]. This Bragg-detuning measurement can also show the grating nonuniformity along the thickness direction [46, 47].

To characterize the performance of the holographic gratings recorded in a photopolymer for HDS, obtained from the setup shown in Figure 5.5, one needs to evaluate other important figures of merit of the material, such as the material recording sensitivity ($S$), dynamic range of the grating until saturation ($M/\#$), and mechanical and thermal stability. Here, $S$ is defined as $(1/I_0 d)\partial\eta^{0.5}/\partial t|_{t=0}$, where $I_0$ is the average recording intensity, and $t$ is the recording time [48]. It is approximated to $\pi\Delta n_{sat}/\lambda I_0 \tau$ for a single exponential growth of $\Delta n$ with a characteristic time constant $\tau$ [49]. $M/\#$ is defined as $\Sigma_{i=1}{}^{N}\,(\eta_i)^{0.5}$, where $\eta_i$ corresponds to the diffraction efficiency of the

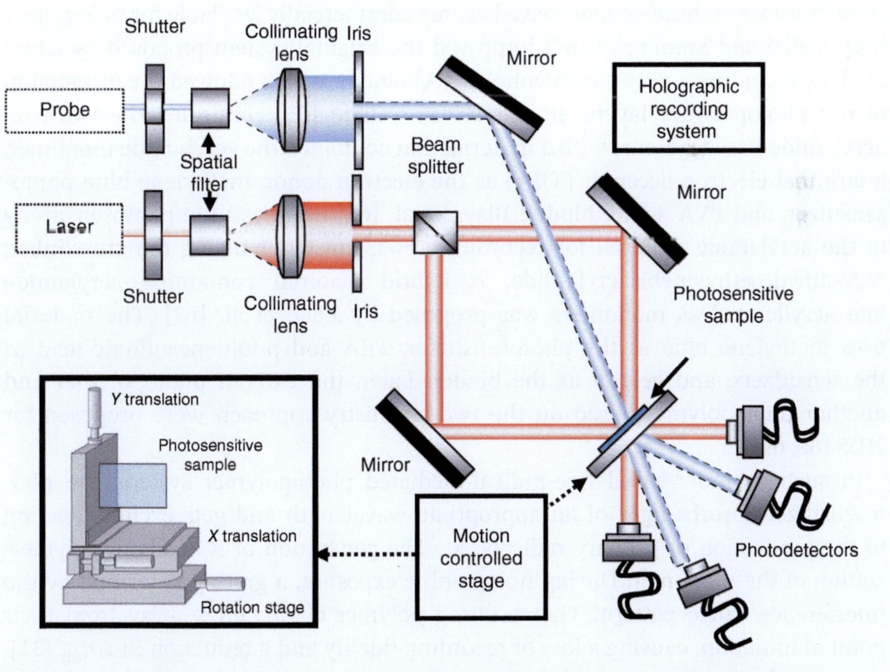


**Figure 5.5** Typical experimental setup for recording unslanted volume transmission holographic gratings with a recording wavelength of $\lambda = 633$ nm, and a probing wavelength of $\lambda = 430$ nm. Source: Jinxin Guo.

$i$th hologram during multiplexing $N$ holograms in one recording area [48, 50]. This system metric may be approximated to $\pi\Delta n_{sat}\ d/\lambda$ for $N$ weak holograms with equal diffraction efficiencies. Generally, for HDS applications, $\Delta n_{sat}$, $S$, and $M/\#$ should be larger than 0.005, 500 cm/J, and 10 for a 500-μm-thick film, respectively, whereas the change in the relative film thickness because of polymerization-induced shrinkage and thermal expansion should be smaller than 0.5% [51].

### 5.2.4 Holographic Photopolymer Materials and Their Improvement for a High-Spatial-Resolution Recording

Thick layers of polymer materials, with their several advantages, are used as true volume gratings, exhibiting high diffraction efficiencies and good angular selectivities. Most photopolymer materials are self-developing and require only simple post-processing, such as an exposure to light or heat treatment. This eliminates the need for wet chemical processes, which makes them suitable for holographic applications, such as 3D display, holographic optical elements for beam manipulations and wearable displays, wavelength filters, hybrid guided-wave optoelectronics [52], and HDS [29, 51, 53–61].

Photopolymer materials were first introduced as holographic recording materials by Close et al. in 1969 [62]. These materials were consisted of a mixture of acrylamide and acrylate monomers as well as a photocatalyst methylene blue. Since then, various photopolymer material systems have been reported, but only a small number of these systems have become commercially available for holography [63]. Sadlej and Smolinska [64] improved the original system proposed by Close et al. by including a poly-vinylalcohol (PVA) binder, which allowed the production of dry photopolymer layers. In the 1980s, Calixto [65] continued the work on acrylamide-based systems with a material that contained the acrylamide monomer, a terminal electron acceptor (TEA) as the electron donor, methylene blue photosensitizer, and PVA as the binder. Blaya et al. [66] improved the photosensitivity of the acrylamide material for recording at 633 nm by changing the crosslinker, *N,N*-dihydroethylenebisacrylamide. A hybrid material containing acrylamide and acrylic acid as monomers was proposed by Zhao et al. [67]. The material uses methylene blue as the photosensitizer, TEA and p-toluenesulfonic acid as the sensitizers, and gelatin as the binder. Later, the DuPont photopolymer and another photopolymer based on the two-chemistry approach were proposed for HDS [68, 69].

In such a binder-based free-radical-mediated photopolymer system, the photosensitizer absorbs light of an appropriate wavelength and gets excited, leading to the production of primary radicals, $R^{\cdot}$. The generation of $R^{\cdot}$ leads to polymerization of the monomer. During holographic exposure, a grating is formed by the interference fringe pattern. The resulting polymer chains grow away from their point of initiation, causing a loss of recording fidelity and a reduction in $\Delta n_{sat}$ [31]. A nonlocal photopolymerization-driven diffusion (NPDD) model was introduced to describe the observed decrease in the grating response of the material at higher spatial frequencies [70, 71]. The grating response of the photopolymer at high spatial frequencies determines the recording resolution, and thus, the data storage capability of the material. The NPDD model predicts that a reduction in the extent of the nonlocal effects (polymer-chain length) within a material improves the spatial frequency response, as experimentally confirmed in a previous study [72–74]. The physical and mechanical characteristics of the polymer chains depend on their molecular weight. The application of CTAs for controlling (reduce) the growth of the nonlocal polymer-chain length has therefore received much attention [38, 39].

To optimize the holographic recording performance by including CTAs, the validity and generality of the extended NPDD model was examined. Guo et al. examined the effect of CTAs on the polymer-chain length and $\Delta n_{sat}$ with the binder-based AA/PVA photopolymer system doped with two types of CTAs, sodium formate (CTA-1) and 1-mercapto-2-propanol (CTA-2) [73]. They theoretically showed that a reduction in $l_{av}$ indeed reduced the nonlocal polymer-chain length. To confirm this hypothesis experimentally, they measured the growth curves and $\Delta n_{sat}$ for each photopolymer material composition in the spatial frequency range from 500 to 3000 lines/mm [74]. Absorption parameters such as the rate of production of the excited-state photosensitizer were obtained from the functions fitted to the normalized transmitted recording beams captured during the recording of the holographic gratings. These parameters determine the absorption behavior of the materials, and therefore, the generation of the initiator radicals. The estimated values were in good agreement with those reported previously [33, 45, 70–72]. The growth curves of $\Delta n$ at each spatial frequency and with each material composition were reproduced from all the repeated experimental datasets and were used during the fitting procedure with appropriate error bars. BY applying the extended NPDD model, the experimental growth curves were fitted in the search ranges of the typical parameter values based on the data reported by other authors [33, 38, 39]. The model was then applied to analyze the temporal growth curves of $\Delta n$ for each material at each spatial frequency. Through this procedure, the monomer diffusion constant, propagation rate constant, initial rate, bimolecular termination rate constant, primary termination rate, chain-transfer rate, reinitiation rate, and nonlocal response length ($\sqrt{\sigma}$), were estimated. For the AA/PVA photopolymer doped with CTA-1, $\sqrt{\sigma} \approx 47.6$ nm was obtained, which corresponds to a reduction of ~22.3% in the mean value for $\sqrt{\sigma}$ compared to that of the undoped one. For the AA/PVA photopolymer doped with CTA-2, $\sqrt{\sigma} \approx 39.8$ nm was obtained, corresponding to a reduction of ~35% in the mean value of $\sqrt{\sigma}$. Therefore, a significant improvement was observed in the high-spatial-frequency responses with the addition of CTAs. Further, it is evident that the shortening of the polymer chains most consistently explains the observed improvement in the spatial frequency response of the recorded gratings as seen in Figure 5.6.

While acrylamide-based free-radical systems received much attention in the past, other photopolymer materials showed several advantages for development of holographic media. In 1996, Oliveira et al. [75] proposed a multicomponent sol–gel-synthesized holographic photopolymer system that included inorganic molecules of zirconium isopropoxide [$Zr(OiPr)_4$] containing a methacrylic acid monomer component at the surface of $Zr(OiPr)_4$. This inorganic compound

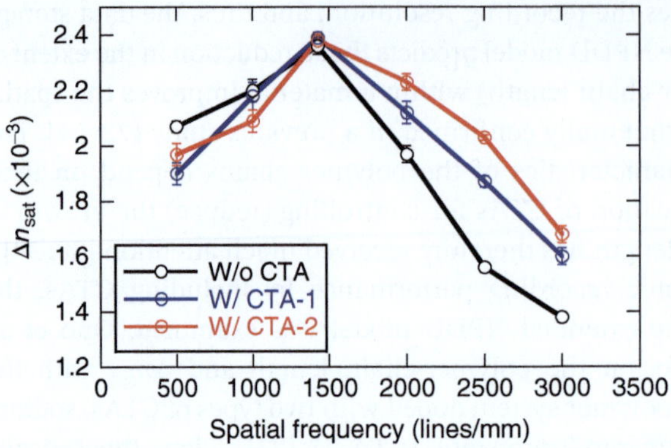


**Figure 5.6** Experimental results of the spatial frequency response corresponding to the saturation refractive index modulation for the standard material, standard material with 0.96 g og CTA-1, and standard material with 0.53 mL of CTA-2. Source: Jinxin Guo.

copolymerized with the organic species of the organo-silane in the matrix material of 3-methacryloxypropyl trimethoxysilane [75, 76]. In 2002, Suzuki et al. proposed the first photopolymerizable NPC system, an organic–inorganic nanocomposite photopolymer incorporating inorganic oxide $TiO_2$ nanoparticles acting as secondary neutral species with high concentrations [77]. Their idea was to use inorganic materials with a wide range of refractive indices for a large increase in $\Delta n_{sat}$ of a recorded volume grating with reduced shrinkage and high thermal stability simultaneously. For example, refractive indices higher than three are even observed in semiconductors [78], compared with polymers, whose refractive indices are of the order of, at most, 1.7. Especially, inorganic oxides, such as $TiO_2$ and $ZrO_2$, are highly transparent in the visible spectral region and have refractive indices higher than 2 in their bulk states. Incorporation of inorganic nanoparticles with (higher or lower) refractive indices into a photopolymer host would give rise to very large $\Delta n_{sat}$ outside the typical range realized by conventional all-organic photopolymer systems. They showed that the incorporation of $TiO_2$ nanoparticles yielded a remarkably large $\Delta n_{sat} \sim 5 \times 10^{-3}$ at a recording wavelength of 532 nm and also suppressed the polymerization-induced shrinkage. Since then, various types of photopolymerizable NPC systems have been reported to show their high performance in volume holographic recording. These include inorganic (such as $SiO_2$, $ZrO_2$, and CdSe quantum dots) and organic nanoparticles (such as hyperbranched polymer) [49, 78–87]. HDS with NPCs using free-radical-mediated thiol-ene/yne photopolymerizations has also been demonstrated [88–90] as these NPCs meet all the material criteria for HDS [82–90]. An overview of the recent studies on photopolymerizable NPCs for holographic applications can be found in [32].

Similar to the AA/PVA photopolymer system described above, we discuss the effect of CTAs on the spatial frequency response of the gratings recorded in photopolymerizable NPCs. In a recent study [91, 92], it was shown that the doping of CTAs substantively improved the spatial frequency response. In these experiments, three types of thiols with mono-, di-, and tri-functionalities, acting as CTAs in free-radical-mediated chain-growth polymerization, were employed. These were methyl-3-mercaptopropionate (mono-thiol), 1,4-bis(3-mercaptobutyryloxy) butane (di-thiol), and tris(3-mercaptopropionate) (tri-thiol).

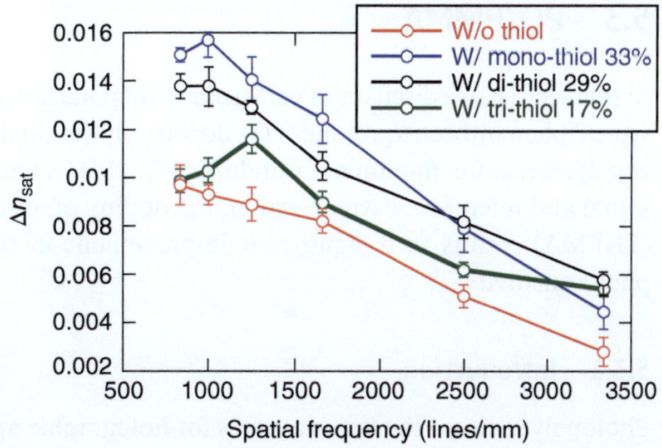


**Figure 5.7** Dependences of $\Delta n_{sat}$ on the spatial frequency of the formed grating for undoped and thiol-doped NPC films. The molar concentrations of mono-, di-, and tri-thiols (with respect to the monomer) are 33, 29, and 17 mol%, respectively. Source: Jinxin Guo.

Each of CTAs was doped into an acrylate monomer, 2-propenoic acid, (octahydro-4, 7-methano-1H-indene-2,5-diyl)bis(methylene) ester, at varied concentrations. $ZrO_2$ nanoparticles (the average size of ~3 nm) were also dispersed in an acrylate monomer host at the optimum concentration of 35 vol% with respect to the monomer for maximizing $\Delta n_{sat}$. To sensitize the NPC at a recording wavelength of 532 nm, they used titanocene (Irgacure784, Ciba) of 1 wt% with respect to the monomer. Figure 5.7 shows the dependences of $\Delta n_{sat}$ on the spatial frequencies of the formed gratings from 833 to 3333 lines/mm, which corresponds to a grating spacing from 1.2 to 0.3 μm, respectively, for undoped and thiol-doped NPC films. It can be clearly seen that the NPC films doped with chain-transferring thiols perform better than that without thiol in the high-spatial-frequency recording condition. The higher concentration of –SH in the acrylate monomer would produce a lower $l_{av}$, although the viscosity increases as the functionality of the thiols increases. This reduction in $l_{av}$ leads to the improvement of the spatial frequency responses as observed similarly in AA/PVA photopolymers [73, 74]. Specifically, the NPC film doped with di-thiol exhibits the highest $\Delta n_{sat}$ ($=6\times10^{-3}$) at a spatial frequency of 3333 lines/mm. This value corresponds to a twofold improvement in $\Delta n_{sat}$ as compared with that of the undoped NPC film.

### 5.2.5 Conclusion

In this section, we reviewed holographic photopolymer materials. We explained the basic kinetic model of free-radical photopolymerization systems and the mechanisms of holographic grating formation in binder-based photopolymer and photopolymerizable NPC systems. Further, the development of these two types of photopolymer systems is described, and experimental investigations of the effects of CTAs on the spatial frequency response of the recorded gratings are outlined for binder-based AA/PVA photopolymer and photopolymerizable NPC systems. Especially, we show the substantive increase in $\Delta n_{sat}$ in the high-spatial-frequency regions with the incorporation of CTAs. Such a simple method based on chemical modifications would be very useful for HDS applications, which require high-density information storage with a high fidelity.

## 5.3 PQ/PMMA

In this section, we discuss the performance of polarization holographic recordings in typical phenanthrenequinone (PQ)-doped poly-(methyl methacrylate) (PMMA) and copolymerization monomer-containing PQ/PMMA with the orthogonally polarized signal and reference waves. Further, the doping of tetrahydrofurfuryl methacrylate (THFMA) results in a significant improvement in the diffraction intensity and photosensitivity.

### 5.3.1 Introduction

Photopolymers are the key materials for holographic applications, like data storage and holographic optical elements, owing to their high optical sensitivity, low fabrication costs, practically unlimited dimensions, and negligible post processing [93].

To improve optical performance, many researchers focused on developing innovative compositions by doping additional functional components into the recording photopolymer materials [94], such as nanoparticles [95], organic–inorganic hybrids [96], lanthanoid organometallic compounds [97], and organic metallic components [98]. However, for some applications, such as data storage, these methods were restricted by their limited thickness, high shrinkage during holographic exposure, relatively high costs, and complex fabrication [99].

Among these photopolymers, PQ-doped PMMA is extensively studied owing to its negligible shrinkage, high thickness, low cost [100, 101], and polarization sensitivity.

However, because of the low solubility (0.7 wt%) of PQ in PMMA solution, the concentration of the photosensitizer is limited. Thus, compared to other materials, the diffraction efficiency and polarization sensitivity of PQ/PMMA are unsatisfactory. To solve these problems, we need to enhance the combination probability of PQ radicals with monomer molecules by increasing the number of PQ molecules.

Liu et al. [102] reported a method to increase the concentration of PQ by raising the prepolymerization reaction temperature. The results showed that approximately 1.0 wt% of PQ was solved into the PMMA solution at 60 °C reaction temperature. Mahilny et al. [103] demonstrated a preparation process, wherein they cast the liquid solution on a substrate and dried it to a solid state. Using their process, the concentration of PQ can be raised up to 4 mol% in the photosensitive layer, but the application of this method is limited by the relatively lower thickness (50–180 μm) of the deposited layer.

In a previous work, we introduced a co-monomer benzyl methacrylate (BzMA) to improve the performance of the holographic recording material PQ/PMMA [105]. However, it is still undefined what kind of the regularity the co-monomer should possess. In this section, we concentrate on the discussion of the photoreaction mechanism and conclude the rules for choosing co-monomer with a certain structure. Compounds with similarities are easily solvable in each other, and accordingly, we suspect that the dopant structure contains an aromatic group that may have a high solubility for PQ molecules. Additionally, the dopant can be efficiently introduced

**Table 5.2** Chemical structures of the dopants for our photopolymer.

| Name | Structure | Experimental result |
|---|---|---|
| Benzyl methacrylate (BzMA) | 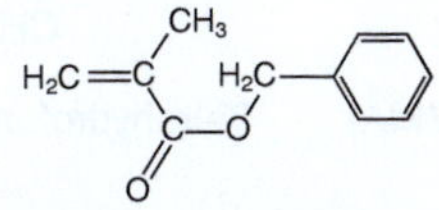 | Experiment succeed |
| Tetrahydrofurfuryl methacrylate (THFMA) | 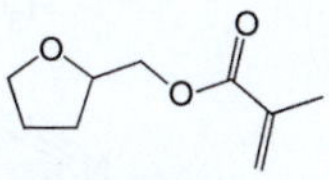 | Experiment succeed |
| Styrene (St) | 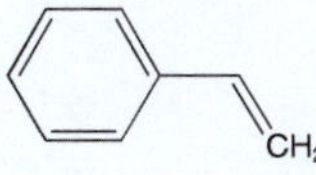 | Haven't succeed yet |
| 2-Phenoxyethyl acrylate (PEA) | 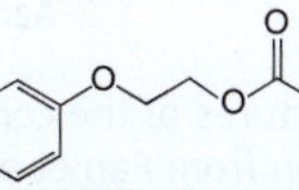 | Haven't succeed yet |

Source: Reproduced with permission from Fan et al. [104]/OSA.

into the PQ/PMMA photopolymer system because of the unsaturated C=C bond in the dopant. In accordance with these requirements, we found the dopants listed in Table 5.2.

We prepared new materials by introducing the aforementioned dopants into PQ/PMMA material.

Among these new materials, two kinds of dopants (St and PEA) have not succeeded since the viscosity of these two dopants is too much higher than that of MMA solution, resulting in a large number of bubbles in material. However, PQ-doped poly (MMA-co-THFMA) photopolymer was successfully prepared via free-radical copolymerization. THFMA is a very interesting monomer as its polymer has a wide range of applicability in the development of biomaterials [106]. By introducing THFMA, the concentration of PQ was improved from 0.7 to 1.3 wt% compared to that in PQ/PMMA, indicating that the material was endowed with a higher diffraction intensity and photosensitivity compared to the typical PQ/PMMA system.

### 5.3.2 Material Preparation

All the original materials were purchased from Shanghai Macklin Biochemical Company and used as received. In this study, THFMA and MMA were selected as the co-monomers, 2,2-azobisisobutyronitrile (AIBN) was employed as the thermo-initiator and PQ as the photosensitizer. Figure 5.8 shows their chemical structures.

In our fabrication process, the monomers MMA and THFMA were mixed first with a weight ratio of 7 : 3, because of the better solubility of PQ in THFMA than in MMA. We introduced THFMA into our system for increasing the dissolved concentration of the PQ molecule. The PQ and AIBN (0.8 wt%) powders were dissolved in

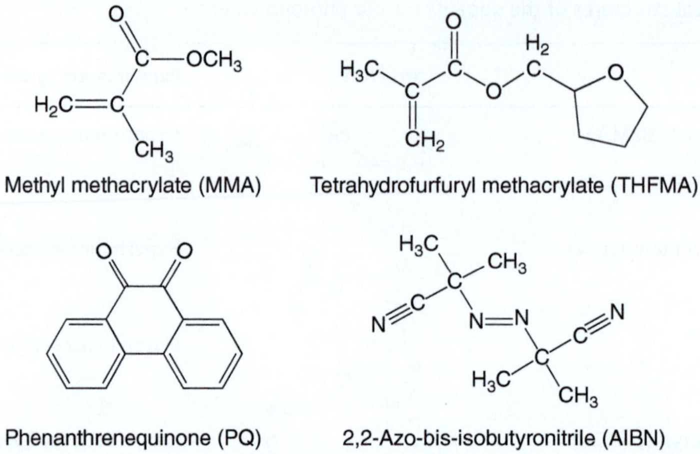


**Figure 5.8** Chemical structures of the components in our photopolymer. Source: Reproduced with permission from Fan et al. [104, 105], Optical Society of America.

the monomer solvent, and the doping concentration of PQ in the monomer mixture was determined by measuring their saturation concentrations (the concentration of PQ can reach 1.3 wt%). The resulting solution was stirred using a magnetic stirrer in a glass bottle at 65 °C for an appropriate duration until it became homogeneously viscoid. This syrup was poured into a glass mold with a 1.5-mm-thick spacer. Then, the mold was then baked at 45 °C for 10 days to solidify the mixture. This procedure ensures that most of the monomers are polymerized, and yellow bulk samples with good optical quality are obtained. This material was thereafter named as PQ/P(T3M7).

In this study, we also prepared PQ/PMMA (PQ: 0.7 wt%) and PQ/P(T3M7) (PQ: 0.7 wt%) photopolymers for comparison. Furthermore, we also compared it with the previous system of PQ/P(B3M7) photopolymer [105].

### 5.3.3 Photochemical Reaction

Before the illumination, the PQ molecule presents highly conjugate coplanar structure [107]. When exposed to a linearly polarized wave, the PQ molecules become oriented depending on the polarization state. For example, Figure 5.9 shows that the PQ molecules that are oriented parallel to the polarization direction of the incident light better react with the monomer [108, 109].

Under the exposure, photons excite the double bond on the carbonyl functional group of the PQ molecules, and the PQ molecules become radicals. Then, the radical reacts with the carbonic double bond on the vinyl functional group on the monomer molecule to form the photoproduct [110, 111]. The polarization anisotropy of the photoproduct is better than that of the PQ molecules, which only process a slight polarization anisotropy. Thus, the photoreaction causes polarization, and the more PQ molecules we introduce into the material, the more PQ molecules are excited by the light to participate into the photoreaction. As described above, the PQ molecule

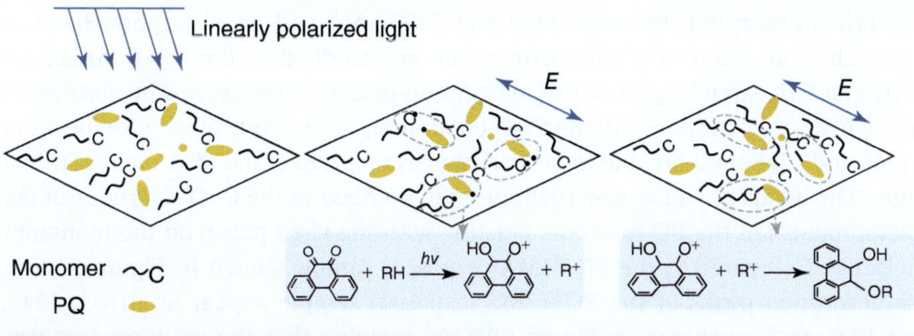


**Figure 5.9** Scheme of PQ photoproduct attachment to the monomer. Source: Reproduced with permission from Fan et al. [104, 105], Optical Society of America.

plays an important role in the PQ-containing polymeric polarization holographic recording material.

### 5.3.4 Results and Discussions

#### 5.3.4.1 Fourier Transform Infrared Spectroscopic Measurements

Fourier transform infrared (FTIR) spectroscopic analysis was performed to evaluate the photochemical reaction of the sample. We compared the IR spectrum of the reactants (PQ and monomers) before and after (bottom in Figure 5.10) the photoirradiation. The IR spectra of the PQ, MMA, THFMA and exposed PQ/MMA, and PQ/P(T3M7) samples were acquired using a Thermo-IS5 FTIR spectrophotometer as shown in Figure 5.10. Before the measurement, KBr plates were dipped in the exposed PQ/MMA and PQ/P(T3M7) solutions. The plate samples were baked at 50 °C for three hours such that the unreacted monomer molecules were removed and the photoproducts were left on the KBr plates. This method is the same as Hsiao's method [112].

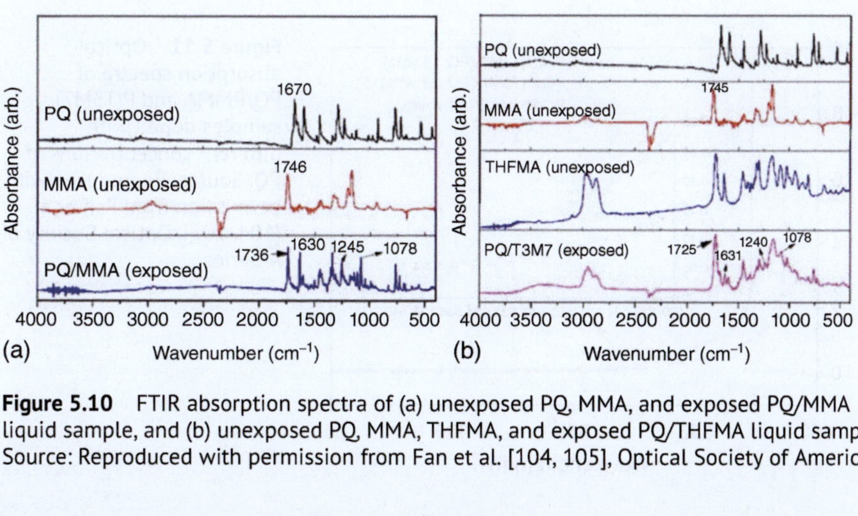


**Figure 5.10** FTIR absorption spectra of (a) unexposed PQ, MMA, and exposed PQ/MMA liquid sample, and (b) unexposed PQ, MMA, THFMA, and exposed PQ/THFMA liquid sample. Source: Reproduced with permission from Fan et al. [104, 105], Optical Society of America.

A broad absorption between 2800 and 3300 $cm^{-1}$ and an absorption band at ~2360 $cm^{-1}$ are seen in Figure 5.10, which are ascribed to the C—H stretching ($CH_3$ and $CH_2$ stretching) and $CO_2$ absorption peaks, respectively. The absorption at ~1670 $cm^{-1}$ corresponds to the C=O bond in PQ. The strong absorption peaks at ~1745 and 1725 $cm^{-1}$ are due to the carbonyl ester group of the MMA and THFMA units. The bands at ~1245 and 1078 $cm^{-1}$ are ascribed to the C–O–C groups of the photoproducts of the PQ molecule reacting with the vinyl group on the monomer molecule. Compared to the PQ/MMA (exposed) sample shown in Figure 5.10(a), the absorption peaks of the PQ/T3M7 (exposed) sample appear at ~1631, 1245, and 1078 $cm^{-1}$ as shown in Figure 5.10(b), implying that the photoproduct was produced by the light exposure [112]. These phenomena show that the THFMA did not affect the photochemical reaction of the PQ and monomer molecules.

#### 5.3.4.2 Ultraviolet–Visible Spectroscopic Measurements

Optical absorption was measured using the U-1901 double-beam UV–Vis spectrophotometer. The plot of absorbance versus wavelength is presented in Figure 5.11, showing a strong absorption for $\lambda < 475$ nm and close to zero for $\lambda > 600$ nm.

We can see from Figure 5.11 that the absorption peak appears redshifted with the increasing content of PQ. It proves that we succeeded in our strategy to increase the concentration of PQ and strengthen the optical absorption. The absorbance of PQ/P(T3M7) changes negligibly with wavelength when the content of PQ is the same as that in PQ/PMMA. This indicates that THFMA has no effect on the light absorption. In our experiments, we chose 532 nm as the recording wavelength, at which the absorbance of PQ/PMMA (PQ: 0.7 wt%), PQ/P(T3M7) (PQ: 0.7 wt%), and PQ/P(T3M7) (PQ: 1.3 wt%) before the exposure were $\alpha = 0.061$, 0.069, and 0.135 $cm^{-1}$, respectively.

#### 5.3.4.3 Photoinduced Birefringence

The photoinduced birefringence was investigated with a continuous He—Ne laser ($\lambda = 632.8$ nm; power, $P = 2.6$ mW/$cm^2$) as the probe light, and a diode-pumped

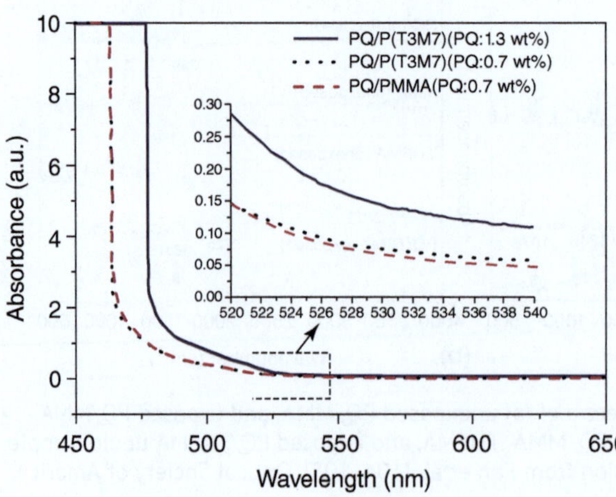


**Figure 5.11** Optical absorption spectra of PQ/PMMA and P(T3M7) samples doped with different concentrations of PQ. Source: Reproduced with permission from Fan et al. [104, 105], Optical Society of America.

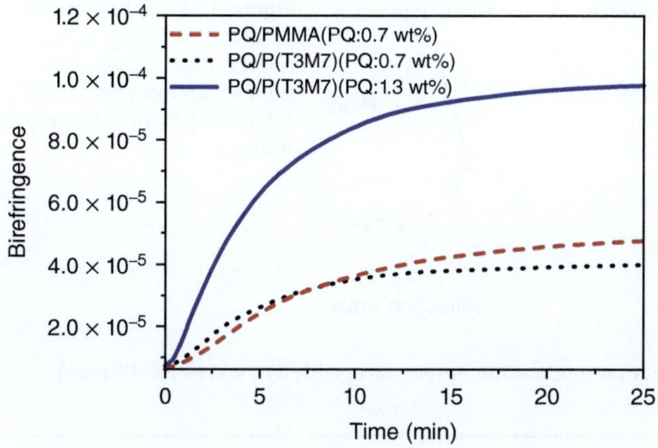


**Figure 5.12** Photoinduced birefringence versus exposure time. Source: Reproduced with permission from Fan et al. [104, 105], Optical Society of America.

solid-state Nd:YAG laser ($\lambda = 532\,\text{nm}$, $P = 52\,\text{mW/cm}^2$) as the pump light. The experimental setup and corresponding calculation formula of photoinduced birefringence were adopted from [105].

Figure 5.12 shows the experimental results of photoinduced birefringence analysis. We can observe that the photoinduced birefringence of PQ/P(T3M7) (PQ: 1.3 wt%) rises much more quickly than that of PQ/P(T3M7) (PQ: 0.7 wt%). Additionally, a higher PQ content in the sample leads to a larger birefringence value. For the PQ/P(T3M7) and PQ/PMMA samples with the same concentration of PQ (PQ: 0.7 wt%), the curve of the photoinduced birefringence remained almost unchanged with only negligible variations.

#### 5.3.4.4 Orthogonal Linearly Polarized Holography

Polarization holography is a promising technology with its unique ability of recording amplified phase and polarization. In contrast to traditional holography, in which the information is recorded via refractive index modulation, a polarization hologram is recorded in polarization-sensitive materials by the modulation of photoinduced anisotropy, such as birefringence, optical dichroism, and optical activity.

Figure 5.13 shows the schematic of polarization holography, which employs two orthogonal polarization waves for recording information. Compared to traditional hologram, where an intensity inference pattern exists, a polarization hologram shows periodically modulated polarization states with a constant intensity distribution in the resultant area. The details of the periodic polarization state modulation of orthogonal linear waves are listed in Table 5.3. As the recording material is polarization sensitive, the hologram is formed by the photoinduced birefringence gratings, and the birefringence modulation originates from the selective photoreaction of the reoriented molecules and structure rearrangement by the photochemical reaction. In the readout stage as shown in Figure 5.13, the polarization hologram is reconstructed by the reference wave, which satisfies the Bragg condition, similar to intensity holography.

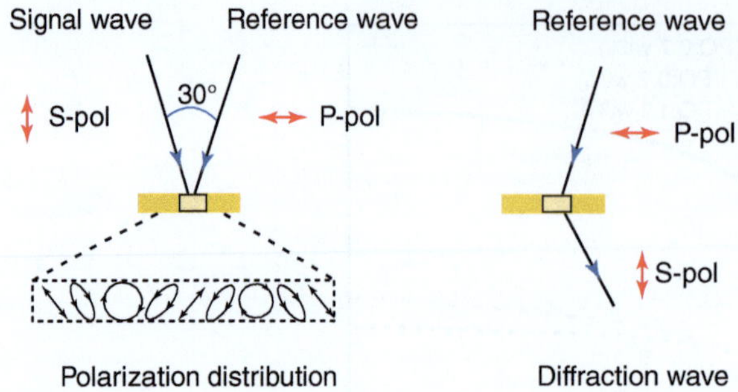


**Figure 5.13** Schematic of polarization holography. Source: Reproduced with permission from Fan et al. [104, 105], Optical Society of America.

**Table 5.3** Periodic polarization modulation in recording with two waves for different polarization geometries.

| Writing beams | Configuration of the interference pattern | | | | | | | | |
|---|---|---|---|---|---|---|---|---|---|
| | $\delta = 0$ | $\delta = \pi/4$ | $\delta = \pi/2$ | $\delta = 3\pi/4$ | $\delta = \pi$ | $\delta = 3\pi/4$ | $\delta = 3\pi/2$ | $\delta = 7\pi/4$ | $\delta = 2\pi$ |
| $s, p$  | | | | | | | | | |

*Note:* δ is the phase shift between the two beams.
Source: Reproduced with permission from Fan et al. [104, 105], Optical Society of America.

The material sensitivity $S$ is defined as [49]:

$$S = (\partial\sqrt{\eta}/\partial t\big|_{t=t_{\text{ind}}})/I_0 d \tag{5.10}$$

where $\eta$ is the diffraction efficiency ($\eta = I_\eta/I_0$); $I_\eta$ and $I_0$ are the diffraction and incident intensities, respectively; and d is the thickness of the sample.

The experimental setup for the measurement of orthogonal linearly polarized holography is shown in Figure 5.14. A 532-nm (green) laser was used for the recording and reconstruction. The green laser beam reference waves, with horizontal polarization (P-pol), and signal waves, with vertical polarization (S-pol), by a polarization beam splitter (PBS) with the same intensity of $P = 52\,\text{mW/cm}^2$. These beams were then symmetrically focused onto the sample with a crossing angle of 30°. The diameter of the light spot on the sample was 6 mm.

In the recording process, we opened shutters 1 and 2, and shutter 3 was kept closed. The recording waves interfered for 4 seconds with the crossing angle of 30°. In the reconstruction process, we closed shutter 2 and opened shutter 3. We used the original reference wave to read the recorded hologram for 0.4 s.

Figure 5.15(a) shows the typical dependence of the diffraction intensity on the exposure time. The diffraction intensity of each sample was elevated after an increase in the exposure time. When the exposure time reached a sufficient value, the diffraction intensity of each sample was comparatively stable.

It can be seen that the diffraction intensity of PQ/P(T3M7) increased with the increasing concentration of PQ. The data indicated that the diffraction intensity of PQ/P(T3M7) can be doubled when the concentration of PQ is increased from 0.7 wt% to an optimum value of 1.3 wt%. As shown in Figure 5.15(a), under the same

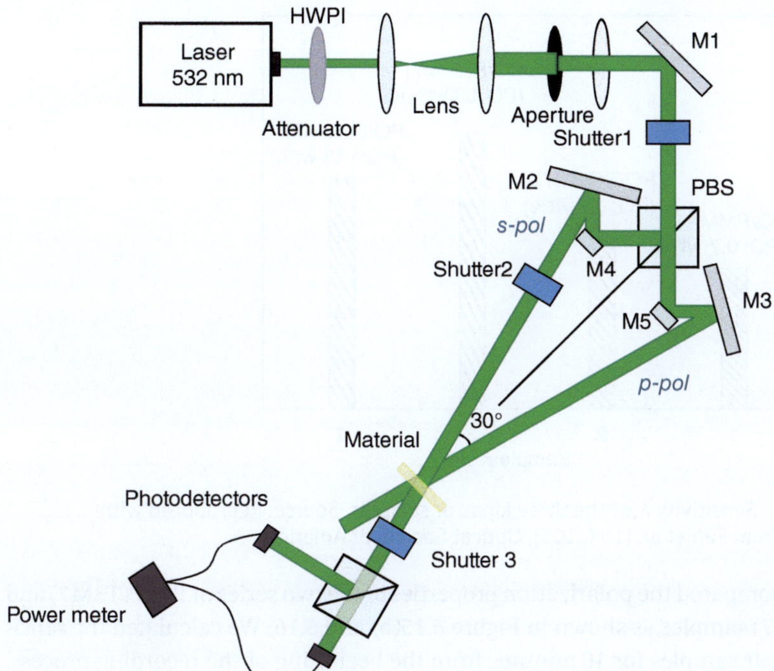


**Figure 5.14** Schematic of the experimental setup. M, mirror; HWP, half-wave plate; PBS, polarization beam splitter. Source: Reproduced with permission from Fan et al. [104, 105], Optical Society of America.

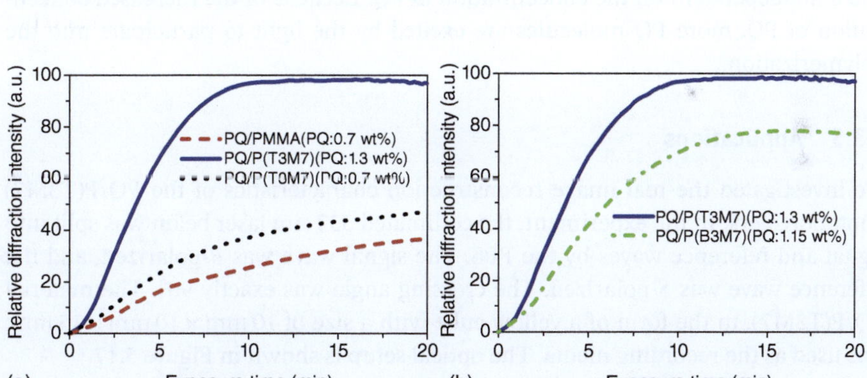


**Figure 5.15** Evolution of diffraction intensity as a function of exposure time. Source: Reproduced with permission from Fan et al. [104, 105], Optical Society of America.

concentration of PQ (0.7 wt%), the diffraction intensity of PQ/P(T3M7) showed a slight increase compared to that of the typical sample of PQ/PMMA.

Thus, the incorporation of the dopant THFMA into the system leads to the formation of a highly concentrated PQ photopolymer material, in which a higher diffraction intensity can be maintained.

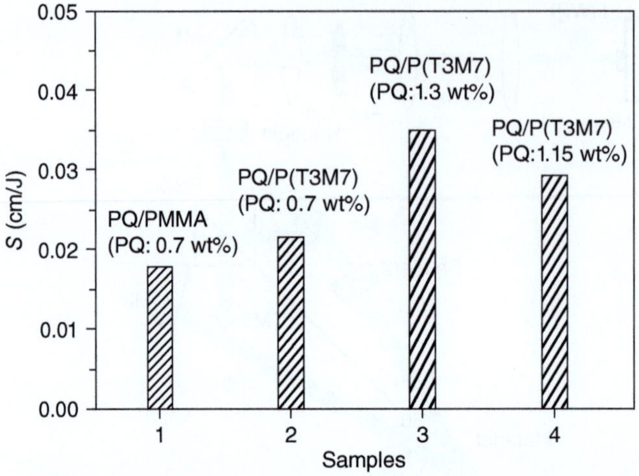


**Figure 5.16** Sensitivity $S$ of the three kinds of samples. Source: Reproduced with permission from Fan et al. [104, 105], Optical Society of America.

We also compared the polarization properties of the two series of PQ/P(T3M7) and PQ/P(B3M7) samples as shown in Figure 5.15(b) and 5.16. We calculated the sensitivity $S$ of our samples for 10 minutes from the beginning of the recording process. The experimental results revealed that PQ/P(T3M7) exhibits a higher diffraction intensity and sensitivity. It can be seen that the diffraction intensity and $S$ increase with the increasing concentration of PQ, demonstrating that the diffraction intensity and $S$ are dependent on the concentration of PQ. Because of the increased concentration of PQ, more PQ molecules are excited by the light to participate into the polymerization.

### 5.3.5 Applications

We investigated the real image reconstruction characteristics of the PQ/P(T3M7) photopolymers. In the experiment, the collimated 532-nm laser beam was split into signal and reference waves by the PBS. The signal wave was $P$-polarized, and the reference wave was $S$-polarized. The crossing angle was exactly 90°. The material PQ/P(T3M7), in the form of a yellow cube with a size of 10 mm × 10 mm × 15 mm, was used as the recording media. The optical setup is shown in Figure 5.17.

First, we received the original image (Figure 5.18(a)) transmitted by a complementary metal-oxide semiconductor sensor. The transmitted image was formed by the original image (on the SLM) when it directly passed through the material without holographic recording and reconstruction. In the holographic recording process, the signal wave carried the original image through the SLM. Then, we opened shutter 1 so that the reference waves can interfere with the signal waves in our material. In the reconstruction process, we closed shutter 2 and opened shutter 1 to read the recorded hologram. The reconstructed image is shown in Figure 5.18(b), indicating a very good contrast and high fidelity, sufficient for the application of volume polarization holographic storage.

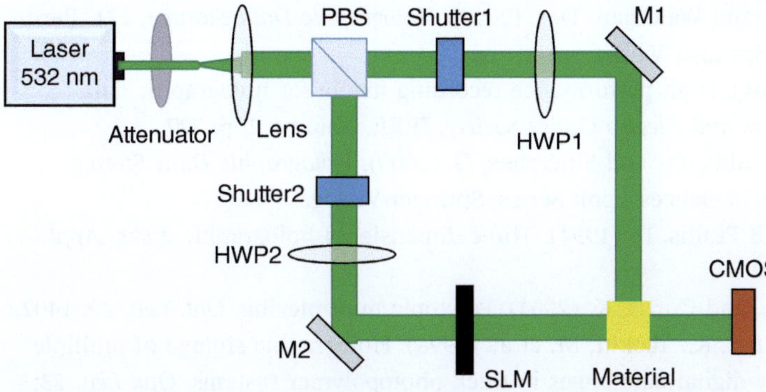


**Figure 5.17** Experimental setup for recording the image. PBS, polarization beam splitter; HWP, half-wave plates; M, mirrors; and SLM, spatial light modulator. Source: Reproduced with permission from Fan et al. [104, 105], Optical Society of America.

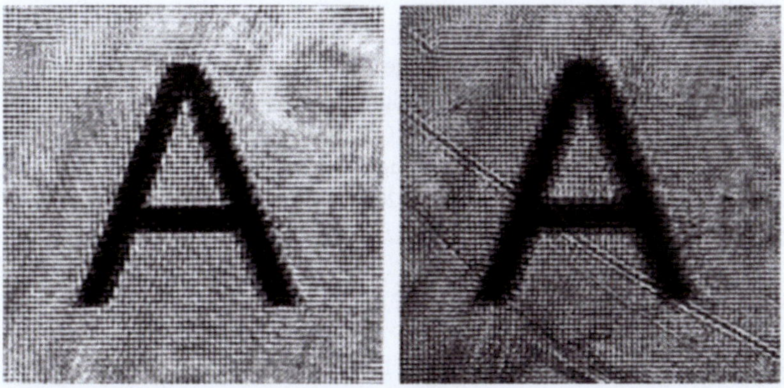

(a) (b)

**Figure 5.18** Image reconstruction in a polarization holography system: (a) original transmitted image and (b) reconstructed image. Source: Fan et al. [104, 105], Optical Society of America.

## References


**1** Gallego, S., Neipp, C., Ortuño, M. et al. (2008). Analysis of multiplexed holograms stored in a thick PVA/AA photopolymer. *Opt. Commun.* 281: 1480.

**2** Curtis, K., Wilson, W., and Dhar, L. (2002). Commercialization of holographic storage at InPhase Technologies. *International Symposium on Optical Memory and Optical Data Storage Topical Meeting*. IEEE, p. 6.

**3** Omori, M., Kondo, H., Miyata, T. et al (2010). Enhancement of a tunable blue laser for holographic data storage. *Proceedings of SPIE*, Volume 7730, 77300T.

**4** Anderson, K., Ayres, M., Sissom, B., and Askham, F. (2014). Holographic data storage: rebirthing a commercialization effort. *Proceedings of SPIE*, Volume 9006, 90060C.

5 Ingwall, R.T. and Waldman, D.A. (2000). *Holographic Data Storage*, 171. Berlin, Heidelberg: Springer-Verlag.
6 Dhar, L. (2004). High performance recording media for holographic data storage. *Lasers and Electro-Optics Society*. IEEE, Volume 2, p. 727.
7 Coufal, H., Psaltis, D., and Sincerbox, G. (2000). *Holographic Data Storage*, Part 5, Optical Sciences Book Series. Springer-Verlag.
8 Li, H.Y.S. and Psaltis, D. (1994). Three-dimensional holographic disks. *Appl. Opt.* 133: 3764.
9 Anderson, K. and Curtis, K. (2004). Polytopic multiplexing. *Opt. Lett.* 29: 1402.
10 Dhar, L., Curtis, K., Tackitt, M. et al. (1998). Holographic storage of multiple high-capacity digital data pages in thick photopolymer systems. *Opt. Lett.* 23: 1710.
11 Stöckel, N., Bruder, F.K., and Askham, F.R. (2011). Advantageous recording media for holographic applications. US Patent 8, 053, 147.
12 Cole, M.C. and Ayres, M.R. (2010). Latent holographic media and method. US Patent 7, 678, 507.
13 Wilson, W.L., Curtis, K.R., Anderson, K.E. et al. (2003). Realization of high-performance holographic data storage: the InPhase technologies demonstration platform. *Proceedings of SPIE*, Volume 5216, p. 178.
14 Dhar, L., Schnoes, M.G., Katz, H.E. et al. (2000). *Holographic Data Storage*, Part 5, Optical Sciences Book Series. Springer-Verlag.
15 Villafranca, A.B. and Saravanamuttu, K. (2009). Diffraction rings due to spatial self-phase modulation in a photopolymerizable medium. *J. Opt. A Pure Appl. Opt.* 11 (12): 125202.
16 Schnoes, M., Ihas, B., Dhar, L. et al. (2003). Photopolymer use for holographic data storage. *Proceedings of SPIE*, Volume 13, p. 68.
17 Wilson, W., Anderson, K., Curtis, K., and Dhar, L. (2004). Toward the commercial realization of high-performance holographic data storage. *Proceedings of SPIE*, Volume 1, p. 4.
18 Curtis, K., Wilson, W.L., Tackitt, M.C. et al. (2000). *High Density, High Performance Optical Data Storage: Via Volume Holography: The Lucent Technology Hardware Platform*, Holographic Data Storage, 359. Berlin Heidelberg: Springer-Verlag.
19 Waldman, D.A., Ingwall, R.T., Dhal, P.K. et al. (1996). Cationic ring-opening photopolymerization methods for volume hologram recording. *Proceedings of SPIE*, Volume 2689, p. 127.
20 Waldman, D.A., Li, H.-Y.S., and Horner, M.G. (1997). Volume shrinkage in slant fringe gratings of a cationic ring-opening holographic recording material. *J. Imaging Sci. Technol.* 41: 497.
21 Shelby, R.M., Waldman, D.A., and Ingwall, R.T. (2000). Distortions in pixel-matched holographic data storage due to lateral dimension change of photopolymer storage media. *Opt. Lett.* 25: 713.
22 Waldman, D.A., Li, H.-Y.S., and Cetin, E.A. (1998). Holographic recording properties in thick films of ULSH-500 photopolymer. *Proceedings of SPIE*, Volume 3291, p. 83.

**23** Waldman, D.A., Butler, C.J., and Raguin, D.H. (2003). CROP holographic storage media for optical data storage at greater than 100 bits/$\mu m^2$. *Proceedings of SPIE*, Volume 5216, p. 10.

**24** Curtis, K., Pu, A., and Psaltis, D. (1994). Method for holographic storage using peristrophic multiplexing. *Opt. Lett.* 19: 993.

**25** Waldman, D.A., Kolb, E.S., and Wang, C. (2007). DHD™ CROP holographic storage media for advanced optical data storage. *Optical data storage*, WDPDP4.

**26** Curtis, K., Dhar, L., Murphy, L., and Hill, A. (2010). Future developments. In: *Holographic Data Storage: From Theory to Practical Systems* Chapter 5, 105–119. Wiley.

**27** Coufal, H.J., Psaltis, D., and Sincerbox, G.T. (2000). *Holographic Data Storage.* New York: Springer.

**28** Ye, C. and McLeod, R.R. (2008). GRIN lens and lens array fabrication with diffusion-driven photopolymer. *Opt. Lett.* 33 (22): 2575–2577.

**29** Sato, A., Scepanovic, M., and Kostuk, R. (2003). Holographic edge-illuminated polymer Bragg gratings for dense wavelength division optical filters at 1550 nm. *Appl. Opt.* 42 (5): 778–784.

**30** Manivannan, G. and Lessard, R.A. (1994). Trends in holographic recording materials. *Trends Polym. Sci.* 2: 282–290.

**31** Lawrence, J.R., O'Neill, F.T., and Sheridan, J.T. (2001). Photopolymer holographic recording material. *Optik* 112 (10): 449–463.

**32** Tomita, Y., Hata, E., Momose, K. et al. (2016). Photopolymerizable nanocomposite photonic materials and their holographic applications in light and neutron optics. *J. Mod. Opt.* 63: S1–S31.

**33** Odian, G. (1991). *Principles of Polymerization, 4e*, 4e. New York: Wiley.

**34** Gleeson, M.R. and Sheridan, J.T. (2009). Nonlocal photopolymerization kinetics including multiple termination mechanisms and dark reactions. Part I. Modeling. *J. Opt. Soc. Am. B: Opt. Phys.* 26 (9): 1736–1745.

**35** Gleeson, M.R., Liu, S., McLeod, R.R., and Sheridan, J.T. (2009). Nonlocal photopolymerization kinetics including multiple termination mechanisms and dark reactions. Part II. Experimental validation. *J. Opt. Soc. Am. B: Opt. Phys.* 26 (9): 1746–1754.

**36** Gleeson, M.R., Liu, S., Guo, J., and Sheridan, J.T. (2010). Non-local photo-polymerization kinetics including multiple termination mechanisms and dark reactions: part III. Primary radical generation and inhibition. *J. Opt. Soc. Am. B* 27 (9): 1804–1812.

**37** Bruder F-K, Deuber F, Fäcke T, Hagen R, Hönel D, Jurbergs D, Kogure M, Rölle T and Weiser M-S 2010 Reaction-diffusion model applied to high resolution Bayfol®HX photopolymer, in Proc. SPIE: Practical Holography XXIIV: Materials and Applications vol 7919, pp 0I-1-15

**38** Mayo, F.R. (1943). Chain transfer in the polymerization of styrene: the reaction of solvents with free radicals. *J. Am. Chem. Soc.* 65 (12): 2324–2329.

**39** Fevola, M., Hester, R., and McCormack, C. (2003). Molecular weight control of polyacrylamide with sodium formate as a chain-transfer agent: characterization via size exclusion chromatography/multi-angle laser light scattering and determination of chain-transfer constant. *J. Polym. Sci. Part A: Polym. Chem.* 41 (4): 560–568.

**40** Tomita, Y., Suzuki, N., and Chikama, K. (2005). Holographic manipulation of nanoparticle distribution morphology in nanoparticle-dispersed photopolymers. *Opt. Lett.* 30 (8): 839–841.

**41** Tomita, Y., Chikama, K., Suzuki, N. et al. (2006). Two-dimensional imaging of atomic distribution morphology created by holographically induced mass transfer of monomer molecules and nanoparticles in a silica-nanoparticle-dispersed photopolymer film. *Opt. Lett.* 31 (10): 1402–1404.

**42** Kogelnik, H. (1969). Coupled wave theory for thick hologram gratings. *Bell Syst. Tech. J.* 48 (9): 2909–2947.

**43** Moharam, M.G. and Gaylord, T.K. (1981). Rigorous coupled-wave analysis of planar-grating diffraction. *J. Opt. Soc. Am.* 71 (7): 811–818.

**44** Syms, R.R.A. (1990). *Practical Volume Holography*. Oxford: Clarendon.

**45** Gleeson, M.R. and Sheridan, J.T. (2009). A review of the modelling of free-radical photopolymerization in the formation of holographic gratings. *J. Opt. A Pure Appl. Opt.* 11 (2): 024008.

**46** Uchida, N. (1973). Calculation of diffraction efficiency in hologram gratings attenuated along the direction perpendicular to the grating vector. *J. Opt. Soc. Am.* 63: 280–287.

**47** O'Neill, F.T., Lawrence, J.R., and Sheridan, J.T. (2000). Automised testing and recording of holographic optical element arrays. *Optik* 111: 459–467.

**48** Hesselink, L., Orlov, S.S., and Bashaw, M.C. (2004). Holographic data storage systems. *Proc. IEEE* 92: 1231–1280.

**49** Nakamura, T., Nozaki, J., Tomita, Y. et al. (2009). Holographic recording sensitivity enhancement of $ZrO_2$ nanoparticle–polymer composites by hydrogen donor and acceptor agents. *J. Opt. A Pure Appl. Opt.* 11 (2): 024010-1-7.

**50** Mok, F.H., Burr, G.W., and Psaltis, D. (1996). System metric for holographic memory systems. *Opt. Lett.* 21 (12): 896–898.

**51** Dhar, L., Hale, A., Katz, H.E. et al. (1999). Recording media that exhibit high dynamic range for digital holographic data storage. *Opt. Lett.* 24 (7): 487–489.

**52** Sullivan, A.C., Grabowski, M.W., and McLeod, R.R. (2007). Three-dimensional direct-write lithography into photopolymer. *Appl. Opt.* 46 (3): 295–301.

**53** Dhar, L., Hale, A., Kurtis, K. et al. (2000). High Density Holographic Data Storage. *IEEE Conference Digest: Optical Data Storage*, pp. 158–160.

**54** Schultz, S., Glytsis, E., and Gaylord, T. (2000). Design, fabrication, and performance of preferential-order volume grating waveguide couplers. *Appl. Opt.* 39 (8): 1223–1232.

**55** Sheridan, J.T., Kelly, J.V., O'Brien, G. et al. (2004). Generalized non-local responses and higher harmonic retention in non-local polymerization driven diffusion model based simulations. *J. Opt. A Pure Appl. Opt.* 6 (12): 1089–1096.

**56** McLeod, R.R., Daiber, A.J., McDonald, M.E. et al. (2005). Microholographic multilayer optical disk data storage. *Appl. Opt.* 44 (16): 3197–3207.

**57** Milster, T.D. (2005). Horizons for optical data storage. *Opt. Photonics News* 16: 28–33.

**58** Sheridan, J.T., Gleeson, M.R., Kelly, J.V. et al. (2006). Optimized holographic data storage: diffusion and randomization. *J. Opt. A Pure Appl. Opt.* 8 (3): 236–243.

**59** Toisihi, M., Tanaka, T., Watanabe, K., and Betsuyaku, K. (2007). Analysis of photopolymer media of holographic data storage using non-local polymerization driven diffusion model. *Jpn. J. Appl. Phys.* 46: 3438–3447.

**60** Nagy, Z., Koppa, P., Ujhelyi, F. et al. (2007). Modeling material saturation effects in microholographic recording. *Opt. Express* 15 (4): 1732–1737.

**61** Toishi, M., Takeda, T., Tanaka, K. et al. (2008). Two-dimensional simulation of holographic data storage medium for multiplexed recording. *Opt. Express* 16 (4): 2829–2839.

**62** Close, D.H., Jacobson, A.D., Magerum, R.C. et al. (1969). Hologram recording on photopolymer materials. *Appl. Phys. Lett.* 14 (5): 159–160.

**63** Smothers, W.K., Trouts, T.J., Weber, A.M., and Mackish, D.J. (1990). Hologram recording in du Pont's new photopolymer materials. *Proceedings of SPIE*, Volume 1212, pp. 30–39.

**64** Sadlej, N. and Smolinska, B. (1975). Stable photo-sensitive polymer layers for holography. *Opt. Laser Technol.* 7: 175–179.

**65** Calixto, S. (1987). Dry polymer for holographic recording. *Appl. Opt.* 26 (18): 3904–3910.

**66** Blaya, S., Mallavia, R., Carretero, L. et al. (1998). Highly sensitive photopolymerizable dry film for use in real time holography. *Appl. Phys. Lett.* 75: 1628–1630.

**67** Zhao, F., Frietmann, E.E., and Geng, W. (1998). Combinative action of zinc chloride and triethanolamine on improving photosensitivity in red-sensitized dichromated gelatin for optical storage. *Proceedings of SPIE*, Volume 3468, pp. 317–321.

**68** Curtis, K. and Psaltis, D. (1994). Characterization of the DuPont photopolymer for three-dimensional holographic storage. *Appl. Opt.* 33 (23): 5396–5399.

**69** Jurbergs, D., Bruder, F.K., Deuber, F. et al. (2009). New recording materials for the holographic industry. *Proceedings of SPIE*, Volume 7233, pp. 72330K-1–72330K-10.

**70** Sheridan, J.T. and Lawrence, J.R. (2000). Nonlocal-response diffusion model of holographic recording in photopolymer. *J. Opt. Soc. Am. A* 17 (6): 1108–1114.

**71** Lawrence, J.R., O'Neill, F.T., and Sheridan, J.T. (2002). Adjusted intensity nonlocal diffusion model of photopolymer grating formation. *J. Opt. Soc. Am. B: Opt. Phys.* 19 (4): 621–629.

**72** Gleeson, M.R., Sabol, D., Liu, S. et al. (2008). Improvement of the spatial frequency response of photopolymer materials by modifying polymer chain length. *J. Opt. Soc. Am. B: Opt. Phys.* 25 (3): 396–406.

**73** Guo, J., Gleeson, M.R., Liu, S., and Sheridan, J.T. (2011). Non-local spatial frequency response of photopolymer materials containing chain transfer agents: I. Theoretical modelling. *J. Opt.* 13 (9): 095601.

**74** Guo, J., Gleeson, M.R., Liu, S., and Sheridan, J.T. (2011). Non-local spatial frequency response of photopolymer materials containing chain transfer agents: I. Theoretical modelling. *J. Opt.* 13 (9): 095602.

**75** Oliveira, P.W., Krug, H., Müller, P., and Schmidt, H. (1996). Fabrication of GRIN-materials by photopolymerization of diffusion- controlled organic-inorganic nanocomposite materials. *Mater. Res. Soc. Symp. Proc.* 435: 553–558.

**76** Krug, H. and Schmidt, H. (1994). Organic-inorganic nanocomposites for micro-optical applications. *New J. Chem.* 18: 1125–1129.

**77** Suzuki, N., Tomita, Y., and Kojima, T. (2002). Holographic recording in TiO[sub 2] nanoparticle-dispersed methacrylate photopolymer films. *Appl. Phys. Lett.* 81 (22): 4121–4123.

**78** Suzuki, N. and Tomita, Y. (2004). Silica-nanoparticle-dispersed methacrylate photopolymers with net diffraction efficiency near 100%. *Appl. Opt.* 43 (10): 2125–2129.

**79** Suzuki, N., Tomita, Y., Ohmori, K. et al. (2006). Highly transparent $ZrO_2$ nanoparticle-dispersed acrylate photopolymers for volume holographic recording. *Opt. Express* 14 (26): 12712–12719.

**80** Omura, K. and Tomita, Y. (2010). Photopolymerization kinetics and volume holographic recording in $ZrO_2$ nanoparticle-polymer composites at 404 nm. *J. Appl. Phys.* 107: 023107-1–023107-6.

**81** Liu, X., Tomita, Y., Oshima, J. et al. (2009). Holographic assembly of semiconductor CdSe quantum dots in polymer for volume Bragg grating structures with diffraction efficiency near 100%. *Appl. Phys. Lett.* 95: 261109-1–261109-3.

**82** Hata, E. and Tomita, Y. (2010). Order-of-magnitude polymerization-shrinkage suppression of volume gratings recorded in nanoparticle-polymer composites. *Opt. Lett.* 35: 396–398.

**83** Hata, E., Mitsube, K., Momose, K., and Tomita, Y. (2011). Holographic nanoparticle-polymer composites based on step-growth thiol-ene photopolymerization. *Opt. Mater. Express* 1 (2): 207–222.

**84** Hata, E. and Tomita, Y. (2011). Stoichiometric thiol-to-ene ratio dependences of refractive index modulation and shrinkage of volume gratings recorded in photopolymerizable nanoparticle-polymer composites based onstep-growth polymerization. *Opt. Mater. Express* 1 (6): 1113–1120.

**85** Mitsube, K., Nishimura, Y., Nagaya, K. et al. (2014). Holographic nanoparticle-polymer composites based on radical-mediated thiol-yne photopolymerizations: characterization and shift-multiplexed holographic digital data page storage. *Opt. Mater. Express* 4 (5): 982–996.

**86** Tomita, Y., Furushima, K., Ochi, K. et al. (2006). Organic nanoparticle (hyperbranched polymer)-dispersed photopolymers for volume holographic storage. *Appl. Phys. Lett.* 88: 071103-1–071103-3.

**87** Tomita, Y., Urano, H., Fukamizu, T. et al. (2016). Nanoparticle-polymer composite volume holographic gratings dispersed with ultrahigh-refractive-index hyperbranched polymer as organic nanoparticles. *Opt. Lett.* 41 (6): 1281–1284.

**88** Momose, K., Takayama, S., Hata, E., and Tomita, Y. (2012). Shift-multiplexed holographic digital data page storage in a nanoparticle-(thiol–ene) polymer composite film. *Opt. Lett.* 37 (12): 2250–2252.

**89** Takayama, S., Nagaya, K., Momose, K., and Tomita, Y. (2014). Effects of symbol modulation coding on readout fidelity of shift-multiplexed holographic digital data page storage in a photopolymerizable nanoparticle-(thiol-ene)polymer composite film. *Appl. Opt.* 53: B54–B59.

**90** Nagaya, K., Hata, E., and Tomita, Y. (2016). Readout fidelity of coaxial holographic digital data page recording in nanoparticle–(thiol–ene) polymer composites. *Jpn. J. Appl. Phys.* 55: 09SB03-1-4.

**91** Fujii, R., Guo, J., Klepp, J. et al. (2014). Nanoparticle polymer composite volume gratings incorporating chain transfer agents for holography and slow-neutron optics. *Opt. Lett.* 39 (12): 3453–3456.

**92** Guo, J., Fujii, R., Ono, T. et al. (2014). Effects of chain-transferring thiol functionalities on the performance of nanoparticle-polymer composite volume gratings. *Opt. Lett.* 39 (23): 6743–6746.

**93** Aswathy, G., Rajesh, C.S., Sreekumar, K. et al. (2016). Improving the performance of methylene blue sensitized photopolymer by doping with nickel ion. *Opt. Mater.* 55: 27–32.

**94** Goourey, G.G., Chung, P.W.W., Jestin, F.D. et al. (2015). Photostability of acrylate photopolymers used as components in recording materials. *Polym. Degrad. Stab.* 119: 208–216.

**95** Suzuki, N. and Tomita, Y. (2007). Holographic scattering in $SiO_2$ nanoparticle-dispersed photopolymer films. *Appl. Opt.* 46 (27): 6809–6814.

**96** Li, C., Li, X., Xue, X., and Huang, M. (2014). Holographic properties of $Fe_3O_4$ nanoparticle-doped organic–inorganic hybrid photopolymer. *Optik (Stuttg.)* 125 (21): 6509–6512.

**97** Chen, Y.F., Hsiao, Y.N., Lin, S.H. et al. (2009). Effect of lanthanoid organometallic compounds on holographic storage characteristics of doped PQ/poly(hydroxyethyl methacrylate- co-methyl methacrylate) hybrids. *J. Opt. A Pure Appl. Opt.* 11 (12): 1–6.

**98** Yu, D., Liu, H., Wang, J. et al. (2011). Study on holographic characteristics in ZnMA doped PQ-PMMA photopolymer. *Opt. Commun.* 284 (12): 2784–2788.

**99** Zang, J., Kang, G., Li, P. et al. (2017). Dual- channel recording based on the null reconstruction effect of orthogonal linear polarization holography. *Opt. Lett.* 42 (7): 1377–1380.

**100** Steckman, G.J., Solomatine, I., Zhou, G., and Psaltis, D. (1998). Characterization of phenanthrenequinone-doped poly(methyl methacrylate) for holographic memory. *Opt. Lett.* 23 (16): 1310–1312.

**101** Lin, S.H., Hsu, K.Y., Chen, W.Z., and Whang, W.T. (2000). Phenanthrenequinone-doped poly(methyl methacrylate) photopolymer bulk for volume holographic data storage. *Opt. Lett.* 25 (7): 451–453.

**102** Liu, H., Yu, D., Li, X. et al. (2010). Diffusional enhancement of volume gratings as an optimized strategy for holographic memory in PQ-PMMA photopolymer. *Opt. Express* 18 (7): 6447–6454.

**103** Mahilny, U.V., Marmysh, D.N., Stankevich, A.I. et al. (2005). Holographic volume gratings in a glass-like polymer material. *Appl. Phys. B Lasers Opt.* 82 (2): 299–302.

**104** Fan, F., Liu, Y., Hong, Y., Zang, J., et al. (2018). Improving the polarization-holography performance of PQ/PMMA photopolymer by doping with THMFA. *Opt. Exp.* 26 (14): 17794–17803.

**105** Fan, F.L., Liu, Y., Hong, Y.F. et al. (2018). Volume polarization holographic recording in phenanthrenequinone doped poly(MMA-Co-BzMA) photopolymer. *Chem. Lett.* 47 (4): 520–523.

**106** Rajendrakumar, K. and Dhamodharan, R. (2009). Ambient temperature atom transfer radical copolymerization of tetrahydrofurfuryl methacrylate and methyl methacrylate: reactivity ratio determination. *Eur. Polym. J.* 45 (9): 2685–2694.

**107** Li, C., Cao, L., Wang, Z., and Jin, G. (2014). Hybrid polarization-angle multiplexing for volume holography in gold nanoparticle-doped photopolymer. *Opt. Lett.* 39 (24): 6891–6894.

**108** Mahilny, U., Trofimova, A., Nazarov, S. et al. (2016). Highly concentrated phenanthrenequinone-polymethylmethacrylate composite for thick reflection holograms recording at 532 nm. *Opt. Mater. Express* 6 (11): 3427–3437.

**109** Kawatsuki, N. (2011). Photoalignment and photoinduced molecular reorientation of photosensitive materials. *Chem. Lett.* 40 (6): 548–554.

**110** Lin, S.H., Chen, P.L., and Lin, J.H. (2009). Phenanthrenequinone-doped copolymers for holographic data storage. *Opt. Eng.* 48 (3): 035802.

**111** Steckman, G.J., Shelkovnikov, V., Berezhnaya, V. et al. (2000). Holographic recording in a photopolymer by optically induced detachment of chromophores. *Opt. Lett.* 25 (9): 607–609.

**112** Hsiao, Y.N. (2004). Analyses on physical mechanism of holographic recording in phenanthrenequinone-doped poly(methyl methacrylate) hybrid materials. *Opt. Eng.* 43 (9): 1993–2002.

# 6

# Data Format

## 6.1 3 : 16 Modulation

As already described, two-dimensional page data are composed of a ring pattern formed by the reference light and an information-light pattern in the inner part of the ring. Figure 6.1 shows a specific pattern example of page data; here, the ON pixels used for recording are shown in black color, and the OFF pixels are shown in white color. Figure 6.1(a) shows a pattern displayed on the spatial light modulator (SLM) during the recording. A reference-light pattern and an information-light pattern are simultaneously displayed. Figure 6.1(b) shows a pattern displayed on the SLM during the reproduction, and only the annular reference light pattern same as that observed during the recording is displayed. The reference light that appears as a radial line is also a pixel-by-pixel pattern when enlarged. Thus, the key to realizing high-quality holographic recording and reproduction by the collinear method depends on the design of the reference- and information-light patterns.

First, an information-light pattern display rule, that is, a data page format is described. User data are digitized in advance, and after preprocessing, such as adding error correction codes and random arrangement as necessary, the data page is coded in units of 8 bits according to a predetermined rule, and sequentially arranged on a two-dimensional page. Figure 6.2 shows an example of the page format and coding.

A data page (or simply called a page) is composed of one-page sync mark in the upper left and 51 sub-pages, and each sub-page has a sync mark in the middle. There are 32 symbols per sub-page, and each symbol is composed of 16 pixels.

Considering the high storage-density demand, usually an aperture in the media plane should be used to block the high frequencies. The size of the aperture could be 1.1–1.3*x* the Nyquist size. Therefore, the image quality of the reproduced beam is comparatively poor because of other complex noises.

In our coding strategy, a kind of sparse modulation coding [2], called 3 : 16 modulation coding, is used to resist the image decay. That means in one symbol only three pixels are ON, and the remaining pixels are OFF. The 3 : 16 modulation coding always displays only three of the 16 pixels, and according to the coding rule, three ON pixels cannot be adjacent. Figure 6.3 shows the examples of forbidden combination style of three ON pixels in one symbol. This rule is also for resisting

*Collinear Holography: Devices, Materials, Data Storage*, First Edition.
Xiaodi Tan, Hideyoshi Horimai, Tsutomu Shimura, and Xiao Lin.

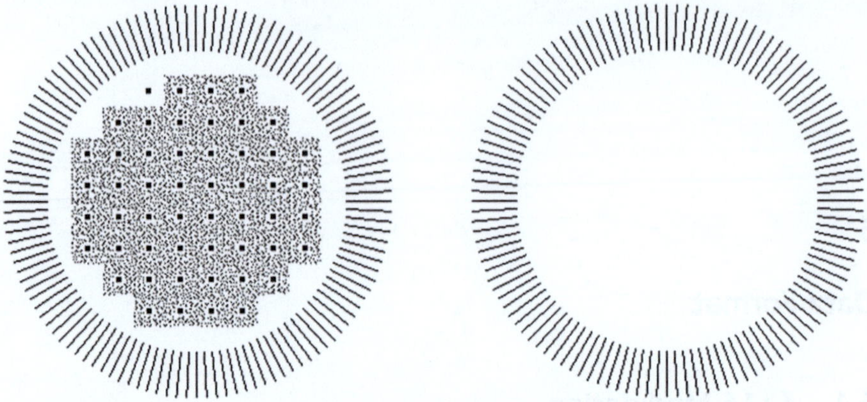

**Figure 6.1** Example of two-dimensional page data in collinear holographic memory. Source: Reproduced with permission of Liu et al. [1], Optical Society of America.

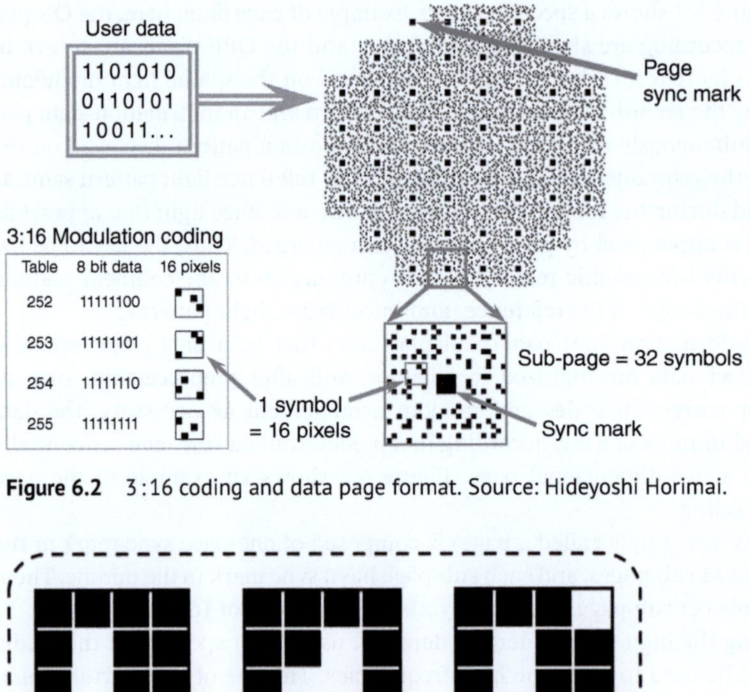


**Figure 6.2** 3 : 16 coding and data page format. Source: Hideyoshi Horimai.

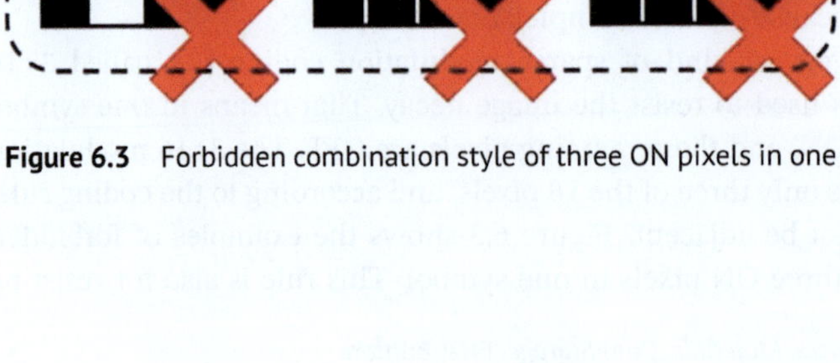

**Figure 6.3** Forbidden combination style of three ON pixels in one symbol. Source: Xiao Lin.

**Table 6.1** Comparison of different number of ON pixels coding.

| ON pixels/symbol | 2 | 3 | 4 | 5 |
|---|---|---|---|---|
| Number of symbol patterns | 96 | 275 | 403 | 265 |
| Amount of data/symbol | 6 bit | 8 bit | 8 bit | 8 bit |

the image decay. After considering the rules, 275 combination patterns are observed in one symbol, and 256 symbol patterns are picked to create a one-to-one look-up table (LUT). The user data are encoded and decoded based on this LUT. The other 19 symbol patterns can provide situations for the correct code. Because 256 symbol patterns = 8 bit, 1 symbol can store an 8 bit data; additionally, we obtain, code rate = 8 bit/16 pixel = 0.5. The amount of information per data page is 51 sub-page × 32 symbol × 1 byte = 1.632 kB.

Based on the nonadjacent coding rule, we can prove that the 3 : 16 modulation coding is most suitable in the case of 4 × 4 pixels as one symbol. We calculated the situations of 2 : 16, 3 : 16, 4 : 16, and 5 : 16 modulation coding and compared them as shown in Table 6.1. Obviously, 3 : 16 modulation coding can store 8 bit data with less ON pixels consumption. Further, the 4 : 16 and 5 : 15 modulation coding will consume more media and exhibit a higher symbol error probability.

In the recording process, the light intensity of the data page remains constant, and in the reading process, we choose the three brightest pixels to decode; this method can further resist the uneven energy distribution of reproducing.

## 6.2 Sync Format

Sync mark, short form of synchronization mark, plays a crucial role of two-dimensional page data format in data storage system. The data page format of a collinear holographic data storage system (CHDSS), based on the sub-page, is used to eliminate the problems of illuminated intensity distribution in a data page, distortion and aberration of the optical system, tilting, and the estimation error caused by amplification. As shown in Figure 6.4, there are two kinds of sync marks: page and sub-page sync marks. The page sync mark is located at the center of 24 × 24 pixels to aid in easy identification of the coordinate of the whole page. Further, the sync mark of a sub-page, i.e. the size of the sub-page (24 × 24 pixels), depends on the parameters and degree of inhomogeneity of a system. In each sub-page, there are 32 byte data symbols (8 × 8 pixels) and a sync mark (4 × 4 pixels) at its center. The synchronous symbol, which includes a rectangular block of 4 × 4 pixels, is used to locate the sub-page and provide the necessary coordinate information for data decoding.

Traditionally, a sync mark is a solid whiteboard pattern. The size of the hologram is dominated by the diffraction pattern of the digital micromirror device (DMD). The sync mark and data information exist at the same diffraction points on the Fourier

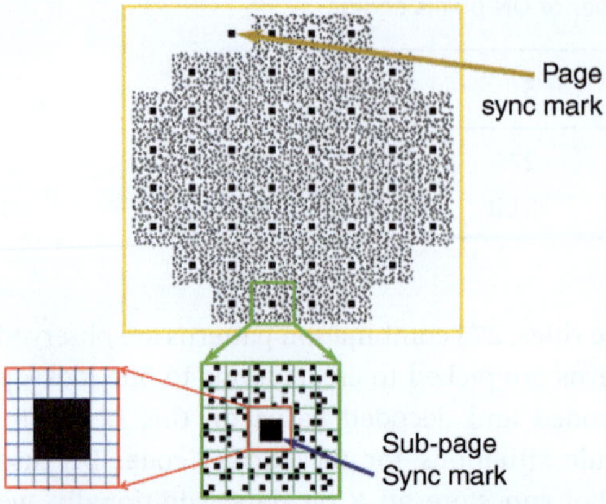


**Figure 6.4** Page and sub-page sync mark in page format. Source: Reproduced with permission of Liu et al. [1], Optical Society of America.

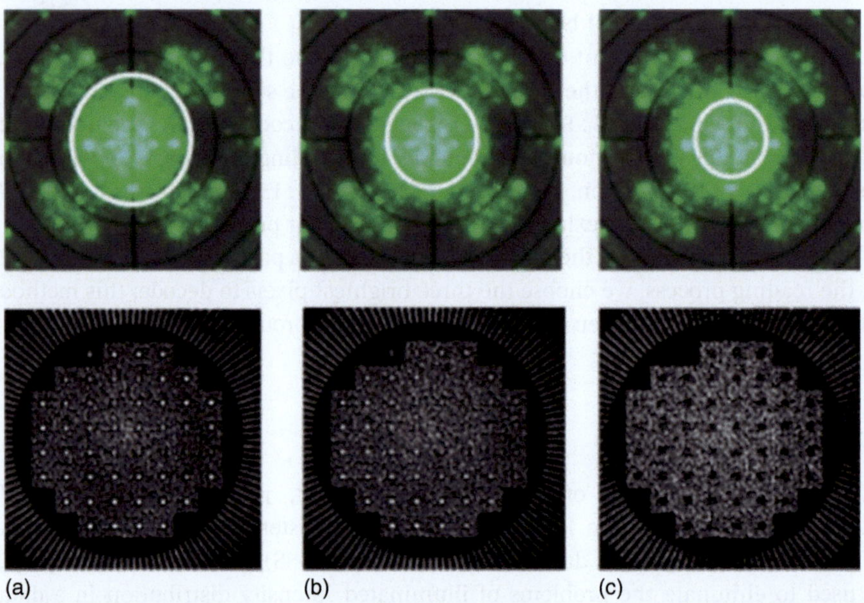


**Figure 6.5** Reconstructed pattern with different aperture sizes at the Fourier plane; white circle represents the aperture; the diameter of the aperture is (a) 8.0 mm, (b) 6.5 mm, and (c) 5.0 mm. Source: Liu et al. [1], Optical Society of America.

plane. Small apertures are often desirable in CHDSS to reduce material consumption and noise. However, aperture reduction decreases the intensity of the sync mark. It is impossible to recover the data without the sync mark, even if the data points are well regenerated.

As shown in Figure 6.5, in a traditional situation, where the sync mark is a solid whiteboard pattern, the energy of holograph is dominated by the four basic points. As a result, the sync mark reconstructed in the reading process is degraded if the aperture is smaller than the range of the four basic points. When the aperture is too small, the sync mark is lost in the reconstructed image. The information data cannot be reconstructed if the sync mark is missing, leading to whole data page loss.

Hence, we need to find a way to keep the sync mark in a good condition, irrespective of the reduction in the aperture size.

To reduce the decrease in the intensity of the reconstructed sync mark with spatial aperture reduction, we tried to locate the sync mark information in the middle of the diffraction pattern and found that sync marks with chessboard and random patterns can retain their initial intensities during the reconstruction process, independent of the aperture size reduction. However, compared to the chessboard pattern, the random pattern generated a larger energy distribution area at the Fourier plane, which is not suitable for the reduced aperture size. Thus, we designed a chessboard pattern to replace the solid whiteboard pattern of sync mark. Changing the solid whiteboard pattern into chessboard pattern can increase the frequency of sync mark points on the Fourier plane. As shown in Figure 6.6, a new diffraction point appears in the middle of the previous four points. Notably, the chessboard pattern design is based on the optimal tilt angle of the micromirror of the DMD described in Section 4.2 of Chapter 4.

Using the new diffraction point to carry the sync mark information can reduce the influence of aperture reduction on the sync mark reconstruction. Figure 6.7 compares the bit error rate (BER) at different aperture sizes for solid whiteboard and chessboard patterns, based on numerical simulation. We can clearly see the improved BER performance of the white chessboard pattern.

Then, the performance of the chessboard sync mark design was experimentally investigated. Figure 6.8 shows a portion of the reconstructed data page with the chessboard sync mark at different aperture sizes. From Figure 6.5, we can see that the sync mark disappears when the aperture diameter is smaller than 5.0 mm. Then, we tested the performance of the newly designed chessboard sync mark for an aperture diameter < 5.0 mm, and found good agreement between the experimental and simulation results. Evidently, the sync mark remains unchanged for an aperture diameter of 2.5 mm. It also shows that we can still reconstruct the sync mark signal even if the data points are missing due to the reduced aperture size.

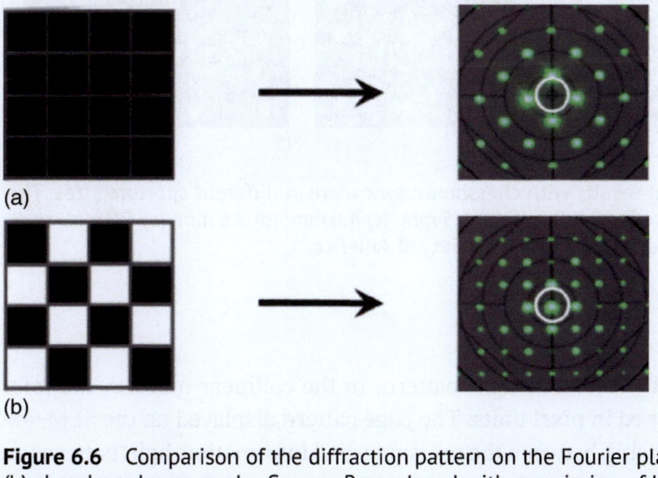


**Figure 6.6** Comparison of the diffraction pattern on the Fourier plane: (a) whiteboard and (b) chessboard sync marks. Source: Reproduced with permission of Liu et al. [1], Optical Society of America.

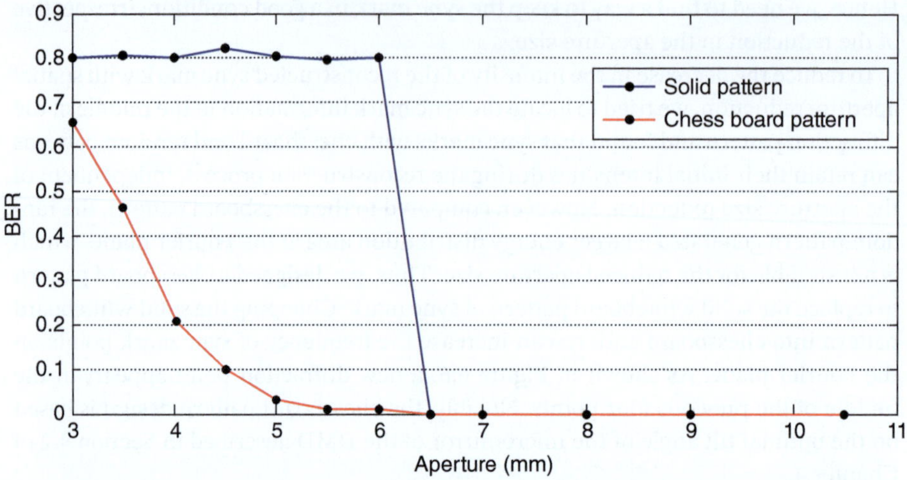


**Figure 6.7** BER comparison using a solid pattern and chessboard of sync mark in the CHDSS, based on simulation. Source: Reproduced with permission of Liu et al. [1], Optical Society of America.

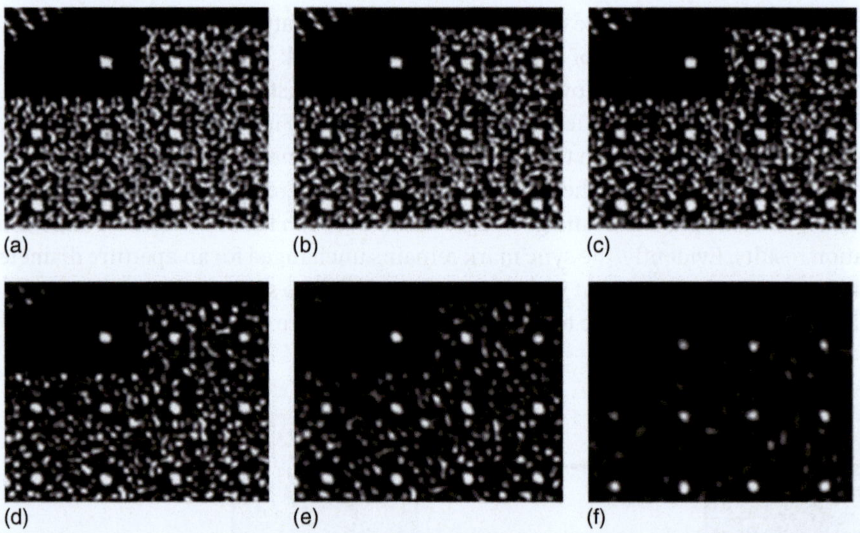


**Figure 6.8** Experiment results with chessboard sync mark at different aperture sizes. The diameter of the aperture is (a) 5.0 mm; (b) 4.5 mm; (c) 4.0 mm; (d) 3.5 mm; (e) 3.0 mm; and (f) 2.5 mm. Source: Liu et al. [1], Optical Society of America.

## 6.3 Reference Pattern

Here, we elucidate the reference-light pattern. In the collinear method, the reference light is also defined in pixel units. The page pattern displayed on the SLM, and the positional relationship between the reference- and information-light patterns are uniquely defined at the pixel level.

(a) (b) (c)

**Figure 6.9** Reference patterns: (a) radial lines, (b) random, and (c) multiple rings. Source: Tsutomu Shimura.

In theory, the reference pattern can be arbitrary; however, no arbitrary reference pattern shows good performance in the holographic data storage. Basically, there are three types of reference-light patterns: radial lines, random, and multiple rings as shown in Figure 6.9. The reference-light patterns greatly affect the shift selectivity.

In Section 3.1, the pixel spread function (PxSF) has been introduced. PxSFs are calculated as a function of the media shift from the original position for various reference and media conditions. The number of pixels in the SLM were 321 × 321.

### 6.3.1 Dependence on the Reference Patterns

As shown in Figure 6.10, the difference in the decrease in the relative intensity of the signal is small, and there is no significant difference in the shift selectivity with the reference pattern (media thickness was 0.5 mm).

### 6.3.2 Dependence on the Width of the Reference Pattern

The signal decays with media shift are compared with the width of the reference pattern (Figure 6.11). A radial reference pattern was used (Figure 6.9(a)) with inner radii (a) 99%, (b) 84%, and (c) 78% of the outer radius. The reference pattern for the last case was the same as that shown in Figure 6.9(a). A wider reference pattern exhibits a better shift selectivity.

Figure 6.12 shows the same result as that shown in Figure 6.6(a), but on a linear scale. For the thin-ring reference pattern, we can calculate the signal decay with the media shift as,

$$\int_0^{2\pi} \exp(i\varphi)\mathrm{d}\theta = \int_0^{2\pi} \exp(i\Delta x K_x)\mathrm{d}\theta$$
$$= \int_0^{2\pi} \exp\left(i\Delta x\frac{kR}{f}\sin\theta\right)\mathrm{d}\theta = J_0\left(\Delta x\frac{kR}{f}\right) \tag{6.1}$$

where $R$ is the radius of the ring reference pattern. From Eq. (6.1), the first zero is calculated as $\Delta x = 0.29\ \mu\mathrm{m}$, for $\lambda = 408\ \mathrm{nm}$ and $R = 160 \times 13.7\ \mu\mathrm{m}$. The result shows good agreement with that presented in Figure 6.12.

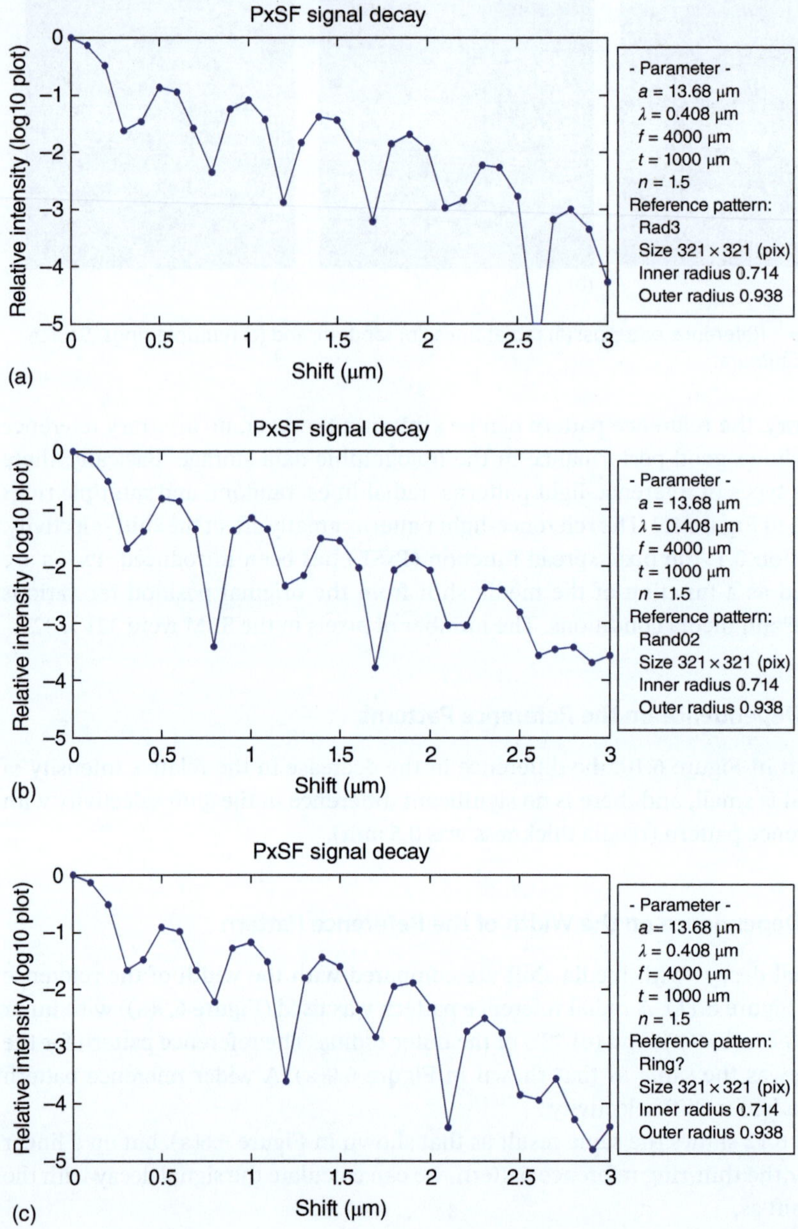


**Figure 6.10** Peak signal intensity for a single pixel for the three reference patterns as functions of media shifts. (a) radial, (b) random, and (c) multiple ring reference patterns. Source: Tsutomu Shimura.

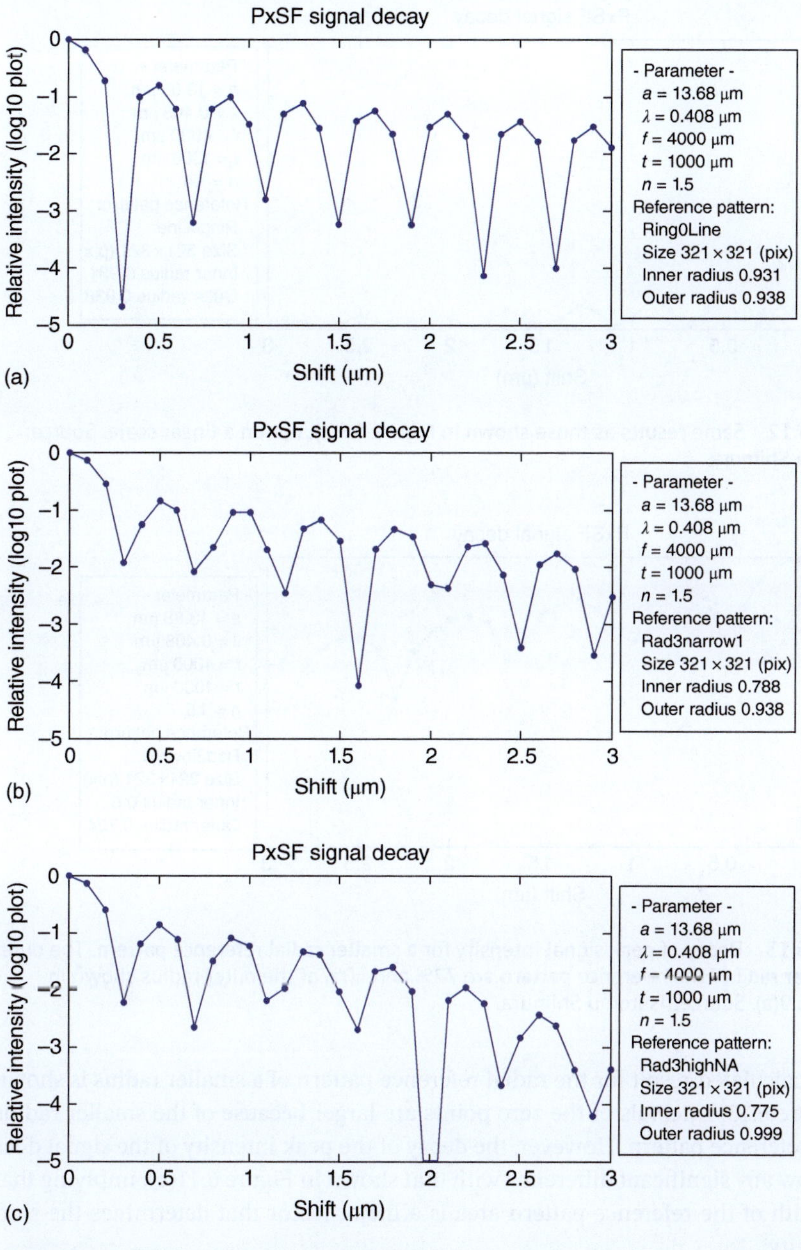


**Figure 6.11** Decay of the signal peak intensity of a single pixel for the radial reference patterns as function of the media shifts. Inner radii of the reference patterns are (a) 99%, (b) 84%, and (c) 78%. Source: Tsutomu Shimura.

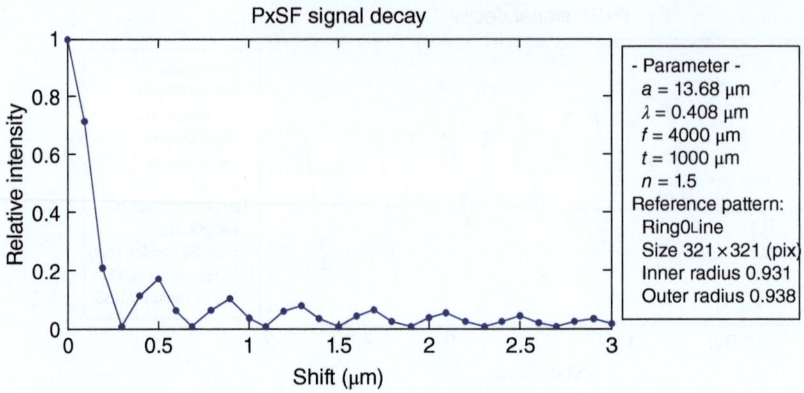


**Figure 6.12** Same results as those shown in Figure 6.6(a), but on a linear scale. Source: Tsutomu Shimura.

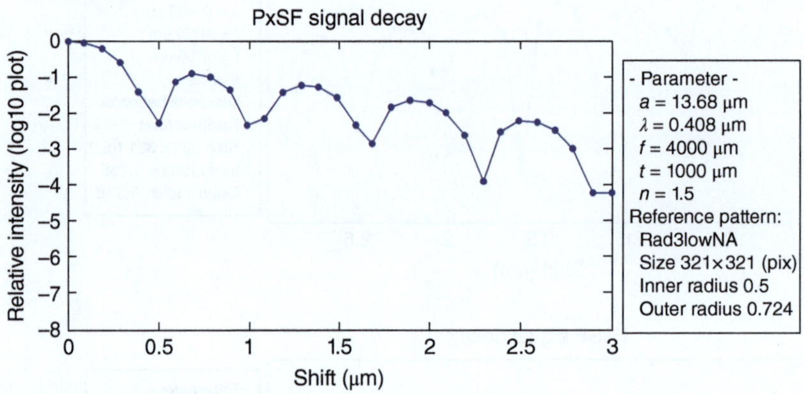


**Figure 6.13** Decay of peak signal intensity for a smaller radial reference pattern. The outer and inner radii of the reference pattern are 72% and 50% of the outer radius shown in Figure 6.9(a). Source: Tsutomu Shimura.

The calculated result for the radial reference pattern of a smaller radius is shown in Figure 6.13. Intervals of the zero points are larger because of the smaller radius of the reference pattern. However, the decay of the peak intensity of the signal does not show any significant difference with that shown in Figure 6.11(c), implying that the width of the reference pattern area is a major factor that determines the shift selectivity.

### 6.3.3 Adjustment Light Energy Ratio

In hologram recording, control of the light energy ratio between the reference light and the information light is also important. In the collinear method, as shown in Figure 6.14, by changing the number of pixels of the reference light, it is possible to define the light energy ratio in a robust and simple manner.

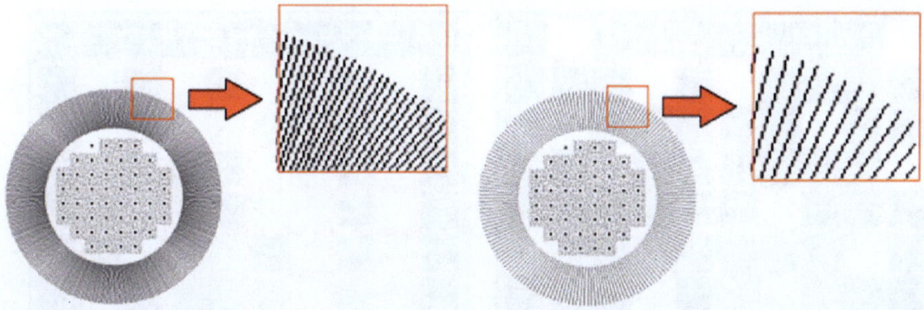

**Figure 6.14** Adjustment of light energy ratio between the reference light and the object light in collinear method. Source: Hideyoshi Horimai.

## 6.4 Phase Modulation

### 6.4.1 Introduction

In the hologram optical memory, a large capacity and high transfer rate have been realized by recording and reproducing two-dimensional page data on a three-dimensional volume recording medium. To perform more efficient recording/reproduction, research to improve the coding method and obtain a code rate as close to 1.0 as possible [3–6] has been conducted. In this case, one pixel of the recording/reproduction data is matched with one pixel of complementary metal-oxide semiconductor (CMOS) sensor (i.e. 1 : 1). Research on the Pixel Matching method is also reported [7, 8]. In addition, attempts have been reported to achieve further ultrahigh density and ultrahigh transfer rate. One of them is a half-tone multilevel hologram memory research conducted at IBM [9]. This is an attempt to dramatically increase the recording density and data transfer rate by converting one pixel, which has been binary digital data, into multiple values by halftone display to increase the amount of information. However, since recording with high signal-to-noise ratio (SNR) is necessary to accurately record and reproduce halftones, it is theoretically concluded that the recording density can only be improved by about 30% at the maximum [10]. In addition, many researchers, including Denz et al., studied holographic optical memory that records and reproduces "light phase," one of the major features of holograms, as information data [2, 11–19]. However, since air fluctuations in the optical path and minute vibrations of the recording medium cause phase disturbance and affect recording and reproduction, phase information recording and reproduction experiments are difficult, and the reported contents remain within the basic proof of principle.

On the other hand, in collinear holography, information light and reference light are recorded and reproduced while traveling along the same optical path, so that the optical system can be configured with a common path. Therefore, it is possible to construct a hologram recording/reproducing environment that is hardly affected by air fluctuation in the optical path and that is resistant to vibration of the recording medium. In particular, when the optical servo technology is applied, the relative

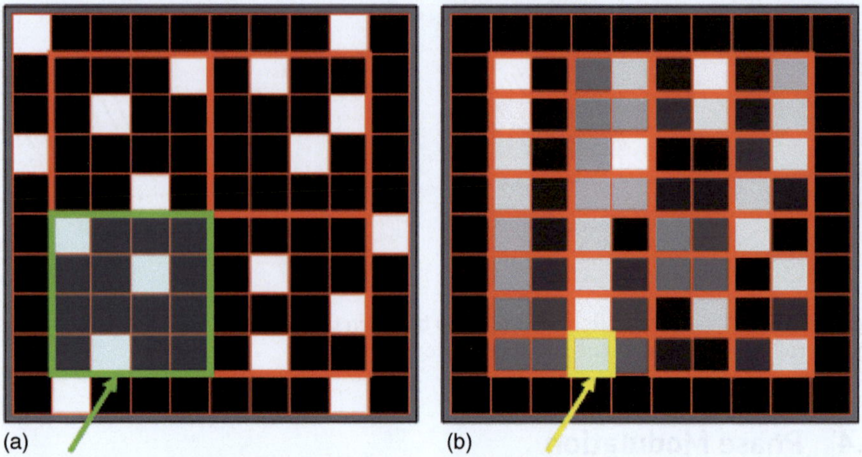


**Figure 6.15** The comparison between (a) basic collinear coding method and (b) phase lock collinear coding method. Source: Hideyoshi Horimai.

position with respect to the recording medium can be dynamically reproduced on the order of 1/20 or less (several tens of nanometers) of the wavelength.

Therefore, basic research and development of a next-generation hologram recording/reproducing method in which phase information is added to two-dimensional page data was conducted based on the basic technology established so far by the collinear method. Up to now, as shown in Figure 6.15 a, 8 bit recording/reproduction has been performed with 16 pixels using digital data that has been subjected to intensity modulation of binary black and white in the SLM. Effectively utilizing the phase information of light originally possessed, as shown in Figure 6.15(b), the possibility of realizing collinear holography with phase multilevel information of up to 8 bits with only one pixel ultimately is seen. Like the phase locked loop (PLL) circuit used in electrical circuits, it can be modified to reduce the phase detection error while applying feedback, so it is called "optical phase-locked collinear method" or simply "optical phase-locked method."

In this chapter, first, the basic principle of the optical phase-locked collinear system is described, followed by the configurations of the experimental optical system and page data. Next, the results of a recording/reproduction experiment of a phase multilevel gradation performed using the optical phase lock method are presented. The phase value can be readout at a low error rate even when the information of 32 gradations (5 bits) is added to one pixel. The experimental results are discussed in detail. Furthermore, the basic principle of a method for phase-coding the reference light and multiplex recording a plurality of phase multilevel page data at the same position are demonstrated.

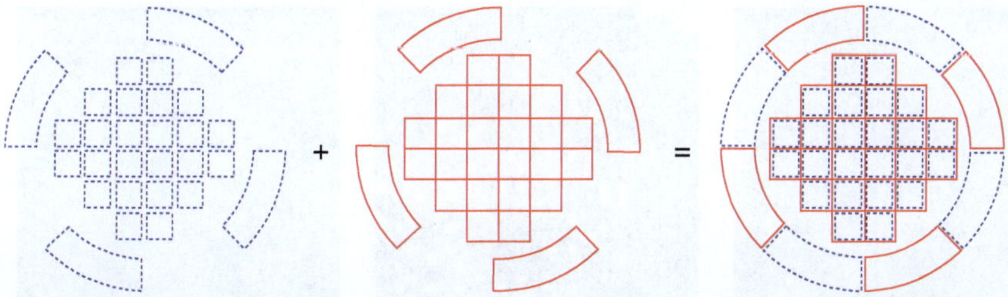


**Figure 6.16** Basic principle of optical phase-locked collinear system. Source: Hideyoshi Horimai.

## 6.4.2 Fundamental Principle of Optical Phase-Locked Collinear System

### 6.4.2.1 Fundamental Principles and Experimental Results

Figure 6.16 is an explanatory diagram of the basic principle of the "optical phase-locked collinear method" for recording/reproducing phase multilevel gradation data. The annular reference light used in the basic collinear method is divided into two parts. Figure 6.16 shows a case of eight divisions, and four of them are used to generate a "lock light page pattern" as a phase reference. Then, the other four are used to generate a "data light page pattern" having phase multilevel gradation information as user data. The two page patterns are recorded independently at the same place. Data reproduction is performed by irradiating the recording position with reference light of both "lock light" and "data light," that is, an annular reference light in which eight are connected. As a result, both the "information light pattern of the lock light" and the "information light pattern of the data light" are reproduced at the same time, and the pixels at the same position in both of them interfere with each other to detect the intensity-modulated signal. Since the phase of the lock light is known, some methods are used to strengthen the data light when the phase is the same as that of the lock light, and weaken the data light when the phase is opposite. The phase of the absolute value of each pixel of the data light can be obtained. This is the basic principle of the optical phase lock collinear method.

Figure 6.17 shows the result of actually confirming this basic principle through a recording/reproducing experiment. Figure 6.17(a) shows a reproduced image when data light is recorded and reproduced with only the reference light of the data light. Figure 6.17(b) is an enlarged image of a part thereof. Figure 6.17(c) is phase information added during recording. As can be seen from the image in Figure 6.17(b), the phase information added at the time of recording is not detected at this point, and the reproduced image has a uniform intensity.

Next, Figure 6.17(d,e) is reconstructed images and magnified images when the lock light is overlapped and recorded at the same position where the data

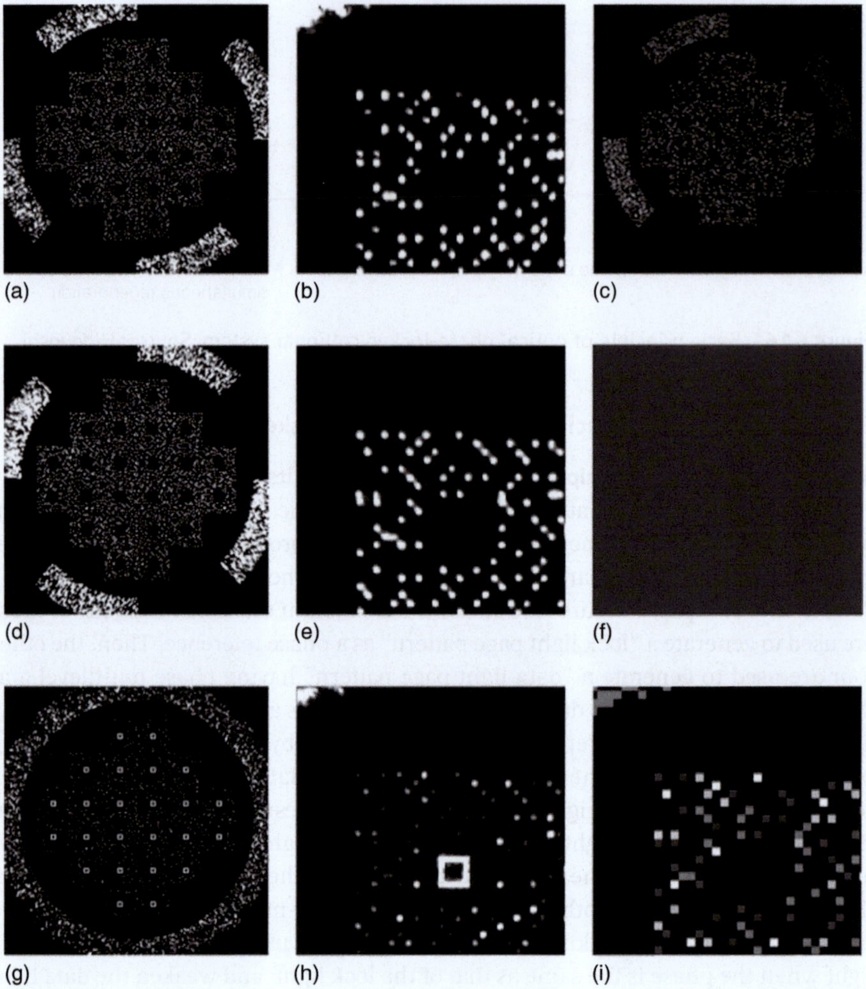


**Figure 6.17** Demonstration of optical phase-locked collier system. Source: Hideyoshi Horimai.

light is recorded and then reproduced with only the reference light of the lock light. Although no pattern appears in Figure 6.17(f), this is because the phase of the lock light is the same for all pixels. Even when the magnified observation image Figure 6.17(e) is viewed, at this point, the reproduced image has a uniform intensity.

Next, reproduction was performed by simultaneously irradiating both "data light" and "lock light" reference lights, and the reproduced image shown in Figure 6.17(g) was obtained. Figure 6.17(h) shows a magnified observation of this, but it has been confirmed that the brightness of each pixel varies from bright to dark differently. Therefore, when phase information added to "data light" (gray-tone image displayed on the phase SLM) is shown side by side in Figure 6.17(i), it was found that the luminance distributions are almost the same.

Through this experiment, it was confirmed that the reproduced image corresponding to the division pattern of the reference light was reproduced independently, despite the fact that "data light" and "lock light" were multiplexed and recorded at exactly the same position. It is also noteworthy that multilevel detection is performed by halftone by phase interference without sacrificing the SNR of each reproduced image. Furthermore, it was confirmed that the multivalue gradation image was displayed very stably, and reproducibility was high due to the benefit of the common path optical system.

These results experimentally validate the basic principle of the optical phase-locked collinear method.

#### 6.4.2.2 Optical Recording and Reproducing System for Optical Phase Locking

Figure 6.18 schematically shows the configuration of an optical phase-lock recording/reproducing experimental optical system. Figure 6.19(a) shows external views of the actual optical system and the sample holder section of the hologram recording medium sample (Figure 6.19(b,c)).

The main features of the optical phase-locked recording/reproducing optical system are the use of two devices: an SLM for intensity modulation (SLM-1) and an SLM for phase modulation (SLM-2). The Liquid Crystal on Silicon (LCOS) manufactured by Hamamatsu Photonics can be used for intensity modulation and phase modulation by changing the polarization direction of linearly polarized light incident on the device by 90°. For this reason, pixel matching is performed by combining the same device with a 1 : 1 telecentric 4f relay lens using L1-1 and L1-2. It became possible to control both phases. The resolution of display gradation is 8 bits (256 gradations) for both intensity and phase. The maximum value of the phase modulation amount is 0 to $2\pi$, and the phase gradation is displayed corresponding to the 8 bit gray gradation input to the drive unit of the device.

The optical system after SLM-2 has exactly the same configuration as the normal collinear holographic optical system.

#### 6.4.2.3 Page Data Structure of Optical Phase-Locked System

As described above, the optical phase-locked optical system is basically different from the optical system used for normal collinear holography, only the function of the SLM is changed to handle the phase. It has the same configuration. That is, it is very easy to understand if only the structure of the page data is considered to have evolved. Next, a page data structure in the optical phase lock method is described with reference to the drawings.

Figure 6.20 schematically shows the optical phase-locked page data broken down into intensity and phase. The intensity modulation pattern is generated by SLM-1, but the 3 : 16 modulation code used in holographic versatile disc (HVD) is used. At this point, the code rate is 0.5, which may seem wasteful, but to make a quantitative comparison with the normal collinear method, we decided to adopt this intensity modulation pattern for the basic principle experiments. As a result, of the 16 pixels, 3 pixels are turned ON, and phase information is added to each of the ON pixels in SLM-2. Figure 6.15 shows a case where a quaternary phase gradation (2 bit phase

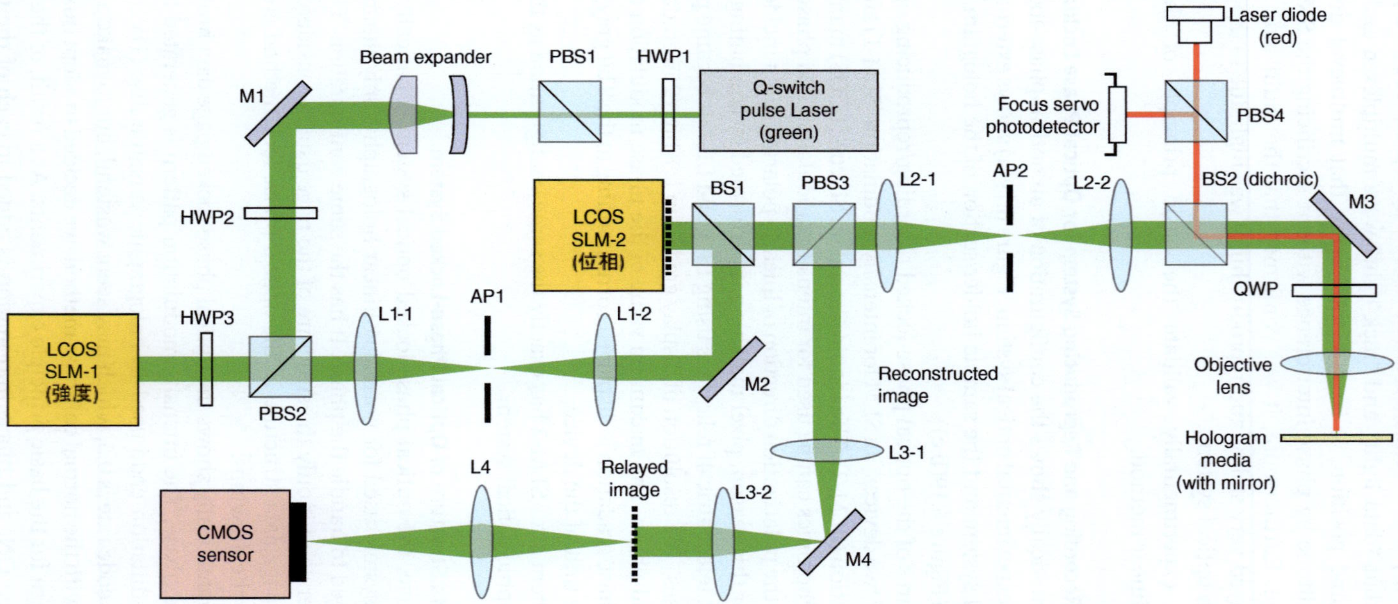


**Figure 6.18** Optical system for recording and reproducing experiment of optical phase-locked system. Source: Hideyoshi Horimai.

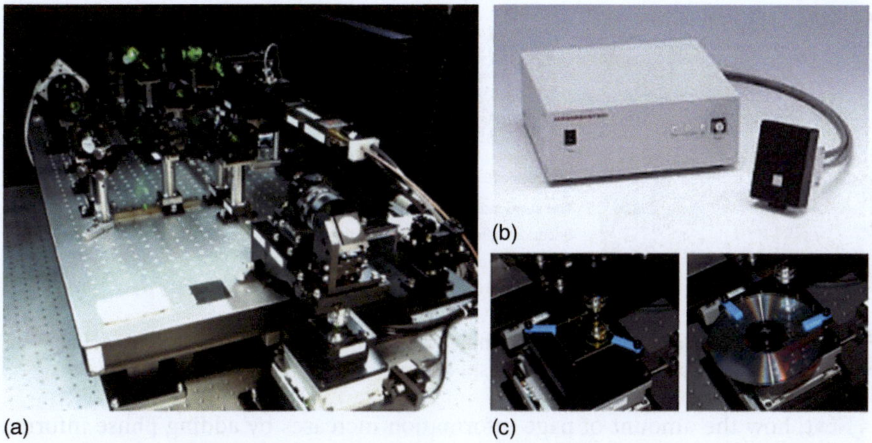


**Figure 6.19** (a) Appearance photograph of the experimental optical system, (b) the spatial light modulator made by Hamamatsu photonics incorporated into the apparatus: LCOS and drive unit, (c) the appearance photograph of the sample holder portion of the hologram recording medium. Source: Hideyoshi Horimai.

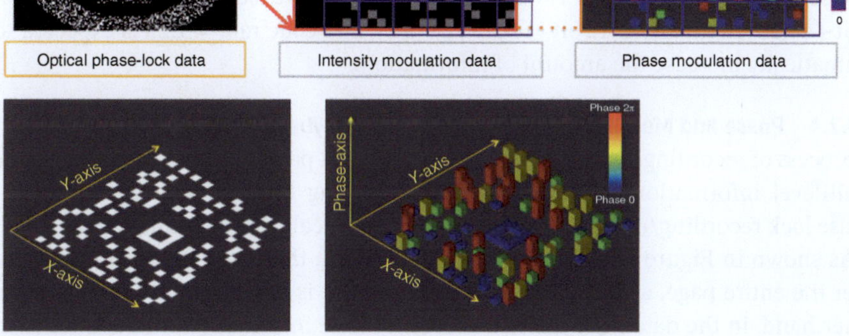


**Figure 6.20** Construction of page data in optical phase-locked collier system. Source: Hideyoshi Horimai.

information) in which 0 to $2\pi$ is divided into four is added. When detected by a CMOS sensor due to phase interference, the phase value is detected as an intensity value as shown in the bar graph in the Figure 6.20 (lower right).

In this way, the amount of information per page can be increased by adding phase modulation to intensity modulation.

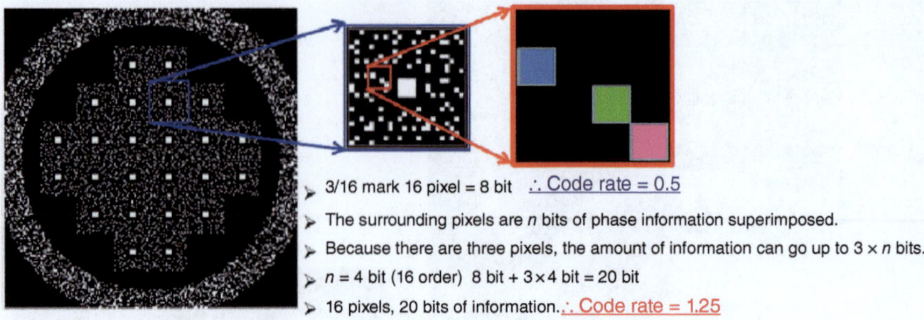


**Figure 6.21** Method for adding phase multivalued information in the case of realizing HVD upper compatibility. Source: Hideyoshi Horimai.

Next, how the amount of page information increases by adding phase information is described with reference to Figure 6.21. In addition, since the SLM (LCOS) incorporated in the experimental apparatus has a pixel size larger than 20 and 13.7 $\mu m^2$ square of DMD, the number of pixels that can be displayed on the pupil diameter of the objective lens is reduced. Therefore, the data page for the optical phase-locked experiment is displayed by trimming the four corners of the $6 \times 6$ sub-page by 3 sub-pages, so that there are 24 sub-pages ($24 \times 32 = 768$ symbols); notably, the shape is smaller. First, when one symbol (= 16 pixel) is extracted, the amount of information carried by the intensity modulation is 8 bits, which is the same as the normal collinear method, and the code rate is 0.5. Next, if n-bit phase information can be incorporated into each ON pixel, the information amount of $3 \times n$ bits can be improved. Although it depends on the characteristics of the recording material, if $n = 4$ bit, that is, 16-level grayscale multilevel recording/reproduction can be realized, even if this HVD format is used as a base, intensity modulation 8 bit + phase modulation 12 bit = 20 bit/symbol, and code rate = 1.25 is expected to dramatically increase the amount of information.

#### 6.4.2.4 Phase and Multipage Data Recording and Playback Process

A process of recording/reproducing phase multilevel page data and detecting phase multilevel information by phase interference using the aforementioned optical phase lock recording/reproducing experimental optical system is elucidated.

As shown in Figure 6.22, in the lock light pattern, the phase is a uniform phase over the entire page, and the divided reference light is labeled with EFGH. On the other hand, in the data light pattern, different phase information is added for each pixel in the information light area (see enlarged view). ABCD is also added to the divided reference light of the data light. Here, in the case of data light, it can be seen that different phase information is incorporated in the divided reference light, but this is for later use when performing multiplex recording by phase-coding the data light page. However, as the details of this method are presented separately, it is omitted here.

The intensity pattern and phase pattern data prepared in this way are displayed on SLM-1 and SLM-2, and as shown in Figure 6.23, the lock light and the data light are sequentially multiplexed and recorded at the same place.

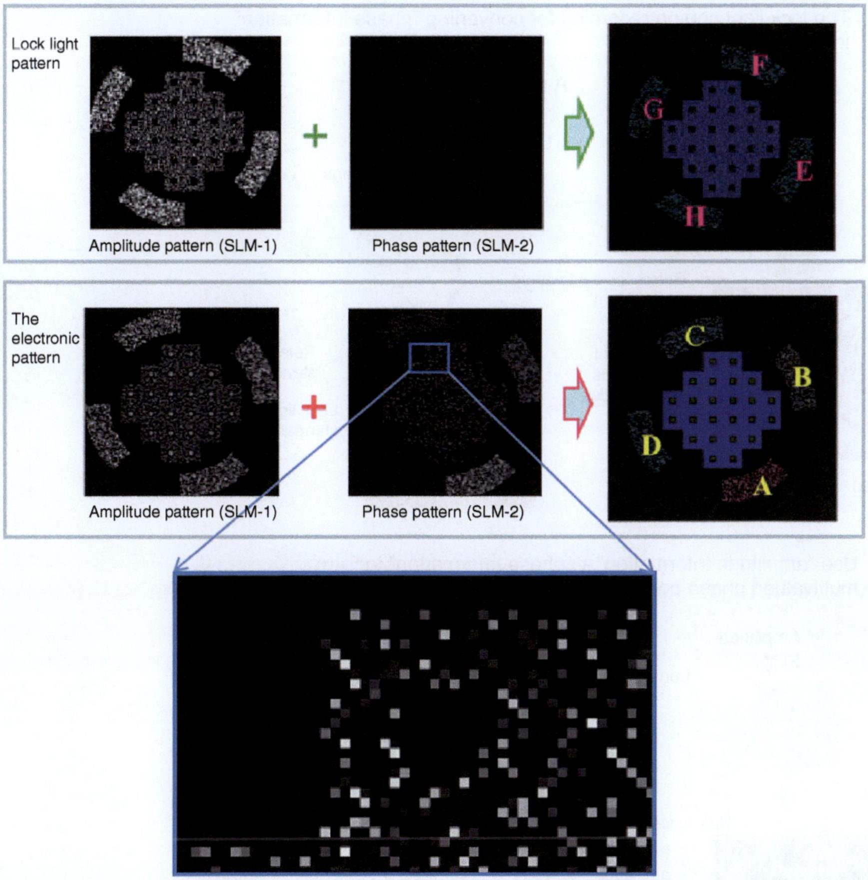


**Figure 6.22** Composition of page data by intensity + phase of locking light and data light. Source: Hideyoshi Horimai.

Next, as shown in Figure 6.24, the two holograms recorded in the same place in this way are combined into a circular shape by combining the divided reference lights and are simultaneously reproduced. At this time, the phase of the page data to be reproduced reflects the phase of the reference light for reproduction. Therefore, as shown on the right side of the figure, the phase of the reproduction image of the lock light that interferes with the reproduced data light is controlled by changing the phase of the lock light portion of the annular reference light.

The multilevel phase information added for each pixel of data light can be accurately obtained by this control. For example, as shown in the figure, four phase interferences are changed to "0," "π/2," "π," and "3π/2" so that the phase of the lock light is different by 90°. When the reproduced images are acquired and the luminance values for each pixel of the reproduced images are $I_A$, $I_B$, $I_C$, and $I_D$, the phase value of the data light is obtained by the following equation.

$$\varphi = \tan^{-1}\left(\frac{I_D - I_B}{I_C - I_A}\right) \quad (6.2)$$

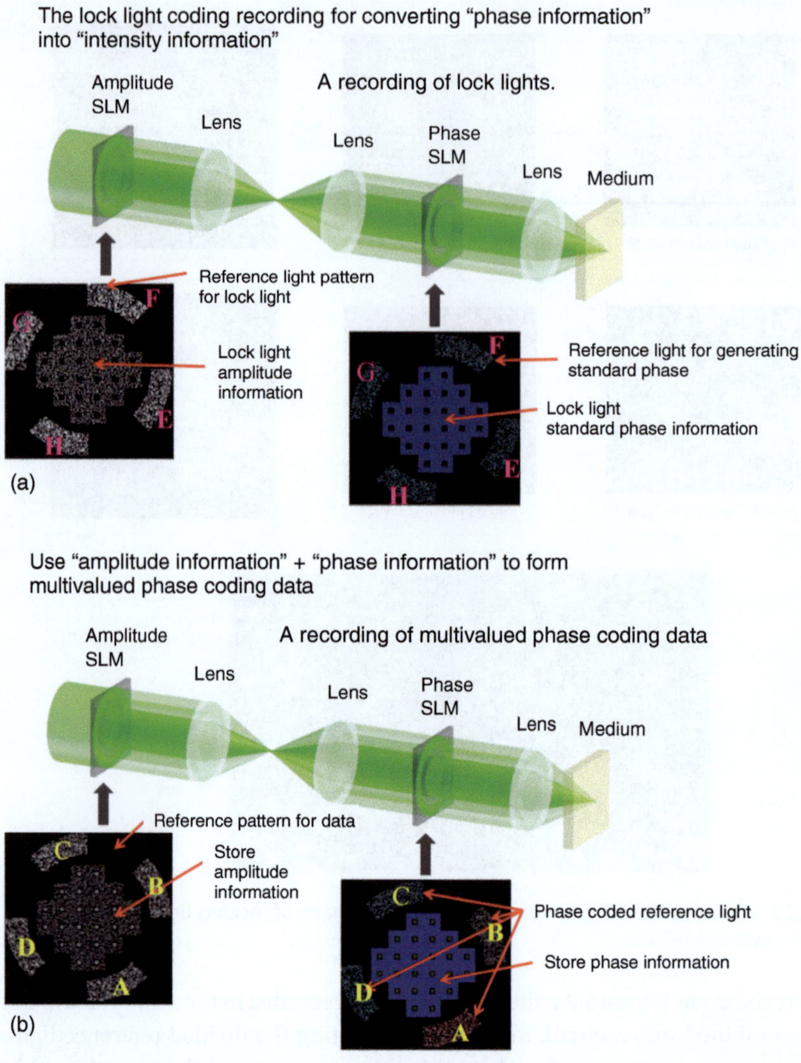


**Figure 6.23** Recording process of locked light and data light. Source: Hideyoshi Horimai.

Alternatively, if a CMOS sensor [20] that can detect a phase that differs by 90° for each pixel, which is used in recent three-dimensional measurement by digital holography, can be used, the phase value can be collectively obtained from one shot of the reproduced image. It is also possible in principle to find. However, at the research stage, to better understand the characteristics of the optical phase-locked collinear method, actually, the phase of the locked light is changed to, for example, 16 values, and more phase interference reproduction images are taken. As shown in Figure 6.25 specifically, a method of obtaining a phase for each pixel by fitting a change in the luminance value with a change in the phase of the lock light with a sine wave was used.

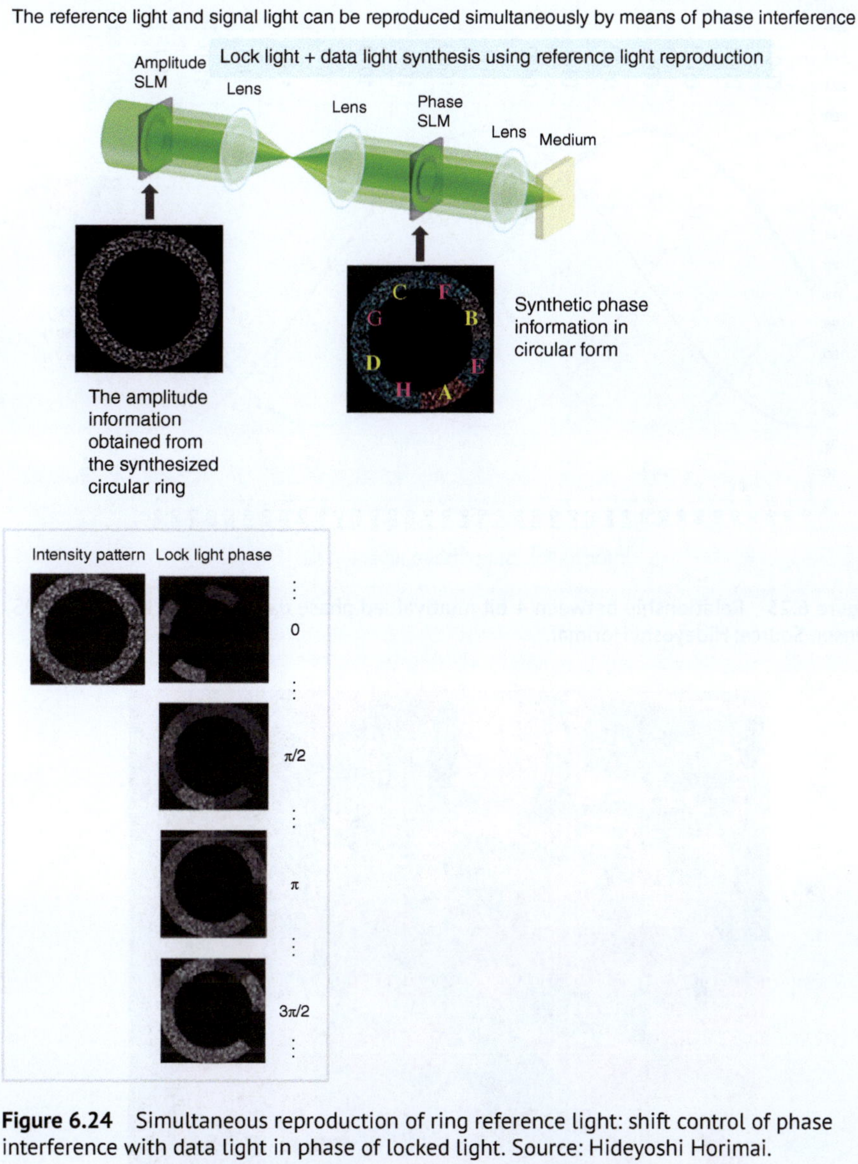


**Figure 6.24** Simultaneous reproduction of ring reference light: shift control of phase interference with data light in phase of locked light. Source: Hideyoshi Horimai.

### 6.4.3 Phase Locked Multi Phase Recording with Optical Phase Locking

#### 6.4.3.1 Fundamental Experiment of Four-Phase Gradation Recording

First, using a data page having a quaternary (2 bit) phase gradation, a basic experiment of recording and reproducing phase multilevel information by the optical phase lock method was performed. The recording medium was developed through the MEXT project (Ministry of Education, Culture, Sports, Science and Technology), and the recording material was "Nanogel Photopolymer" developed at Kyoeisha Chemical Co. Figure 6.26 shows a photograph of the appearance of the hologram

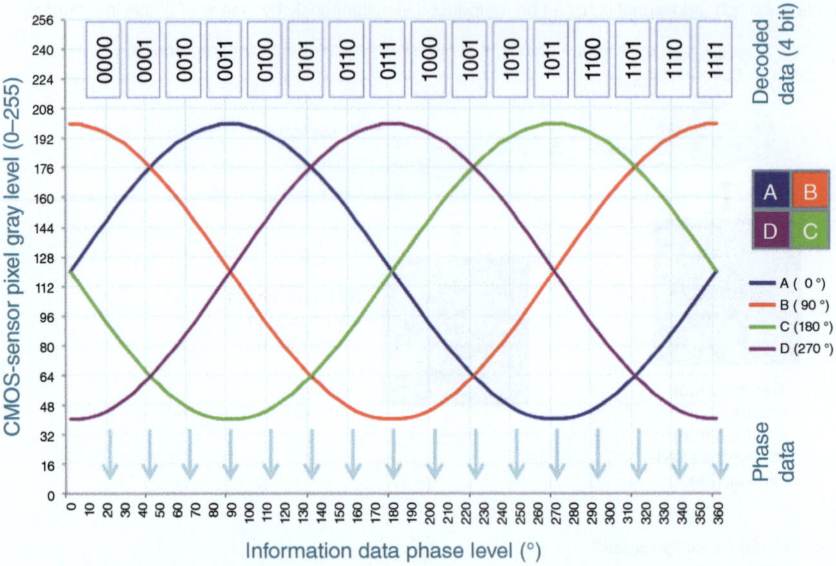


**Figure 6.25** Relationship between 4-bit multivalued phase data and gray level on CMOS sensor. Source: Hideyoshi Horimai.

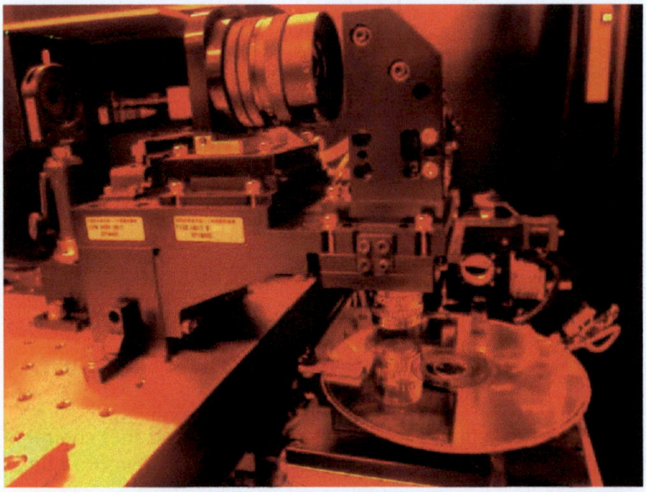

**Figure 6.26** Hologram recording media (HVD) set in optical phase-locked experimental apparatus. Source: Hideyoshi Horimai.

recording medium (HVD) set around the optical head of the experimental apparatus and the sample holder by the XYZ automatic stage. In the optical phase-lock recording/reproducing experiment, only the focus position control by the focus servo was performed, and the recording/reproducing experiment was performed while the optical disk was not rotated and stopped.

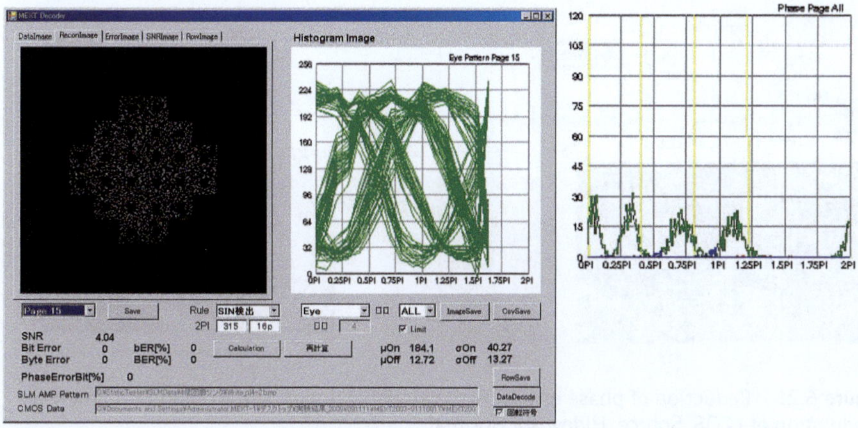


**Figure 6.27** Experimental results of four-phase multivalued data recording and reproduction using optical phase-locked method. Source: Hideyoshi Horimai.

Figure 6.27 shows the result of recording and reproducing the four-level phase multivalue data and detecting the phase value. Figure 6.27(a) is a screen of an analysis tool designed exclusively. The drawn waveform is a plot of luminance values for each pixel connected by broken lines from a reproduced image obtained by changing the phase of the lock light in 16 steps. The data were fitted to a sine curve for each pixel, and the reproduced phase value was estimated. Figure 6.27(b) represents the phase value of each pixel thus obtained as a histogram on the input phase value (horizontal axis). The phase value actually used for recording is indicated by a yellow line in the right histogram. Although the quaternary phases were detected separately, there was a deviation from the center value of the reproduced phase distribution, and it was confirmed that a reproduction phase value error (error) occurred.

#### 6.4.3.2 Factor Analysis of Phase Error

When the cause of this phase value error was analyzed, it was found that the intensity modulation unevenness was a major cause. When the intensity-modulated page data generated by SLM-1 were observed, a relatively large amount of stray light noise that could be generated from other optical components was observed. Further, as shown in Figure 6.28, it was also found that the intensity-modulated image generated by LCOS has an incomplete black level and has a low-contrast intensity-modulated image. Therefore, as shown in the figure, in addition to improvements to replace optical components with less stray light, feedback by direct observation of the image is generated to generate a corrected image that maximizes the intensity modulation amount and is attached to the LCOS driver. Using this function, the input signal is corrected and displayed. As a result, stray light is removed, and an intensity-modulated image having an extremely high quality of 1400 : 1 can be generated.

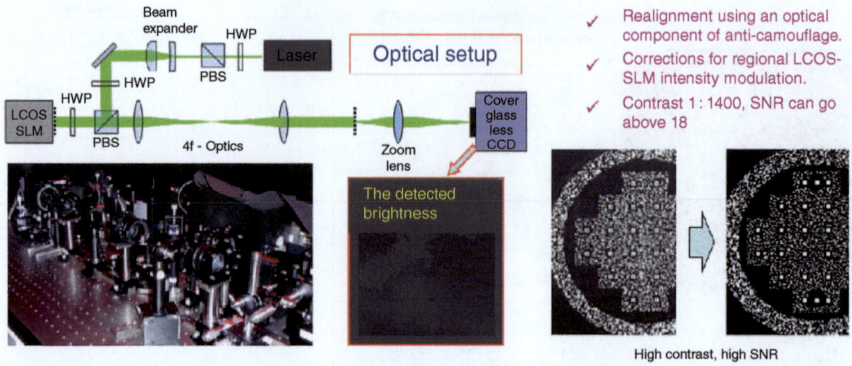


**Figure 6.28** Reduction of phase error due to compensation image with uniform intensity modulation at LCOS. Source: Hideyoshi Horimai.

### 6.4.4 Phase Multi Level Gray Scale Recording and Reproduction Characteristics of Optical Phase-Locked System

After optimization adjustment of the optical system, the phase gradation is changed to 4 values (2 bits), 8 values (3 bits), 16 values (4 bits), 32 values (5 bits), and 256 values (8 bits), and the recording/reproduction characteristics of phase multilevel gradation data by the optical phase lock method were evaluated. The analysis result from 4 values (2 bits) to 32 values (5 bits) is shown in Figure 6.29. As a result of the optimization adjustment of the optical system, it can be seen that the distribution width of the detected phase is smaller than the recording/reproduction result of the quaternary phase gradation in the basic experiment. Notably, the peak of the detected phase distribution becomes smaller as the gradation of the multilevel phase increases, because the number of pixels in the page data is the same as a result of the decrease in the total number.

In Figure 6.30, the horizontal axis indicates the number of multilevel gradations of the phase information added per pixel, and the recorded original phase is compared with the phase detected after recording/reproduction, resulting in a different phase value. Figure 6.30 is a graph showing the experimental results with the phase error bit (%) representing the number of pixels expressed as a ratio on the vertical axis. Since the phase width has a maximum value of 0 to $2\pi$, when this phase is divided by the number of gradations, the window for detecting the phase is narrowed as the number of phase gradations is increased. Therefore, if the phase distribution is large compared to this window, the probability of erroneous detection in the adjacent phase detection window will increase, and phase error bit (%) will deteriorate. From the viewpoint of error correction, if the value of phase error bit (%) is 20% or less, it is within the allowable range. The evaluation results show for the first time the effectiveness of the optical phase-locked collinear method based on 16-level (4 bit) phase multilevel gradations, even if it is more desirably 10% or less.

Furthermore, an optical phase-lock recording/reproducing experiment was performed using page data with 256-phase (8 bit) phase added. The analysis result is

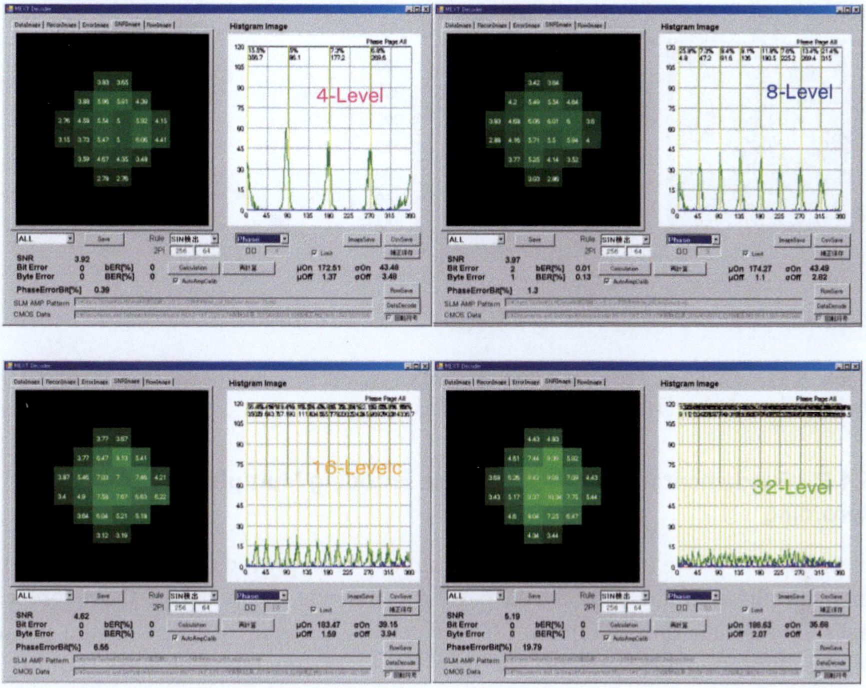


**Figure 6.29** Analysis of phase multilevel grayscale recording and reproduction data by optical phase-locked method. Source: Hideyoshi Horimai.

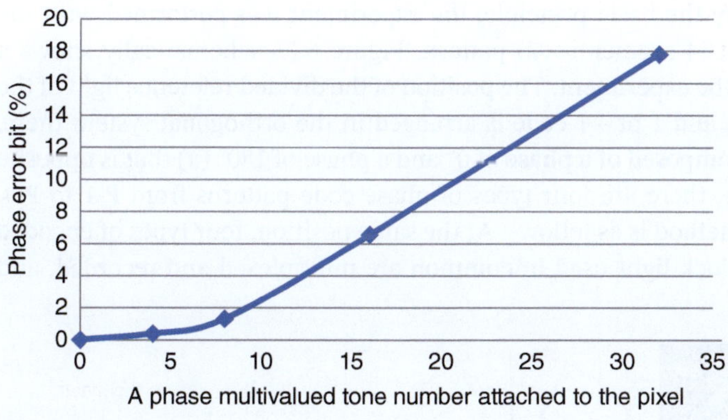


**Figure 6.30** Phase error bit characteristics for phase multiple valued gradations. Source: Hideyoshi Horimai.

shown in Figure 6.31(a). The phase of each pixel was fitted with a clean sine wave by changing the phase of the lock light, but we obtained phase error bit of 92.7%. However, when the relationship between the recorded phase and the reproduced phase was plotted on a graph, it was confirmed that the phase gradation was displayed linearly as shown in Figure 6.31(b).

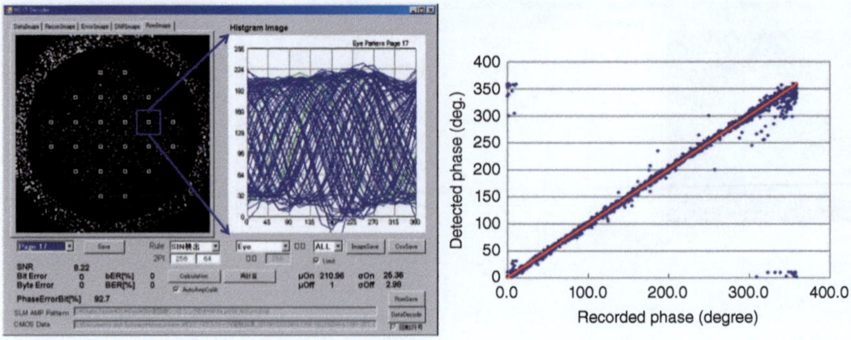


**Figure 6.31** Analysis of the optical phase-locked recording experiment using phase multilevel gradation (8 bits). Source: Hideyoshi Horimai.

### 6.4.5 Fundamental Principles of Phase Encoding Multiplexed Recording

In the optical phase lock method, it has already been proved through experiments that the phase of the reproduced page data reflects the phase of the reference light for reproduction. Therefore, it was considered that by adjusting the phase of the divided reference light for data light, so-called phase-coded multiplex recording/reproduction in which different page data are multiplexedly recorded at the same place can be performed. Depending on the number of page data to be multiplexed, for example, as shown in Figure 6.32, the number of divisions of the annular reference light can be increased.

However, to verify the basic principle, the experiment was performed with the same egiht-segment (4 segments × 2) pattern. Figure 6.33 schematically shows a phase code used in the experiment. The position of the divided reference light of the data light is ABCD, and 1 or − 1 code is arranged in the orthogonal system there. The light phase is composed of a phase of 0° and a phase of 180° (π) that is opposite in phase. Therefore, there are four types of phase code patterns from P-1 to P-4. The experimental method is as follows. At the same position, four types of encoded data light and one lock light used in common are multiplexed and recorded, and

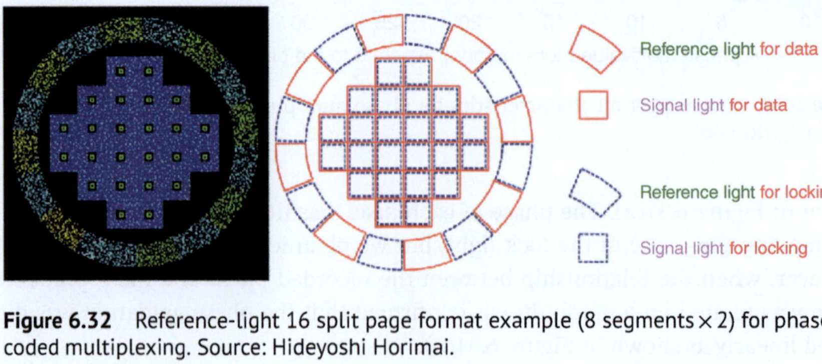


**Figure 6.32** Reference-light 16 split page format example (8 segments × 2) for phase coded multiplexing. Source: Hideyoshi Horimai.

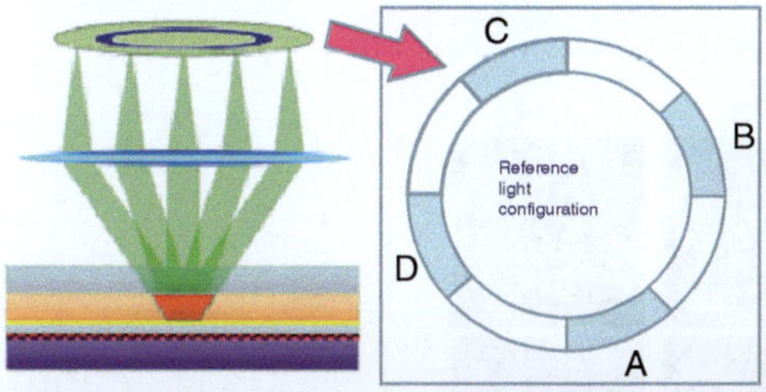


| | P-1 | P-2 | P-3 | P-4 |
|---|---|---|---|---|
| A | 0 | 0 | 0 | 0 |
| B | 0 | 0 | π | π |
| C | 0 | π | 0 | π |
| D | 0 | π | π | 0 |

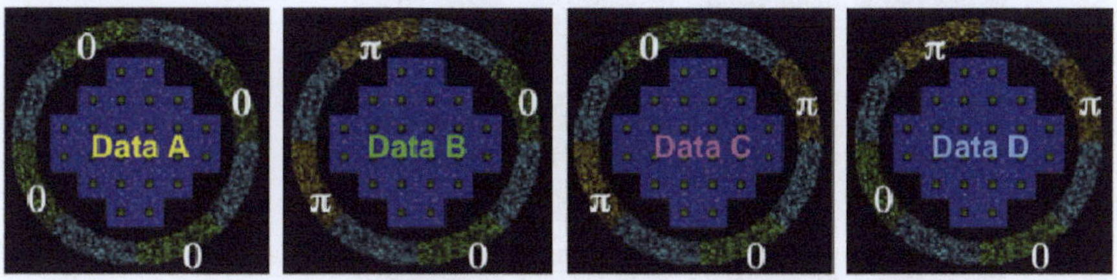


**Figure 6.33** Orthogonal phase codes and page data for multiple recording and playback with phase codes. Source: Hideyoshi Horimai.

reproduction is attempted with each phase code. If only the data corresponding to the phase code are reproduced, the basic principle is proved. Different data (Data A to Data D) were used for the page light, and the error rate of the decoded data was checked. Further, data (Data A to Data D) used were data with four-level phase multilevel gradation added.

First, a basic experiment on phase code reproduction was performed. The experimental results are shown in Figure 6.34. The experimental results are described using the four playback data in the upper stage. First, a hologram was recorded with a recording phase code of P-1 as a single hologram, and a reproduction image when the reproduction phase code of P-1 to P-4 was irradiated on the hologram was read with a CMOS sensor. In the upper row, it was found that a reproduced image can be obtained only with the reference light of the reproduction phase code of P-1, which is the same as the recording. Similarly, as a result of performing recording/reproducing experiments from P-2 to P-4 and arranging the reproduced images, the recording phase code P-2 is slightly inferior in selectivity, but generally has its own phase at the diagonal position. It was confirmed that playback was performed only with chords.

Next, experiments on phase-coded multiplex recording and phase code reproduction by an optical phase-lock method using quaternary phase multivalue gradation page data were performed. The experimental results are shown in Figure 6.35. In the recording process, four types of encoded data light and one commonly used lock light were sequentially multiplexed and recorded at the same position using an optical phase lock optical system. In the reproduction process, the recorded position was post-cured with light-emitting diode (LED) light immediately thereafter, and then the data light was reproduced with an annular synthetic reference light combining the reproduction phase code and the lock light. To select quaternary phase multivalue gradation data accurately, a reproduced image was photographed while changing only the phase of the lock light in 16 steps (while the reproduction phase

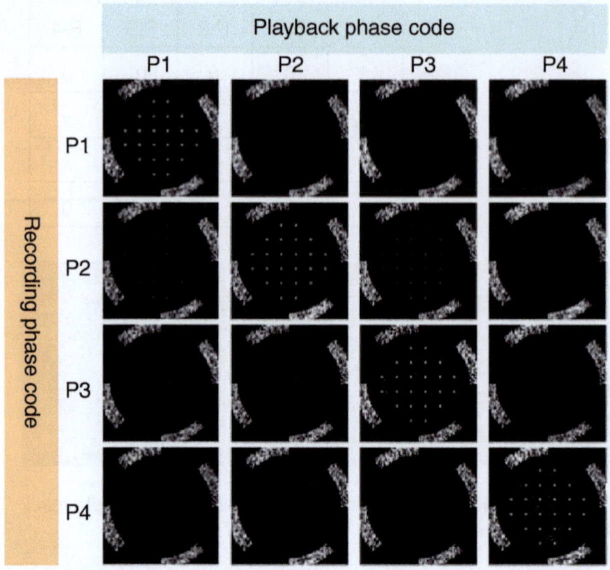


**Figure 6.34** Experimental results of playback selectivity by phase encoding of light. Source: Hideyoshi Horimai.

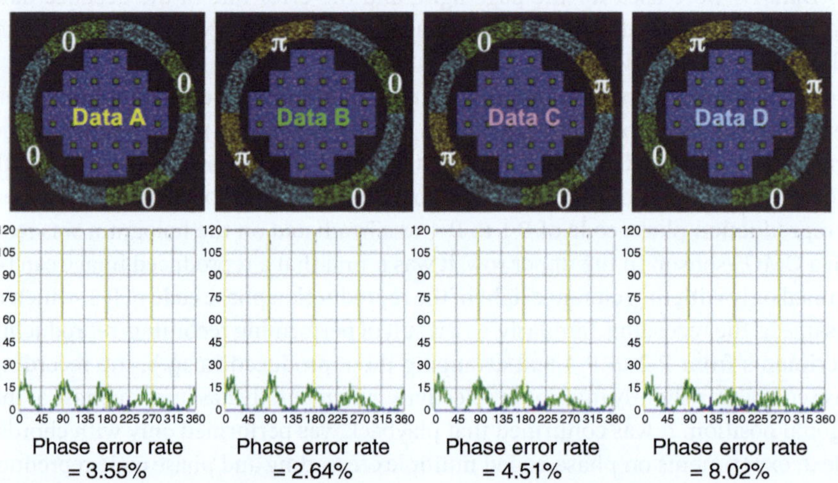


**Figure 6.35** Experimental results of phase coded multiplex recording and reproduction using phase multivalued grayscale page data. Source: Hideyoshi Horimai.

code remains the same). Similarly, for other reproduction phase codes, 16 reproduction images were taken and the phase data were analyzed.

As a result of the analysis, although the four data of the phase multilevel gradation are multiplexed and recorded at the same position, the data read by each reproduction phase code may be Data A to Data D. As shown in the figure, the error rate values were all 10% or less, and it was confirmed that the quaternary phase multi-level gradation was also faithfully reproduced.

As described above, recording and reproduction of "phase-encoded multiplex" using "phase multivalue gradation page data" by "optical phase-locked collinear method" was successful, and all basic principles were proved by experiments. The results of this basic principle proof experiment also proved the upward compatibility of the HVD format. Based on this result, we would like to develop research and development for the practical application of ultra-high speed/large capacity holographic optical data storage.

### 6.4.6 Summary

We devised an "optical phase-locked collinear method" in which "phase of light," one of the major features of holography, is recorded and reproduced. By using the common path optical system, which is a feature of collinear holography, phase multilevel gradation page data can be recorded and reproduced stably while eliminating the effects of phase disturbance caused by air fluctuation and vibration in the experimental system as much as possible. The basic principle of the optical phase-locked collinear method was explained, and how the experimental optical system capable of independently modulating the intensity and phase and the structure of the devised recording/reproducing page data was shown. Actually, a phase multivalue gradation recording/reproducing experiment was performed using an optical phase lock method, and even when the information of 32 gradations (5 bits) is added per pixel, the phase value can be read out with an error rate in a practical range. This was demonstrated for the first time through experimental results. In addition, a recording/reproducing basic experiment was performed in which phase multilevel information of up to 256 gradations (8 bits) was placed per pixel, and it was experimentally confirmed that the relationship between recording and reproducing phase was reproduced linearly. Furthermore, a method has been devised in which the reference beam is phase-coded and a plurality of phase multilevel page data is multiplexed and recorded at the same position. In fact, for the first time, a basic principle proof experiment, in which four pages are recorded and reproduced at the same position using page data having four-level phase multilevel gradations, is performed. In contrast to the conventional collinear method, collinear shift multiplexing is a general multiplexing method, whereas in the optical phase-locked collinear method, in addition to this, there is no change in the recording position even though it is a single light. However, it is significant to show in principle that multiple recording/reproduction is possible. This indicates a new direction for the future development of ultra-high-speed/high-density holographic optical data storage systems.

Notably, this page data structure and phase multilevel information recording/reproduction using the optical phase-locked collinear method can maintain an upward compatibility with the HVD using the collinear method approved as an international standard, making it extremely practical.

## 6.5 Low-Density Parity-Check Code

Low-density parity-check (LDPC) code is a block error correction code with a sparse check matrix proposed by Robert Gallager of the Massachusetts Institute of Technology in his doctoral dissertation in 1963. In 1996, MacKay et al. re-examined and found that the LDPC codes exhibit an excellent performance [21], close to the Shannon limit, and feature low-complexity decoding algorithms and high parallelism hardware implementation architecture. At present, LDPC codes have been considered as one of the best error control codes in the field of modern channel coding. In the holographic data storage system, we can use LDPC code to decrease BER from decoded data to user data.

Assuming each segment of the block code has $k$ symbols in each codeword in the initial situation (i.e. the length of the original information bit is $k$), and it is extended by encoding to each codeword which contains $n$ symbols (i.e. the code length is $n$), if the result of adding any two codewords in the block is still a codeword in the same block, then the block code is called a linear block code [22].

An LDPC code is essentially a linear block code and can be uniquely defined by its check matrix $H$. The dimensionality of the check matrix $H$ of the LDPC code with code length $n$ and information sequence length $k$ is $m \times n$, where $m = n-k$. Each row of the check matrix $H$ can be expressed as a check equation, which corresponds to a constraint relationship. Because one can compare with the length of the rows and columns of $H$, the number of non-zero elements in each row and column of $H$ (i.e. row weight and column weight) is very small, $H$ is called a sparse parity check matrix. In LDPC codes, the complexity of decoding process and the minimum inter-code distance only increase linearly with the increase of the code length. Eq. (6.3) shows a $5 \times 10$ check matrix and its corresponding check equations, where the codeword $c = (c_1, c_2, \ldots, c_{10}) \in C$ satisfies $H \cdot c^T = 0$.

$$H = \begin{bmatrix} 1 & 0 & 1 & 1 & 0 & 0 & 0 & 0 & 0 & 1 \\ 1 & 1 & 0 & 0 & 1 & 1 & 0 & 0 & 0 & 0 \\ 0 & 1 & 0 & 0 & 1 & 0 & 1 & 1 & 0 & 0 \\ 0 & 0 & 1 & 1 & 0 & 0 & 0 & 1 & 0 & 1 \\ 0 & 0 & 1 & 0 & 0 & 1 & 1 & 0 & 1 & 0 \end{bmatrix} \rightarrow \begin{bmatrix} c_1 + c_3 + c_4 + c_{10} = 0 \\ c_1 + c_2 + c_5 + c_6 = 0 \\ c_2 + c_5 + c_7 + c_8 = 0 \\ c_3 + c_4 + c_8 + c_{10} = 0 \\ c_3 + c_6 + c_7 + c_9 = 0 \end{bmatrix} \tag{6.3}$$

According to Tanner's research, the $H$-matrix can be one-to-one correspondingly described in the form of an equivalent graph, which is called the Tanner graph [23] or bipartite graph. Tanner graph can intuitively display the update and propagation process of messages during the iterative decoding process, and is a useful tool for representing and analyzing the performance of LDPC codes. Tanner graph is a bidirectional graph composed of two parts: nodes and edges. The nodes include check nodes and variable nodes. The connection between these two nodes is called edges. A check node represents an equation constraint relationship, and when a variable node is connected to a check node, it indicates that the variable node participates in the corresponding check equation. Figure 6.36 illustrates a Tanner graph which represents the check matrix of Eq. (6.3).

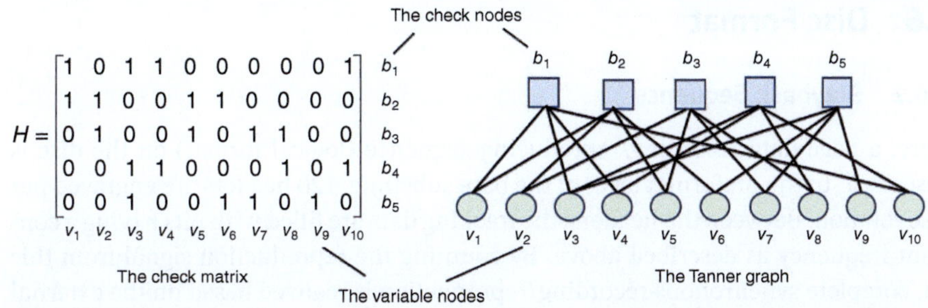


**Figure 6.36** Schematic of correspondence between check matrix and Tanner graph. Source: Xiao Lin.

The ring length is an important concept of LDPC codes. The number of edges that pass through a cycle is called the ring length, where a cycle is formed as that, starting from a certain node, through several variable nodes and check nodes, and returning to the starting node. The closed loop in the Tanner graph is an important factor affecting the performance of LDPC codes. The ring length affects the convergence of the decoding algorithm and also affects the final performance of LDPC codes. Under the similar class of iterative decoding algorithms with Belief Propagation algorithm, different closed loops make LDPC codes show completely different decoding performance.

Research shows that LDPC codes have the following outstanding advantages: (i) excellent decoding performance. Compared with Turbo codes, LDPC codes have better decoding performance and lower error code rate, and have found that its performance is close to the Shannon limit. (ii) Decoding is simple. The sparse check matrix iterative decoding algorithm makes the decoding operation for LDPC code have a low calculation volume, and its decoding can be operated in parallel, which makes the hardware of the decoder simple to implement and can obtain a higher throughput rate. (iii) Simple description and flexible structure. LDPC codes can be simply described by the check matrix and the corresponding Tanner graph, which can flexibly construct the LDPC codes. Based on the above advantages, LDPC codes have been widely used in various fields of modern communications.

LDPC codes have excellent performance and are very convenient to decode, which have a wide range of applications. The applicable scope includes but is not limited to the following areas: high-speed wireless networks, high-speed mobile communications, high-performance space satellites transmission, and high-speed data storage technology [24–26].

For example in the field of holographic storage, the InPhase professional drive uses an LDPC code within each page [27]. Each LDPC code word consists of 2044 bytes of user data together and a single data page contains 38 complete code words. The bytes of a single code word are not written sequentially to rows of pixels across the page. Instead all the code words are spatially interleaved together. Distributing each code word over the whole page provides more uniform performance from the LDPC code over all the code words. Their research indicates that LDPC code provides powerful error correction capabilities in holographic storage [27].

## 6.6 Disc Format

### 6.6.1 Playback Sequence

Here, a page data recording/reproducing sequence (logical format) on the disc is described. In the preformat layer of the base substrate, 120 headers are engraved per disc rotation. Between the headers, the tracking data are filled with pits having a constant frequency as described above. By counting the reproduction signal from this pit, complete synchronous recording/reproduction is realized based on the external clock and address position information generated from the disc. Figure 6.37 is a conceptual diagram thereof. The disc is controlled at a constant angular velocity (CAV). According to the experimental specifications, the rotation speed is explained here at 100 rpm.

The two-dimensional page data includes error correction codes within and between pages, and recording is performed with one pulse. One hologram in which page data are recorded is called a page. Therefore, this page is the minimum recording unit. Similarly, the page is reproduced with one pulse.

Although it is one of the features of HVD, the pages are recorded as jumping in the circumferential direction of the disc. For the sake of explanation, Figure 6.37 shows a case where 120 pages are recorded per round. For convenience of explanation, the data of 120 pages in one round are collected and used as a unit called a section. At this time, when recording the next section, the recording position is slightly shifted in the circumferential direction by slightly adjusting the start timing, and as a result, collinear shift multiplexing is performed in the circumferential direction. Similarly, for example, when holographic recording of 100 sections = 12 000 pages on one track is performed, collinear shift multiplex recording is performed at a pitch of

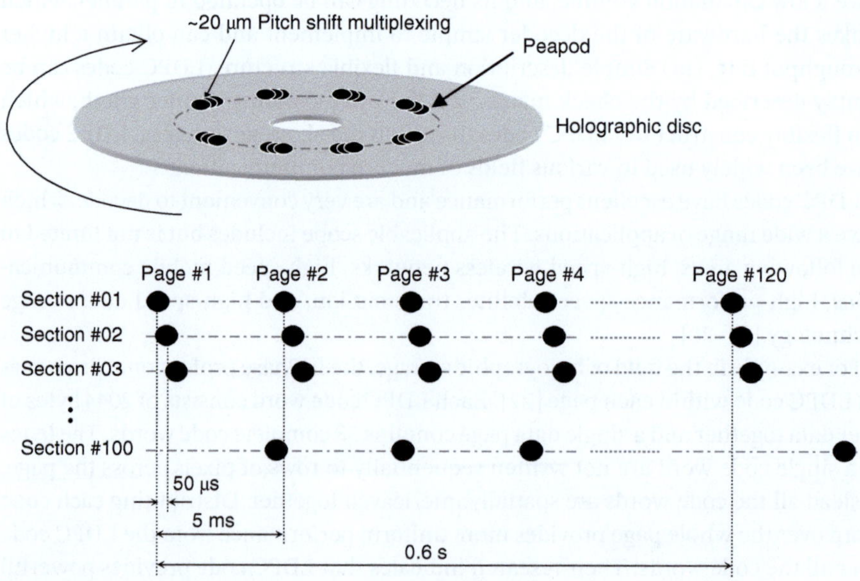


**Figure 6.37** Conceptual diagram of recording/playback sequence in HVD. Source: Hideyoshi Horimai.

about 20 μm in the circumferential direction around a radius of 40 mm (medium circumference) of the disk. In the same track, when the number of sections increases, the collinear shift multiplexed holograms are connected and extended; this is called Peapod.

Under the condition of rotating at 100 rpm as described above, the recording and playback frame rate is 200 fps, and therefore the recording time interval between pages 1 and page 2 is 5 ms. Physically, adjacent pages are arranged in the rotation direction at time intervals of 50 μs, but during playback, the pages are played back based on the clock information from the preformat. It is possible to read out pages in sequence.

Several sections formed on one track are put together, and here, for convenience of explanation, this is a unit called a chapter. For example, one section recorded 100 sections on one track is used as one chapter, and this one chapter can also be formed by collinear shift multiplexing at a predetermined shift pitch in the radial direction. In HVD, since the track pitch of the preformat is 1.6 μm, the shift amount in the radial direction is selected by an integral multiple thereof. And some of these chapters are put together into a unit called a book. The size of each unit can be freely determined to some extent by the application and is ultimately defined as a file format.

Thus, the number of sections per chapter and the number of chapters per book are optimized, including the collinear shift multiplex recording performance in the $x$ and $y$ directions of the recording medium. In the HVD format, these can be selected according to the performance of the recording material. For example, as schematically shown in Figure 6.38, the HVD can handle disc formats, with different storage capacities, only by changing the recording/reproducing pitch. Such a format is called "selectable capacity format." Therefore, in the future, when the performance of recording materials, SLM, and CMOS sensors improves, it is easy to ensure not only upward compatibility but also backward compatibility while improving storage capacity and transfer rate. This is an extremely important design philosophy when trying to store data for more than 50 years in archive applications.

There are several merits that can be obtained by adopting the recording/reproducing sequence as described above, but typical ones are listed below.

(1) Since the rotation speed of the disk can be increased without depending on the frame rate of the device, the disk rotation waiting time for random access can be shortened, and the speed at the time of random access to data can be improved.

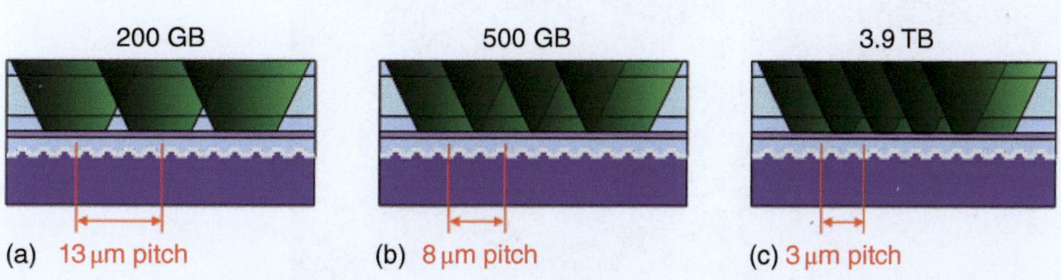


**Figure 6.38** Conceptual diagram of selectable capacity format with different shift multiplexing distances (a) 13 μm, (b) 8 μm and (c) 3 μm Source: Hideyoshi Horimai.

(2) Since continuous page information is recorded while being distributed in the circumferential direction, even when a local defect occurs in the recording medium, it is possible to avoid occurrence of a burst error, and an error correction code between pages is also effective. Therefore, the robustness of stored data is extremely high.
(3) Although the length per round is different between the inner circumference and the outer circumference of the disc, recording and reproduction can be realized while maintaining the highest recording density over the entire disc surface by simply increasing the number of sections per circumference.
(4) Because the frame rate of DMD and CMOS and the repetition frequency of laser pulses can be made constant, the data transfer rate can always be kept at the maximum value.

Notably, the recording/reproduction sequence shown in Figure 6.37 is sequence of the header portion of the version before the international standardization of HVD. For information on the logical format described, refer to the international standard document "ECMA-377" [28, 29].

### 6.6.2 On-the-Fly Recording/Playback Experiment

By integrating various technologies developed with a focus on collinear holography, we succeeded in collinear page data On-the-Fly recording and playback experiments for HVD rotating at 100 rpm. By converting video content into a large number of pages and fusing optical servo technology, pages were recorded and reproduced at any address by continuous recording. Figure 6.39 shows an enlarged photograph of the holographic optical disc that was actually recorded and reproduced, although the moving image content is not shown because of copyright. The pulse width

**Figure 6.39** Enlarged photo of HVD that succeeded in video recording and playback experiment with On the Fly. Source: Hideyoshi Horimai.

of the green laser (532 nm) during recording was about 20 ns, and the energy emitted from the objective lens was about 7 μJ/pulse. The exposure area near the focal plane is approximately $0.0015\,\text{cm}^2$ when the hologram diameter is approximately 220 μm, so the recording energy density per page is $E = 7\times10^{-3}$ (mJ)/ 0.0015 ($\text{cm}^2$) = $4.6\,\text{mJ/cm}^2$.

After recording, the recording medium was removed and chucked once again on the spindle, and data were successfully reproduced in accordance with the physical format at the time of recording. The holographic optical disk in the collinear method was also verified.

In contrast, several phenomena that were not seen before were discovered through dynamic recording and playback experiments using prototype drives.

First of all, during recording/reproduction experiments, it is not observed so prominently, but as time elapses from recording/reproduction experiments, the recording medium used at this time is able to observe the Peapod recorded in the circumferential direction with the naked eye. The phenomenon which changes to is confirmed. In addition, the shape of the observed Peapod seems to be larger than the width of the hologram exposure size actually recorded in multiple recording. The recording energy density per page $E = 4.6\,\text{mJ/cm}^2$ itself is not an extremely large value as compared with the recording sensitivity of the recording material (about $10\,\text{mJ/cm}^2$). However, considering that collinear shift multiplex recording by multiple sections was performed at a pitch of 20 μm in the circumferential direction, at the place where Peapod grew long, at least about 10 multiplex recording was performed. It was found that the amount of irradiation energy per unit area was $46\,\text{mJ/cm}^2$. In addition, the page was read many times with the same amount of irradiation energy.

As a result, multiple recording was performed with a relatively large laser energy density, and excessive energy was also locally concentrated and exposed by reproduction, so stray light and scattered light were also present in the surrounding area of the Peapod, an unrecorded part. It is presumed that the exposure material was gradually exposed by the integration of the above, and the material near the hologram recorded was large enough to be observable with the naked eye, and the refractive index or composition itself changed. Such a large local refractive index change affects the servo error signal detected by the red laser beam, so care must be taken when optimizing the recording/reproduction characteristics when considering commercialization. At the same time, it was found that this is a technical problem that should be fully examined from the viewpoint of the composition of the recording material.

Further, through a continuous recording/reproducing experiment, it was observed that there was an area where the error deteriorated at a predetermined position during one rotation of the disk. Figure 6.40 shows a result of reading and recording/reproducing data for one round, that is, each page data for 1 section/120 pages and performing error evaluation.

In this evaluation example, the error rate of the page data at the positions of the sector numbers 56 to 81 of the recorded page shows a deteriorated value as symbol error rate > $10^{-2}$. Although the image of the captured playback page appears to have been slightly degraded in resolution, no scratches or foreign objects were found on the recording medium even when the disk at that portion was examined with a microscope or the like.

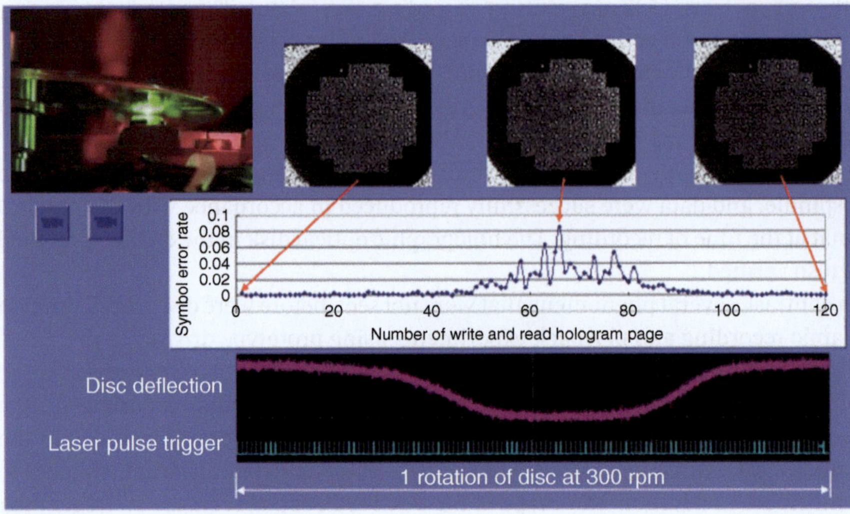


**Figure 6.40** Error analysis results for 1 lap/1 section/120 pages. Source: Hideyoshi Horimai.

Therefore, when the mechanical properties of this HVD were measured, it was found that the amount of surface blur in the portion where the symbol error rate was deteriorated exceeded 300 μm at the maximum amplitude value, as shown in the lower part of the figure. As a result, the symbol error rate deteriorated because the prototype HVD was not flat, and local blurring occurred. Consequently, recording and playback were performed while the objective lens followed the blurring by the focus servo. Thus, it was speculated that this blurring was observed because the specifications of the image plane margin were exceeded.

By combining the above knowledge and experimental know-how built through previous developments, the collinear page by On the Fly under optimized recording and playback conditions for the entire HVD disc was created. The error rate of the entire disk surface during data recording/reproduction was evaluated. According to the new HVD format, the rotation speed was 300 rpm. Figure 6.41 shows the state of the drive system (ES-1) during recording/reproduction. The case was removed and the experiment was performed under pure red illumination where the hologram recording material had no photosensitivity. Q-switched pulsed laser light (green) for hologram recording records page data continuously, so that it can be observed slightly through the back of the HVD as shown in the picture. After recording the entire surface, the HVD was once removed from the drive and post cured with a bleaching device using an LED array.

The HVD, which had been fully recorded and post-cured, was set in the drive again. Here, evaluation is performed as raw data without performing error correction processing.

Since all the data are enormous and the display is crowded and difficult to understand, we decided to sample about every 1000 tracks and visualize the error distribution. The result is shown in Figure 6.42. In Figure 6.42, one circle represents the

**Figure 6.41** Drive system (ES-1) used for the recording and playback (experiment under red illumination). Source: Hideyoshi Horimai.

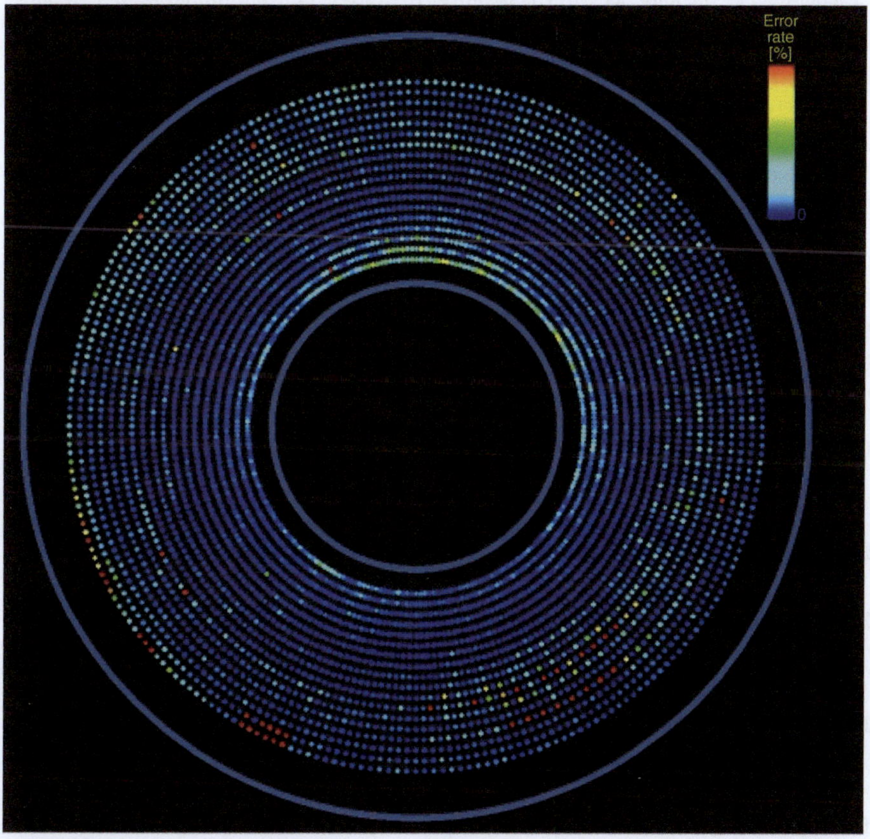


**Figure 6.42** Evaluation results of error distribution due to recording/playback over the entire HVD. Source: Hideyoshi Horimai.

sampled portion, and the color in the circle indicates the error rate. For example, the blue circle represents a low-error-rate state, and the red circle represents a burst error > 0.1. There are red circles in some parts on the disc, but the page format is recorded discretely, so it is possible to recover the data by performing error correction between the pages.

## References


1 Liu, J., Horimai, H., Lin, X., et al. (2017). Optimal micro-mirror tilt angle and sync mark design for digital micro-mirror device based collinear holographic data storage system. *Appl. Opt.* 56 (16): 4779–4784.

2 Denz, C., Pauliat, G., Roosen, G., and Tschudi, T. (1991). Volume hologram multiplexing using a deterministic phase encoding method. *Opt. Commun.* 85: 171–176.

3 Ashley J. and Marcus B. (1997). Constant-weight/lowpass modulation codes for holographic recording. *IBM Research Report RJ-10089 (91905)*, October 1997.

4 Heanue, J., Bashaw, M., and Hesselink, L. (1994). Volume holographic storage and retrieval of digital data. *Science* 265: 749–752.

5 Ashley, J. and Marcus, B. (1998). Two-dimensional lowpass filtering codes for holographic storage. *IEEE Trans. Commun.* 46: 724–727.

6 Burr G.W. and Marcus B. (1999). Coding trade-offs for high-density holographic data storage. *Proceedings of SPIE, Advanced Optical Data Storage: Materials, Systems, and Interfaces to Computers*, Volume 3802, pp. 18-29.

7 Burr, G.W., Ashley, J., Coufal, H. et al. (1997). Modulation coding for pixel-matched holographic data storage. *Opt. Lett.* 22 (9): 639–641.

8 Shelby, R.M., Hoffnagle, J.A., Burr, G.W. et al. (1997). Pixel-matched holographic storage with megabit pages. *Opt. Lett.* 22: 1509–1511.

9 Burr, G., Neifeld, M., Barking, G. et al. (1998). Gray-scale data pages for digital holographic data storage. *Opt. Lett.* 23: 1218–1220.

10 Burr, G.W., Coufal, H., Grygier, R.K. et al. (1998). Noise reduction of page-oriented data storage by inverse filtering during recording. *Opt. Lett.* 23 (4): 289–291.

11 Taketomi, Y., Ford, J.E., Sasaki, H. et al. (1991). Incremental recording for photorefractive hologram multiplexing. *Opt. Lett.* 16: 1774–1776.

12 Bashaw, M.C., Aharoni, A., Walkup, J.F., and Hesselink, L. (1994). Cross-talk considerations for angular and phase-encoded multiplexing in volume holography. *J. Opt. Soc. Am. B: Opt. Phys.* 11: 1820–1836.

13 Denz, C., Pauliat, G., Roosen, G., and Tschudi, T. (1992). Potentialities and limitations of hologram multiplexing by using the phase-encoding technique. *Appl. Opt.* 31: 5700–5705.

14 Lembcke, J., Denz, C., and Tschudi, T. (1995). General formalism for angular and phase-encoding multiplexing in holographic image storage. *Opt. Mater.* 4: 428–432.

**15** Yang, X., Xu, Y., and Wen, Z. (1996). Generation of Hadamard matrices for phase-code-multiplexed holographic memories. *Opt. Lett.* 21: 1067–1069.

**16** Sinzinger, S. and Arrizon, V. (1997). High efficiency detour-phase holograms. *Opt. Lett.* 22: 928–930.

**17** Alves, C., Pauliat, G., and Roosen, G. (1994). Dynamic phase-encoding storage of 64 images in a $BaTiO_3$ photorefractive crystal. *Opt. Lett.* 19: 1894.

**18** Muller, K.-O., Denz, C., Rauch, T. et al. (1998). High capacity holographic data storage based on phase-coded multiplexing. *Opt. Mem. Neural Netw.* 7: 1.

**19** Denz, C., Muller, K.-O., Heimann, T., and Tschudi, T. (1998). Volume holographic storage demonstrator based on phase-coded multiplexing. *IEEE J. Sel. Top. Quantum Electron.* 4: 832–839.

**20** Awatsuji, Y., Fujii, A., Kubota, T., and Matoba, O. (2006). Parallel three-step phase-shifting digital holography. *Appl. Opt.* 45: 2995–3002.

**21** MacKay, D.J.C. and Neal, R.M. (1996). Near Shannon limit performance of low-density parity-check codes. *Electron. Lett.* 32: 1645–1645.

**22** Tanner, R.M., Sridhara, D., Sridharan, A. et al. (2004). LDPC block and convolutional codes based on circulant matrices. *IEEE Trans. Inf. Theory* 50 (12): 2966–2984.

**23** Tanner, R. (1981). A recursive approach to low complexity codes. *EEEE Trans. Inf. Theory* 27 (5): 533–547.

**24** 房明 (Ming Fang) (2014). LDPC码构造方法和编码算法研究[D], (Research on LDPC code construction method and coding algorithm)北京邮电大学. (Beijing University of Posts and Telecommunications)

**25** 王鹏 (Peng Wang) (2013). 低密度奇偶校验码应用于存储系统的关键技术研究[D], (Research on Key Technologies of low density parity check codes applied to storage systems)华中科技大学. (Huazhong University of science and technology)

**26** 燕威 (Wei Yan) (2017) 低密度奇偶校验码在空间通信中的关键技术研究[D], (Research on Key Technologies of low density parity check codes in space communication)中国科学院大学. (University of Chinese Academy of Sciences)

**27** Curtis, K., Dhar, L., Hill, A. et al. (2010). *Holographic Data Storage: From Theory to Practical Systems.* United Kingdom: Wiley.

**28** Standard ECMA-377 (2007), Information Interchange on Holographic Versatile Disc (HVD) Recordable Cartridges, Capacity: 200 Gbytes per Cartridge. http://www.ecma-international.org/publications/standards/Ecma-377.htm (accessed 10 November 2021).

**29** Standard ECMA-378 (2007). Information Interchange on Read-Only Memory Holographic Versatile Disc (HVD-ROM), Capacity: 100 Gbytes per disk. http://www.ecma-international.org/publications/standards/Ecma-378.htm (accessed 10 November 2021).

# 7

# Characteristics of Write and Read

## 7.1 Introduction

In holographic optical memory, the recording density is improved using multiple recording methods, including angle multiplexing [1–6], peritrophic multiplexing [7], wavelength multiplexing [1, 8–10], phase code multiplexing [11], shift multiplexing [12–16], and spatial multiplexing [17, 18]. Further, a polytopic method [19] that applies these methods has been proposed as well. Among these methods, it was considered that the shift multiple recording method, in which the recording position automatically moves according to the rotation of the optical disc, is suitable for the collinear method. In the shift multiplexing method reported by Psaltis et al., the reference light is incident as a spherical wave with respect to the information light that is incident almost perpendicular to the recording medium, such that the reproduction light beam spot is deviated from the recorded position. As a result, the signal strength of the reproduced information pattern declines. This is called regioselectivity or shift selectivity, which is a crucial parameter that determines the theoretical limit of the storage capacity. Because another hologram can be multiply recorded at a position where the reproduction signal from the adjacent hologram becomes negligibly small, the continuous hologram recording/reproduction can be performed by moving (shifting) the recording medium. However, in the method of Psaltis et al., the reference light of this spherical wave is vertically incident, but the information light is incident diagonally, resulting in an asymmetric wavefront state of the interference fringes formed in the recording medium. Unbeknownst to us, a sufficient shift selectivity cannot be obtained depending on the moving direction.

In contrast, for a rotating optical disc-shaped recording medium with a coaxially arranged optical system configuration, Hesselink and Orlov conducted a recording/playback demonstration of page data for the first time [20, 21]. The rotation speed of the optical disc was 300 rpm, and a multiplex method, called, was adopted to realize multiple recording and reproduction in the rotation direction. In correlation shift multiplexing, the light diffused from the random-phase plate is used as the reference light. However, the recording and playback was based on a ±0.1-μm

*Collinear Holography: Devices, Materials, Data Storage*, First Edition.
Xiaodi Tan, Hideyoshi Horimai, Tsutomu Shimura, and Xiao Lin.

precision optical encoder attached to the spindle, and thus, the position of the original data could not be reproduced and access became difficult when the optical disk was attached or detached. Consequently, the same random-phase plate was used as without it, problems such as device incompatibility arose, which hindered its practical use.

This chapter describes the measurement results of the shift selection characteristics of the collinear method. In addition, the results of an experiment, in which shift multiple recording and reproduction were actually attempted, are described. Furthermore, the data page pattern suitable for high-density recording and its effectiveness are elucidated through numerical simulation results, which are compared and verified with those of the recording/reproduction experiment.

## 7.2 Shift Selectivity

### 7.2.1 Evaluation of Playback Signal Strength with Respect to Movement of Playback Position

The collinear method aims to record and play back page data on an on-the-fly optical disk that rotates continuously. Therefore, the shift multiplex recording method, in which multiplex recording is performed while the recording position is automatically moved according to the rotation of the optical disk, is considered effective from the viewpoint of compatibility with the system.

Therefore, the shift selectivity in the collinear method was measured using S-VRD which is mentioned in Section 2.3 of Chapter 2. The experimental procedure is as follows. First, page data are written at a desired position through a collinear method, and post-cure is performed so that the hologram is not lost even if it is bleached by a light-emitting diode (LED) and continuously reproduced. Thereafter, without changing the position, the hologram is irradiated with an annular reference beam to read a reproduced image of the page data, and this is used as the standard data. Next, using a high-precision six-axis stage, the page data at each position are reproduced while being shifted by $\pm 1$ μm from the position where the recording medium is recorded in the $x$ or $y$ direction, and the reproduced image is read.

According to the above sequence, the shift selectivity is actually evaluated. Surprisingly, the brightness of the reconstructed image drops sharply by moving the position by only $\pm 1$ μm in both the $x$ and $y$ directions. The error rate is no longer evaluated, because the brightness of the reproduced image is extremely reduced, such that the sync mark is not detected, and the error rate and signal-to-noise ratio (SNR) calculation cannot be performed. However, in either case, when the recording is returned to the recorded position by the high-precision six-axis stage, the high-quality page data are read out, and the content stored as the reference data is almost the same, validating the reproducibility of the experiment.

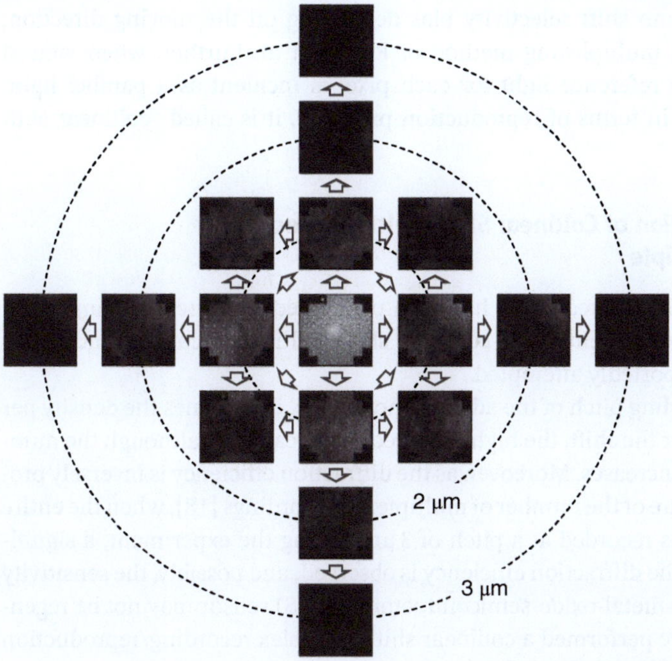


**Figure 7.1** Evaluation result of shift selectivity using the collinear method. Source: Reproduced with permission of Horimai et al. [22], Optical Society of America.

Therefore, when the captured image data are reduced, and the images are further moved and acquired while arranging these reduced images at the moved distance, the image shown in Figure 7.1 is obtained; circles indicated by broken lines indicate the distance moved from the recording position by 1, 2, and 3 μm from the inside. This result indicates that the captured image, reproduced while shifting the position by ±1 μm in the $x$ and $y$ directions from the hologram recording position, exhibits a sharply decreased intensity due to the ±1-μm deviation, and is reproduced until ±2 μm. An image that looks like an image is still left, but disappears almost completely at ±3 μm. According to these experimental results, under the above-mentioned various conditions of the collinear method, if the pitch to the adjacent hologram is 3 μm or more, then the influence of crosstalk from the adjacent hologram is considered to be negligibly small. In addition, since there is no remarkable asymmetry in shift selectivity, it is suggested that in principle multiplex recording can be performed by laying out the holograms at a pitch of 3 μm in the $x$ and $y$ directions. Note that the exposed hologram size is approximately 200 μm at the focal plane of the objective lens; therefore, if it is assumed that multiplex recording/reproduction is realized by spreading it at a 3-μm pitch in the $x$–$y$ direction, then it will be over 4400 multiplex. Consequently, the realization of an ultra-high-density large-capacity holographic optical disk is expected.

This method has no shift selectivity bias depending on the moving direction, as seen in the shift multiplexing method of Psaltis et al. Further, when viewed microscopically, the reference light for each pixel is incident as a parallel light. As it differs greatly in terms of reproduction principle, it is called "collinear shift multiplexing."

### 7.2.2 Demonstration of Collinear Shift Multiplex Recording and Playback Principle

Experimentally, multiplex recording has been performed at a pitch of 3 μm, and a proof-of-principle experiment of the collinear shift multiplex recording and reproduction has been reportedly attempted.

Because the recording pitch of the adjacent holograms determines the density per unit area, the smaller the shift, the higher the recording capacity, although the number of multiplexing increases. Moreover, as the diffraction efficiency is inversely proportional to the square of the number of multiplexed recordings [18], when the entire recording medium is recorded at a pitch of 3 μm during the experiment, a significantly reduction in the diffraction efficiency is observed, and possibly, the sensitivity of a complementary metal-oxide-semiconductor (CMOS) sensor may not be regenerated. Therefore, we performed a collinear shift multiplex recording/reproduction experiment that moved only in one direction.

Figure 7.2 shows a page data pattern used for recording and reproduction. The reference light is radial at intervals of 3°, and the total numbers of reference light pixels and information light pixels are designed to be approximately the same.

After recording 20 holograms with the same power at a pitch of 3 μm and post-curing with an LED light source, one must then precisely move the stage to the position at which the recording media are attached to reproduce the information at the recording position of each hologram.

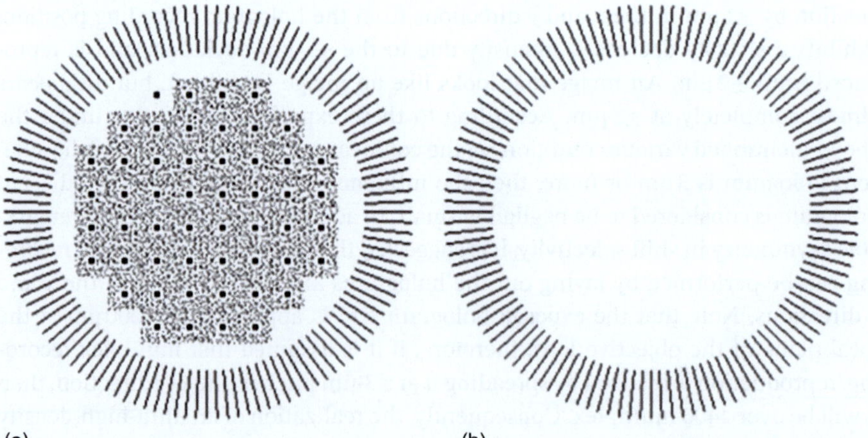


**Figure 7.2** Page data used for (a) recording and (b) reproduction (reference beam: radial pattern at 3° intervals). Source: Reproduced with permission of Horimai et al. [22], Optical Society of America.

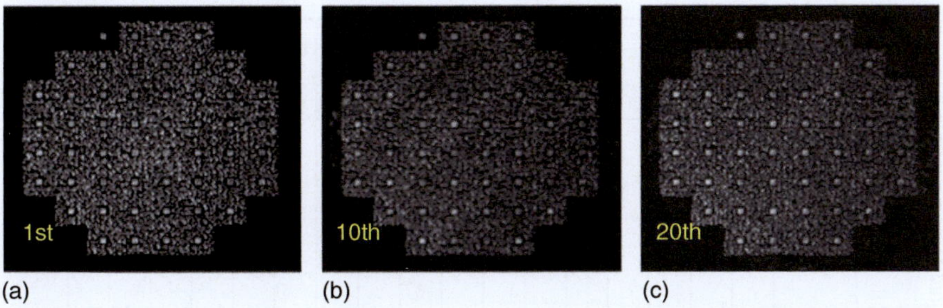


**Figure 7.3** Page data images reproduced after collinear shift multiplex recording: (a) 1st, (b) 10th, and (c) 20th. Source: Horimai et al. [22], Optical Society of America.

As a result, when playback is performed by returning to each recording position, the reproduced images of the page data recorded at all the positions are obtained. Figure 7.3 shows a part of the reproduction page data image detected by the CMOS sensor. The first recorded page data are shown in Figure 7.3a (1st), the last recorded page data are shown in Figure 7.3c (20th), and the page data located at the center with the most multiplexing are shown in Figure 7.3b (10th). All the reproduced images obtained are of good quality and could be analyzed with the S-VRD.

Figures 7.4 and 7.5 show the symbol error rates and SNRs obtained by analyzing the respective reproduced images. The symbol error rate is of the order of $10^{-2}$ or less, and an error-free holographic optical memory can be realized by combining with an error correction code. This experiment demonstrated the practical application feasibility of the collinear shift multiplexing system for the first time.

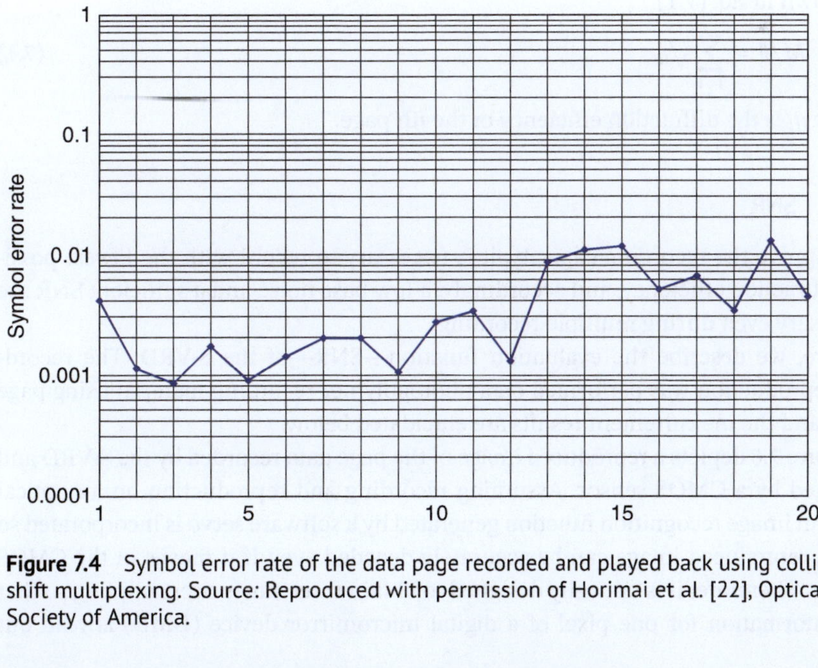


**Figure 7.4** Symbol error rate of the data page recorded and played back using collinear shift multiplexing. Source: Reproduced with permission of Horimai et al. [22], Optical Society of America.

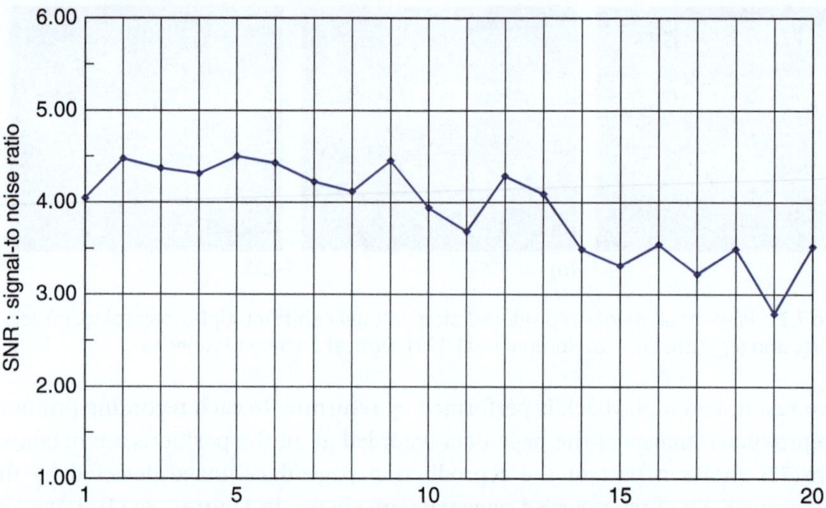


**Figure 7.5** SNR of a data page recorded and played back via collinear shift multiplexing. Source: Reproduced with permission of Horimai et al. [22], Optical Society of America.

## 7.3 Parameters

### 7.3.1 *M*/#

To improve the recording capacity, it is necessary to increase the number of multiplexing recordings. The performance index is called a dynamic range, and a high dynamic range is required. As an index of dynamic range, $M/\#$ ($M$ number) is used as shown in Eq. (7.1).

$$M/\# = \sum_{i=1} \sqrt{\eta_i} \tag{7.1}$$

where $\eta_i$ is the diffraction efficiency of the $i$th page.

### 7.3.2 SNR

To improve the recording capacity, it is necessary to record with the lowest possible diffraction efficiency, and accordingly, a low base noise and a sufficient SNR are necessary even during multiple recording.

Here, we describe the evaluation function—SNR—of the S-VRD. The recording/reproduction was performed on a photopolymer recording material using page data, and the measurement results are elucidated below.

Figure 7.6 depicts a reproduced image of the page data recorded by the S-VRD and detected by a CMOS sensor. Assuming recording and reproduction on an optical disk, an image recognition function generated by a software servo is incorporated so that a reproduced image can be accurately decoded even if it moves in the CMOS sensor. Therefore, it is optically designed to obtain three times oversampling so that the information for one pixel of a digital micromirror device (DMD) is read out

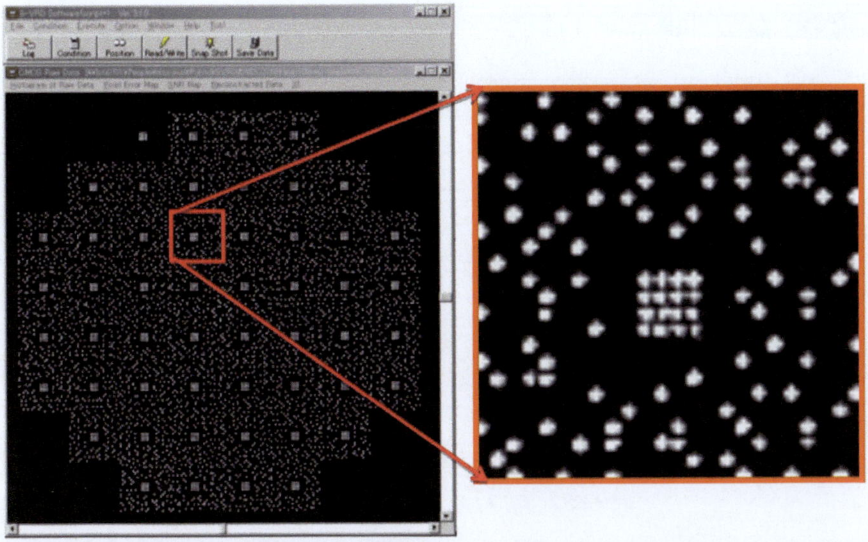

**Figure 7.6** Page data playback image detected by the CMOS sensor (three times oversampling). Source: Hideyoshi Horimai.

by a CMOS sensor using $3 \times 3 = 9$ pixels. An enlarged part of the detected image (Figure 7.6 left) is shown in Figure 7.6 (right), which confirms the effectiveness of the three times oversampling.

The playback image detected in this way is adjusted to correct the position shift using a detection algorithm based on the sync mark generated by the uniquely-developed high-speed software servo, and the pattern is matched to the recording page data format. Thereafter, the luminance information detected by the $3 \times 3 = 9$ pixels of the CMOS sensor is weighted and calculated as the density information for one pixel.

The page data image recalculated in this way becomes an image for the evaluation of the SNR and error rate. Since the evaluation image corresponds to the page data pattern displayed in the DMD, each pixel of the reproduced page data (evaluation image) was originally an ON or OFF pixel. For this reason, the SNR is obtained from the density information for each pixel using Eq. (7.2):

$$\mathrm{SNR} = \frac{\mu_{\mathrm{on}} - \mu_{\mathrm{off}}}{\sqrt{{\sigma_{\mathrm{on}}}^2 + {\sigma_{\mathrm{off}}}^2}} \tag{7.2}$$

where

$\mu_{\mathrm{on}}$ is the ON average pixel density
$\mu_{\mathrm{off}}$ is the OFF average pixel density
$\sigma_{\mathrm{on}}$ is the ON dispersion of pixel density; and $\sigma_{\mathrm{off}}$ is the OFF dispersion of pixel density.

Figure 7.7 shows the SNR evaluation result of the whole page of the detected image shown in Figure 7.6. The ON/OFF distributions of the pixels are clearly separated from each other, and a high SNR value 10.2 is obtained using Eq. (7.2). It is

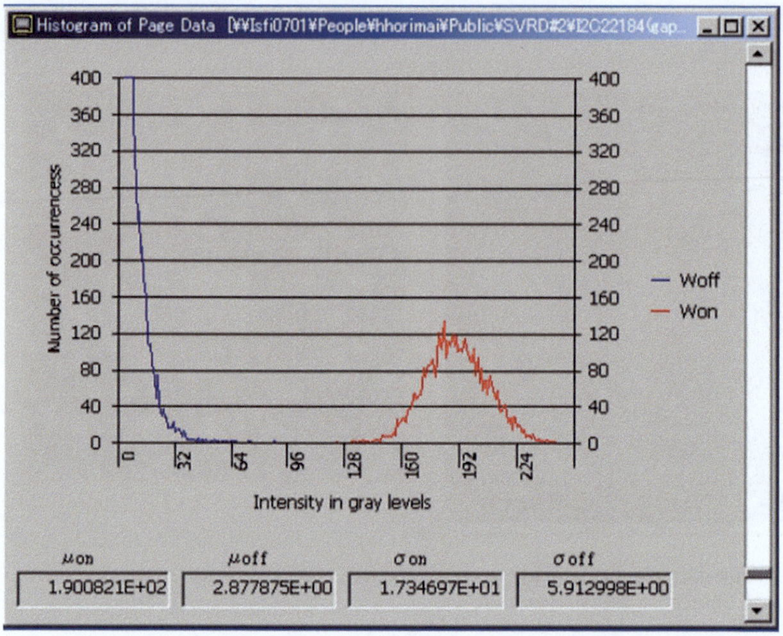


**Figure 7.7** Pixel brightness histogram of the entire data area. Source: Hideyoshi Horimai.

understood that the recording material facilitates the recording of an image with a sufficiently good image quality during the multiple recording experiment.

## 7.4 Symbol Error

### 7.4.1 Burst Error Determination and Image Observation of Error Part

In the collinear method, the 3 : 16 modulation coding is adopted, as it is strong against the random error of the pixel order generated in the data page. In contrast, by incorporating an error correction code in a page and arranging it in a scrambled manner, a correction process is performed even for a burst error that is localized and becomes a large error. However, the amount of information added for the error correction processing reduces the amount of user data per page, affects the high-density recording, and increases the data transfer speed depending on the processing time required for the error correction. To reduce this, it is necessary to consider that the data page can always be recorded and reproduced with a high quality.

Here, we selected several CMOS detection images with noise in the data page, extracted and analyzed the part where the error occurred, and examined how to improve the quality.

Figure 7.8 (left) shows a reproduced image, in which noise, such as haze, appears at the center, obtained when recording/reproducing is performed with a continuous annular reference light pattern. Figure 7.8b (right) is a pixel error map, in which the pixel error is detected from this CMOS image and its distribution is displayed.

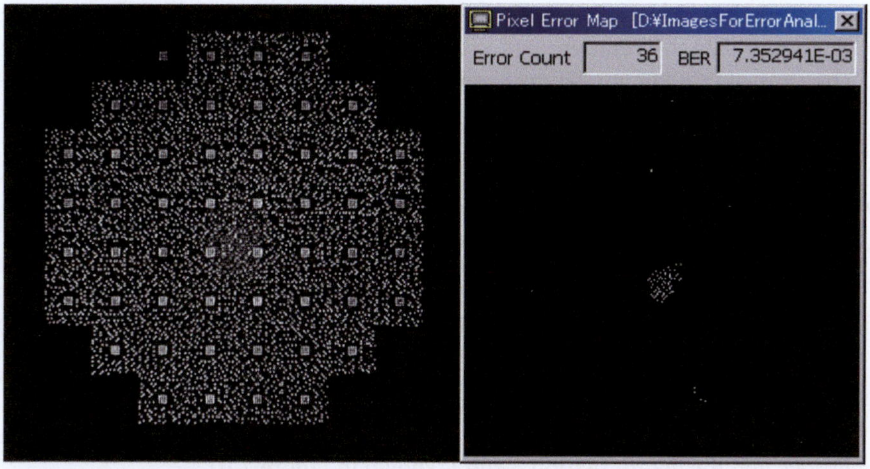


**Figure 7.8** CMOS playback image and pixel error map with noise observed at the center. Source: Hideyoshi Horimai.

On the pixel error map, the portion displayed in white indicates a pixel error. From this map, it seems that a random error has occurred in the center. Therefore, to determine whether the error is a random error or a burst error, the following rules are adopted.

As shown in Figure 7.9, in the S-VRD, the data page detected by the CMOS sensor is divided into 8 × 8 sub-page unit blocks for various types of analysis. Of these, the

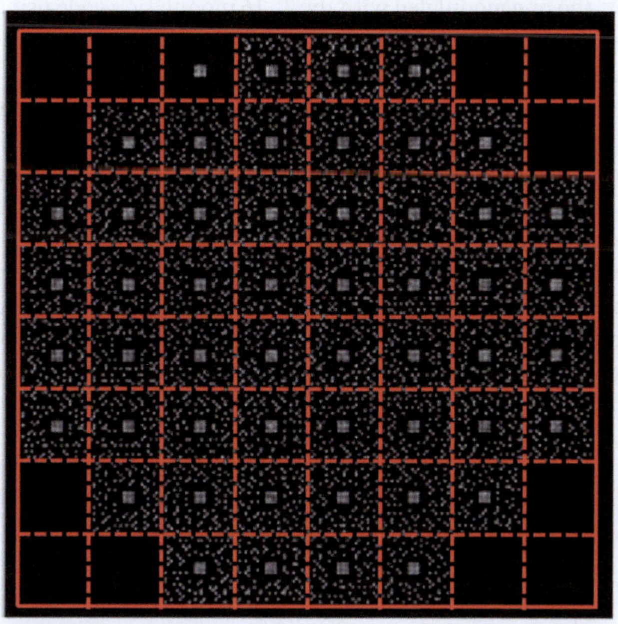

**Figure 7.9** Block diagram of the 8 × 8 sub-page area of the data page. Source: Hideyoshi Horimai.

page sync marks arranged at three corners and near the left corner are the blocks without user data information.

The model for judging the burst errors is as follows:

$$N = \sum_{i} \delta(i) \times \text{subpage}(i), \quad \begin{cases} \delta(i) = 1, & \text{if the } i\text{th subpage has burst error} \\ \delta(i) = 0, & \text{otherwise} \end{cases} \tag{7.3}$$

where

"0": block that is not user data
"1": a block of user data where no burst error is detected (random noise may exist)
"−1": user data block detected as a burst error (if three or more symbol errors continue on the same line, then it is determined as a burst error block. However, not all errors are termed as burst block errors.)

Based on this definition, the above-mentioned CMOS reproduction image was analyzed, and the determination result is shown in Figure 7.10.

The "−1" block at the center of the page is not a random error but a burst error. A burst error block is cut out and extracted from a CMOS image as shown in Figure 7.11, and further expanded by a factor of four to analyze the cause of the error. For comparison and analysis, a block without a burst error was cut out as shown in Figure 7.12.

In the evaluation using an S-VRD, the symbol pattern during the recording is known, and the SNR analysis is performed by comparing with this pattern. In other words, it is possible to evaluate the pixel density at the border position when it is ON and the pixel density at the border position when it is OFF. Therefore, in Figure 7.12, which is a reference block, the one modulated with the 3 : 16 modulation code using the same user data as the burst error part of the selected portion (Figure 7.11). That is, the pattern arrangement of the ON pixels of the arranged symbols is same.

The SNR of the two extracted images with the same data were analyzed using the function of S-VRD, and the corresponding results are shown in Figure 7.13. Figure 7.13a shows the burst error part, and Figure 7.13b presents an error-free part. The histogram showing the luminance distribution of the pixel part indicates the ON and OFF pixels in red and blue colors, respectively.

According to the above analysis result, when the burst error is evaluated, the histograms of the ON pixels of the SNR are distributed in almost the same manner. However, in the block determined to be a burst error, the luminance distribution of the OFF pixel portion is the same as that of the ON pixel, i.e. the histograms overlap.

```
{0, 1, 1,  1,  1, 1, 1, 0},
{1, 1, 1,  1,  1, 1, 1, 1},
{1, 1, 1, -1, -1, 1, 1, 1},
{1, 1, 1, -1, -1, 1, 1, 1},
{1, 1, 1,  1,  1, 1, 1, 1},
{0, 1, 1,  1,  1, 1, 1, 0},
{0, 0, 1,  1,  1, 1, 0, 0}
```

**Figure 7.10** Data page burst error judgment result: "−1" is the burst error block. Source: Hideyoshi Horimai.

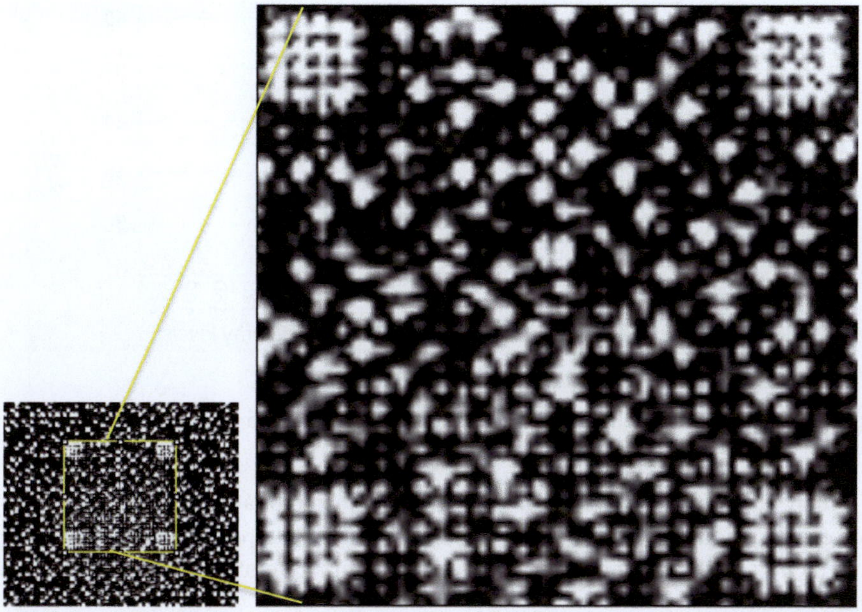

**Figure 7.11** Reproduced CMOS image of the blocked part determined as a burst error. Source: Hideyoshi Horimai.

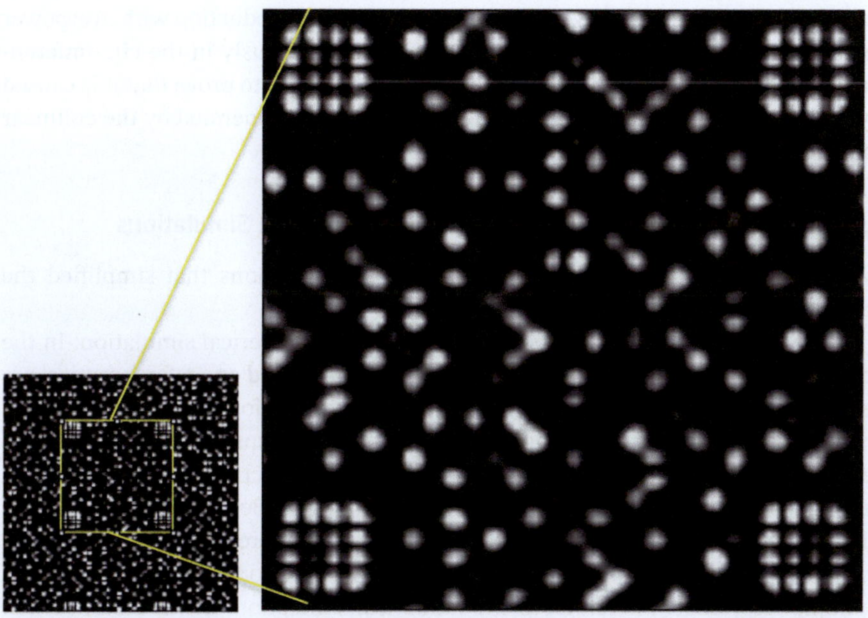

**Figure 7.12** Reproduced CMOS image of the blocked part determined to have no burst error. Source: Hideyoshi Horimai.

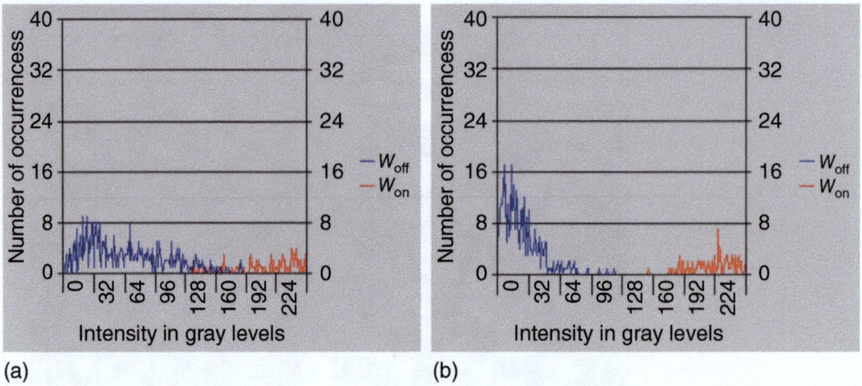


**Figure 7.13** SNR extraction results: (a) Burst error part and (b) no error part. Source: Hideyoshi Horimai.

In other words, the ON pixel mode of the 3 : 16 modulation does not change. However, for some reason, there is no blackening at the position that should be turned OFF; it can be assumed that the decoding function of 3 : 16 fails and causes a burst error.

From the observation of the image, it can be confirmed that the ON pixel of the coded data page is reproduced in the block that caused the burst error. It can also be seen that another noise image overlaps the surrounding area and interferes with it.

The phenomenon that noise is generated in the central portion of the page data as described above is also observed during the recording/reproduction with overpower or when the reference light pixels are arranged continuously in the circumferential direction and recording medium. There is no evidence to prove that it is caused by local defects. Therefore, it is considered that noise is generated by the collinear method.

### 7.4.2 Estimation of Noise Generation Using Numerical Simulations

We estimated the cause of noise via numerical simulations that simplified the collinear recording and playback process.

Figure 7.14 shows the model used for the collinear numerical simulation. In the collinear recording process, when one information pixel and one reference pixel are selected, it may be considered that interference fringes are formed by the two-beam interference method after passing through the objective lens. Therefore, the interference fringe model defined by these two pixels can be described as a simple period grating formed by the plane waves of the two light beams. Because one information pixel actually forms a brilliant grating with the other reference pixels, the optical model is treated as a large number of other single-period gratings recorded in multiple recordings. Therefore, in the numerical simulation, it was assumed that the recording medium could linearly form any number of interference fringes irrespective of the multiplicity.

In the reproduction process, for one grating formed in this way, an image reproduced by a plane wave irradiated from any one reference pixel is obtained, and this operation is performed for all the other reference pixels. Next, the same calculation

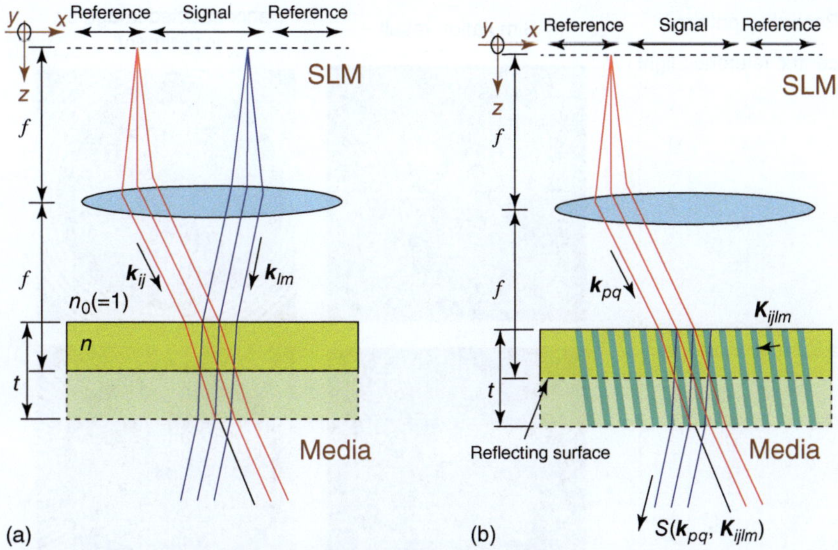


**Figure 7.14** Simplified (a) recording and (b) reproduction of an optical model using the collinear method. Source: Hideyoshi Horimai.

is repeated for another grating, so that a reproduction data page image is finally obtained by simulation based on the sum of the reproduction images of all combinations. To simplify the calculation, it was considered the information and reference pixels did not form a grating.

Although an enormous number of calculations are required in the collinear numerical calculation model, the simulation accurately reflects the individual pixel arrangement of the information and reference lights arranged in pixels. It is considered an optimal method for quantitatively assessing the relationship between the data page pattern, i.e. information light, and the pattern of the reference light. Details of this numerical simulation have been reported by Shimura et al. [23].

Based on this optical model, a reproduction data page was obtained through simulations and compared with the actual recording/reproduction result as shown in Figure 7.15. Although the information light pattern (data page) is the same, it is confirmed that noise is generated at the center of the reproduced data page when a concentric pattern, continuous with the annular reference light pattern, is used. Conversely, when reference light having a radial ring pattern was used, the result was that uniform brightness without noise was reproduced. In addition, when recording and playback experiments were actually performed using both page patterns, the same as those used for the numerical simulation, as shown on the right side of the figure, playback similar to each simulation result was performed. An image was obtained.

As a result, the burst error at the center of the data page is a unique phenomenon that occurs when recording and playback is performed using a continuous concentric reference beam pattern, which is caused by the design of the page data pattern. In addition, this analysis result proved that a radial reference light pattern is selected by trial and error, it is an effective pattern for obtaining a low-noise reproduced image.

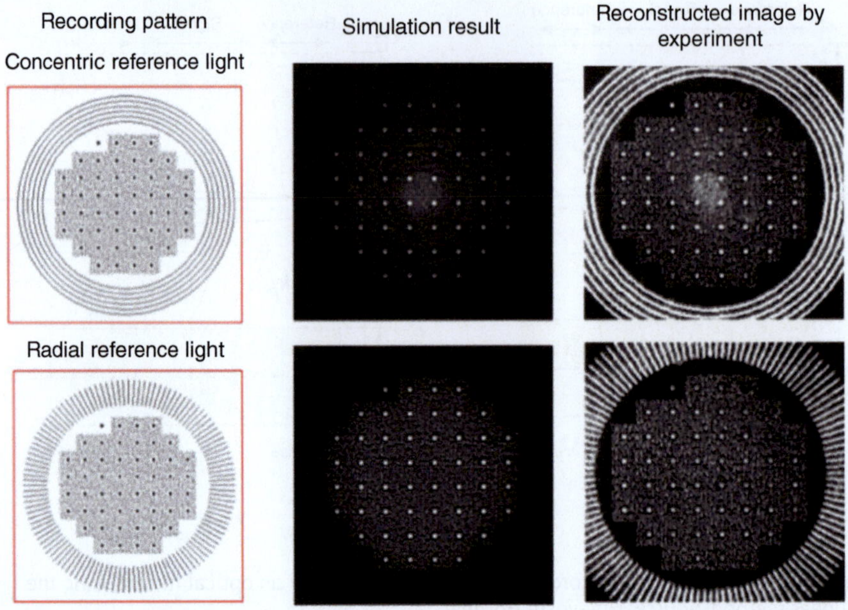


**Figure 7.15** Reproduction quality evaluation of the data page design through numerical simulation. Source: Hideyoshi Horimai.

## 7.5 Numerical Analysis of Collinear Shift Selectivity

As described in the Section 7.4.2, it was found that the cause of noise in recording/reproduction was in good agreement with the actual experimental results, indicating the effectiveness of the numerical simulation. Then, we tried a simulation analysis by a simple numerical modeling of the shift selectivity characteristics in the collinear method.

Figure 7.16 shows the state of the grating and the wavefront when the medium is shifted in-plane from the recording position. Figure 7.16a shows a state in which the medium itself moves in a direction perpendicular to the optical axis ($z$ axis) of the optical system, that is, the in-plane direction of the recording medium. At this time, when observed from the optical axis at the time of recording, it appears that the phase of the grating is shifted from that before the shift. The phase shift of this grating is indicated by $K \cdot r_{\text{shift}}$ where $r_{\text{shift}}$ represents the media shift. On the other hand, when the medium is considered as a reference, the phase of the incident wave is shifted with respect to the medium as shown in Figure 7.16b. In this case, since the optical axis and the CMOS move in the same direction as the incident wave, they can be shown by the same description as in the case of media misalignment.

To understand the behavior of the collinear shift selectivity, it is assumed that only one pixel at the center of the page data is recorded/reproduced by the concentric annular reference light. Although details are omitted, such a typical model can be described as a superposition of reproduced images obtained from the concentric reference beams with different radii. Thus, the change in the reproduction signal,

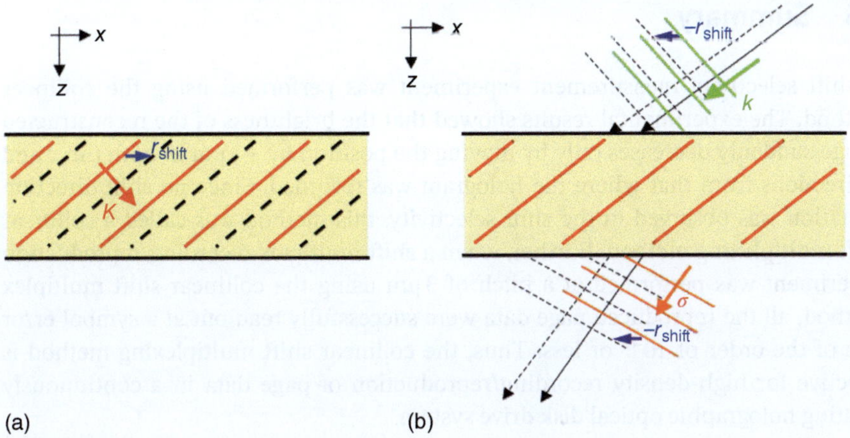


**Figure 7.16** Numerical model with media shifted from recording position in-plane under different object of reference (a) media shifted, (b) incident wave shifted. Source: Hideyoshi Horimai.

$E_{\text{one signal\_shift}}$, accompanying the shift is given by the superposition of zeroth-order Bessel functions for each concentric circle radius.

$$E_{\text{onesignal\_shift}} \approx \int_{r\_\text{inner}}^{r\_\text{outer}} S(r_{pq}) J_0(K(r_{pq}) r_{\text{shift}}) \mathrm{d}r \tag{7.4}$$

The zeroth-order Bessel function is also used to describe the amplitude of the Fraunhofer diffracted light (Airy disk) produced by a circular aperture [24].

Figure 7.17 shows a result obtained by actually substituting the parameters into the Eq. (7.4). This result indicates that the behavior seen in the actual shift selectivity measurement result shown in Figure 7.1 can be explained qualitatively via numerical simulations.

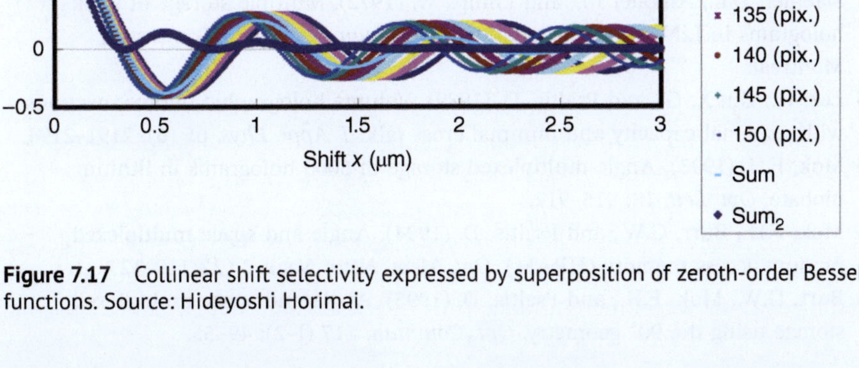


**Figure 7.17** Collinear shift selectivity expressed by superposition of zeroth-order Bessel functions. Source: Hideyoshi Horimai.

## 7.6 Summary

A shift selectivity measurement experiment was performed using the collinear method. The experimental results showed that the brightness of the reconstructed image suddenly decreases only by moving the position by $\pm 1$ µm in both the $x$ and $y$ directions from that where the hologram was recorded. Since no shift direction deviation was observed in the shift selectivity, this method was called a collinear shift multiplexing method. Further, when a shift multiplex recording/reproduction experiment was performed at a pitch of 3 µm using the collinear shift multiplex method, all the reproduced page data were successfully read out at a symbol error rate of the order of $10^{-2}$ or less. Thus, the collinear shift multiplexing method is effective for high-density recording/reproduction of page data in a continuously rotating holographic optical disk drive system.

Moreover, to record and reproduce low-noise and high-quality data pages suitable for high-density recording, data page error analysis was performed. When a burst error determination criterion was set, and a burst error occurrence portion was extracted and an image was observed, it was found that an error occurred overlapping the central portion of the data page. For this reason, collinear recording/reproduction is used as a simple model, and as a result of analysis by numerical simulation, it was found that the noise is peculiar to a reference light pattern that is continuous with a ring shape. From this result, it was found that the radial reference light pattern effectively suppressed the generation of noise, confirming the effectiveness of the pattern.

Furthermore, a numerical simulation-based analysis was performed to evaluate the shift selectivity feature of a collinear system. It was shown that the change in the reproduction signal accompanying the shift can be expressed by the superposition of zeroth-order Bessel functions by the concentric reference light. Thus, the experimentally observed shift selectivity can be qualitatively explained by the numerical simulation-based analysis.

## References


1 Leith, E.N., Kozma, A., Upatnieks, J. et al. (1966). Holographic data storage in three-dimensional media. *Appl. Opt.* 5 (8): 1303–1311.

2 Staebler, D.L., Amodei J.J., and Philips W. (1972). Multiple storage of thick holograms in $LiNb0_3$. *VII International Quantum Electronics Conference*, Montreal.

3 Lee, H., Gu, X.-G., and Psaltis, D. (1989). Volume holographic interconnections with maximal capacity and minimal cross talk. *J. Appl. Phys.* 65 (6): 2191–2194.

4 Mok, F.H. (1993). Angle-multiplexed storage of 5000 holograms in lithium niobate. *Opt. Lett.* 18: 915–917.

5 Mok, F.H., Burr, G.W., and Psaltis, D. (1994). Angle and space multiplexed random access memory (HRAM). *Opt. Mem. Neur. Netw.* 3 (2): 119–127.

6 Burr, G.W., Mok, F.H., and Psaltis, D. (1995). Angle and space multiplexed storage using the 90° geometry. *Opt. Commun.* 117 (1–2): 49–55.

**7** Curtis, K., Pu, A., and Psaltis, D. (1994). Method for holographic storage using peristrophic multiplexing. *Opt. Lett.* 19 (13): 993–995.

**8** Rakuljic, G.A., Levya, V., and Yariv, A. (1992). Optical data storage by using orthogonal wavelength-multiplexed volume holograms. *Opt. Lett.* 17 (20): 1471–1473.

**9** Yin, S., Zhou, H., Zhao, F. et al. (1993). Wavelength-multiplexed holographic storage in a sensitive photorefractive crystal using a visible-light tunable diode-laser. *Opt. Commun.* 101 (5–6): 317–321.

**10** Lande, D., Heanue, J.F., Bashaw, M.C., and Hesselink, L. (1996). Digital wavelength-multiplexed holographic data storage system. *Opt. Lett.* 21: 1780–1782.

**11** Denz, C., Pauliat, G., Roosen, G., and Tschudi, T. (1991). Volume hologram multiplexing using a deterministic phase encoding method. *Opt. Commun.* 85: 171–176.

**12** Barbastathis, G., Levene, M., and Psaltis, D. (1996). Shift multiplexing with spherical reference waves. *Appl. Opt.* 35: 2403–2417.

**13** Paltis, D., Levene, M., Pu, A. et al. (1995). Holographi storage using shift multiplexing. *Opt. Lett.* 20 (7): 782–784.

**14** Dhar, L., Curtis, K., Tackitt, M. et al. (1998). Holographic storage of multiple high- capacity digital data pages in thick photopolymer systems. *Opt. Lett.* 23 (21): 1710–1722.

**15** Barbastathis, G., Pu, A., Levene, M., and Psaltis, D. (1995). Shift-multiplexed holographic 3d disk. *Optical Data Storage Proceedings,* San Diego, CA, USA (1995, SPIE).

**16** Pu, A. and Psaltis, D. (1996). Shift-multiplexed holographic 3-D disk system. *International Symposium on Optical Memory and Optical Data Storage*, Maui, Hawaii.

**17** Li, H.-Y.S. and Psaltis, D. (1994). Three dimensional holographic disks. *Appl. Opt.* 33 (17): 3764–3774.

**18** Psaltis, D. (1992). Parallel optical memories. *Byte* 17 (9): 179.

**19** Anderson, K. (2004). Polytopic multiplexing. *Technical Digest of Optical Data Storage Topical Meeting 2004*, pp. 255–257, Monterey, California, USA.

**20** Hesselink, L., Orlov, S.S., and Bashaw, M.C. (2004). Holographic data storage systems[J]. *Proc. IEEE* 92 (8): 1231–1280.

**21** Orlov, S.S., Bjornson, E., Phillips, W. et al. (2000). 6 Gbit/sec transfer rate demonstration in a holographic disk digital storage system. *Post Deadline Papers*, Optical Data Storage 2000, p. P31.

**22** Horimai, H., Tan, X., and Li, J. (2005). Collinear holography[J]. *Appl. Opt.* 44 (13): 2575–2579.

**23** Shimura, T., Ichimura, S., Fujimura, R. et al. (2006). Analysis of a collinear holographic storage system: introduction of pixel spread function. *Opt. Lett.* 31 (9): 1208–1210.

**24** 栖原　敏明 (1998). "光波工学 光エレクトロニクス教科書シリーズ". コロナ社 第 1 版.

# 8

# System Margin of a Collinear Holographic Data Storage System

## 8.1 Defocus Margin

The collinear method has an optical servo mechanism similar to that of optical discs such as CD, DVD, and HD-DVD. Therefore, the margin is compared with the DVD standard values [1–4]. First, the focus servo is used to focus on the reference plane of the disc accurately. However, when the recording medium rotates mechanically like a disc, it may not follow the amplitude and frequency of the disc runout. That is, as schematically shown in Figure 8.1, it is assumed that a residual occurs at the focal position. Such a shift of the focal position with respect to the focal plane is referred to as defocus or defocus amount. It is important to quantify the defocus amount that does not affect the hologram reproduction, i.e. the defocus margin, while designing the mechanical characteristics of a holographic optical disk, response characteristics of an actuator, weight of an objective lens, and so on. Therefore, the defocus margin in the collinear method was evaluated based on the reproduction error rate.

First, the optical servo mechanism in the S-VRD was used to record the page data with hologram by applying the focus servo to the reference reflecting surface, and then by moving the stage of the recording media, sample playback was performed at a shifted position (defocused). This position shift is achieved by shifting the recording medium from the focal position, and the amount of movement, that is, the relationship between the defocus amount and the number of occurrences of symbol error was obtained. From the viewpoint of error correction, since it is considered that the error range is acceptable if the symbol error is 100 or less, the defocus amount reaching a symbol error of 100 was obtained as the defocus margin. The evaluation results are shown in Figure 8.2.

According to the evaluation results, the symbol error changes asymmetrically depending on whether the defocus direction approaches or moves away from the disc. However, it was confirmed through experiments that the defocus amount reaching symbol error = 100 is significantly larger than the DVD defocus standard value, with the defocus amount being more than six times that of the wide focus.

The reason for this is that in optical discs, the resolution of the light spot is reduced owing to defocusing, and the quality of the pit playback signal is degraded, resulting in an increase in errors. In contrast, in the hologram regeneration based on the collinear mode, the regeneration deterioration is caused by the mismatch of the wave

*Collinear Holography: Devices, Materials, Data Storage*, First Edition.
Xiaodi Tan, Hideyoshi Horimai, Tsutomu Shimura, and Xiao Lin.

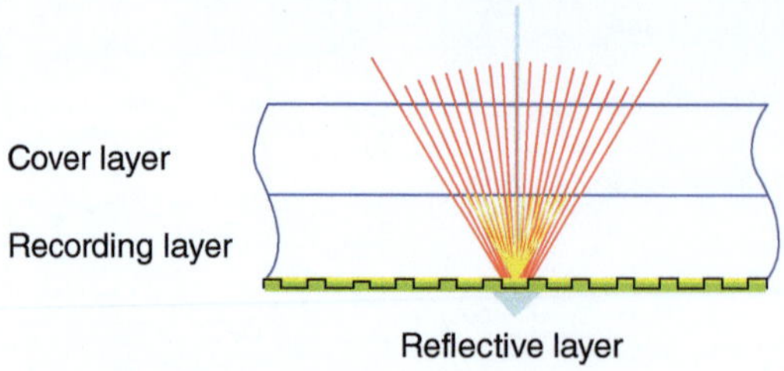


**Figure 8.1** Schematic of the state of defocusing in the collinear method. Source: Hideyoshi Horimai.

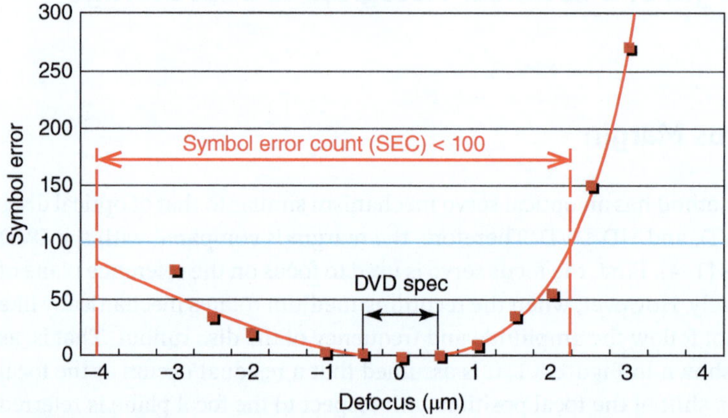


**Figure 8.2** Results of the collinear defocus margin evaluation, SEC: Symbol error count. Source: Hideyoshi Horimai.

surface of the interference fringe. Therefore, even if the focus is shifted in the optical axis direction, the range in which the wavefronts are matched is relatively wide.

In the actual drive system, the focus servo is applied in the same way as in a DVD. Therefore, when the servo is functioning, no defocus errors occur as long as the focus servo is functioning within the DVD specification range.

## 8.2 Detrack Margin

When the disk rotates, the so-called eccentricity occurs, in which the movement of the regenerative position is recorded in the radial direction owing to the centering of the spindle motor, mechanical accuracy of the chucking mechanism of the disk, and deviation of the central hole due to manufacturing variations of the disk itself. In the collinear method, to determine the position of a regenerated hologram to be recorded, a tracking servo that follows a preformatted phase pit is used in the same manner as in a DVD. However, as with the focus servo, the eccentricity of the disk is not monotonous, and if the amplitude and frequency cannot be tracked, then it is assumed that a tracking residual occurs as shown in Figure 8.3. Thus, the amount of deviation of the position in the radial direction with respect to the track is called detrack or detrack amount.

To perform collinear shift multiplex recording regeneration with a higher reliability, a sufficiently small detrack amount is necessary. Therefore, the recorded

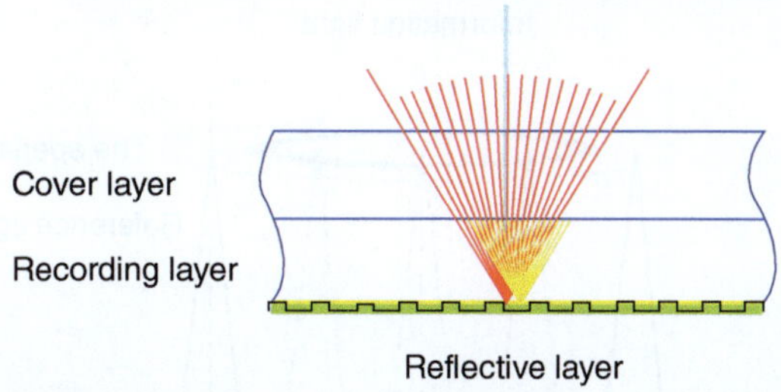


**Figure 8.3** Schematic of the detracking state in the collinear method. Source: Hideyoshi Horimai.

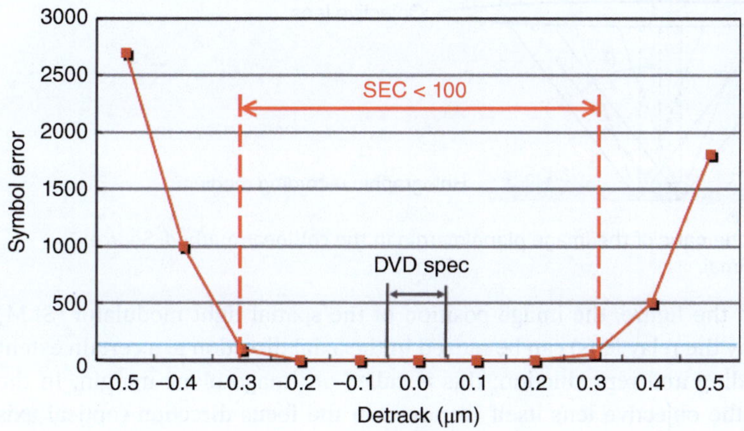


**Figure 8.4** Detrack margin in the collinear method. Source: Hideyoshi Horimai.

position was slightly moved in the plane of the disc, and how the movement amount increased the error of the reproduced page data was evaluated. Figure 8.4 shows the evaluation result of the collinear detrack margin. The evaluation results indicated that the holographic recording regenerated using the collinear method had a margin of six times or more than the DVD detrack standard value when compared with 100 or less symbol errors. In the case of collinear shift multiplex recording, all the errors occurred when the recording position shifted by 1 μm in the plane. From this result, it can be seen that there is no increase in the error up to a shift amount of about ±0.2 μm. Therefore, as long as the tracking servo is functioning within the DVD specification range, there is no risk of errors due to detracking. Further, if recording/reproduction based on the address is performed, then the page data recorded by the collinear shift multiplex recording method can be accessed accurately and reproduced.

## 8.3 Image Plane Margins

The collinear approach is based on the concept that a holographic disc can be attached to or removed from a drive device to enable media exchange. Therefore, it is necessary to ensure compatibility of recording and reproduction between different drive devices as well as recording and reproduction with the same drive device. From this perspective, as shown in Figure 8.5, the image position of the

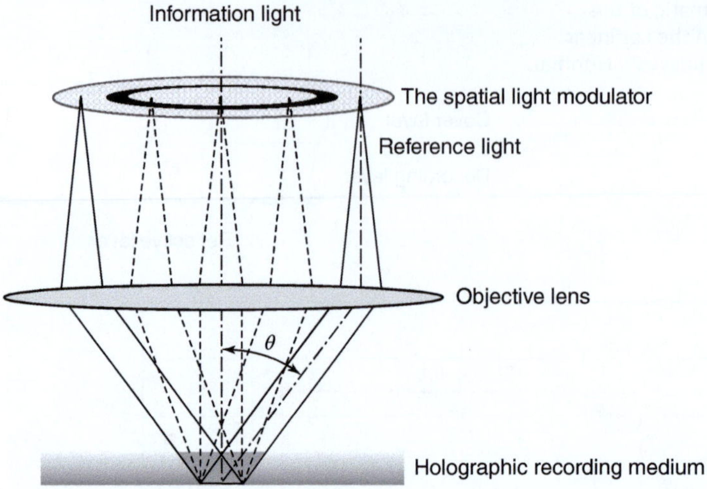


**Figure 8.5** Schematic of the image plane margin in the collinear method. Source: Hideyoshi Horimai.

page data (in the figure, the image position of the spatial light modulator (SLM) transmitted by the relay lens) can be moved in the axial direction to a certain extent during recording and reproduction; this is called an image plane margin. In the experiment, the objective lens itself is moved in the focus direction (optical axis direction) by moving the recording medium away or closer using the stage, while maintaining the state where the focus servo is accurately applied. The symbol error was measured when the lens was shifted relative to the entrance pupil position of the objective lens.

Figure 8.6 shows the evaluation results, indicating that the symbol error increases rapidly from around ±200 μm. Due to the mechanical constraints of the optical

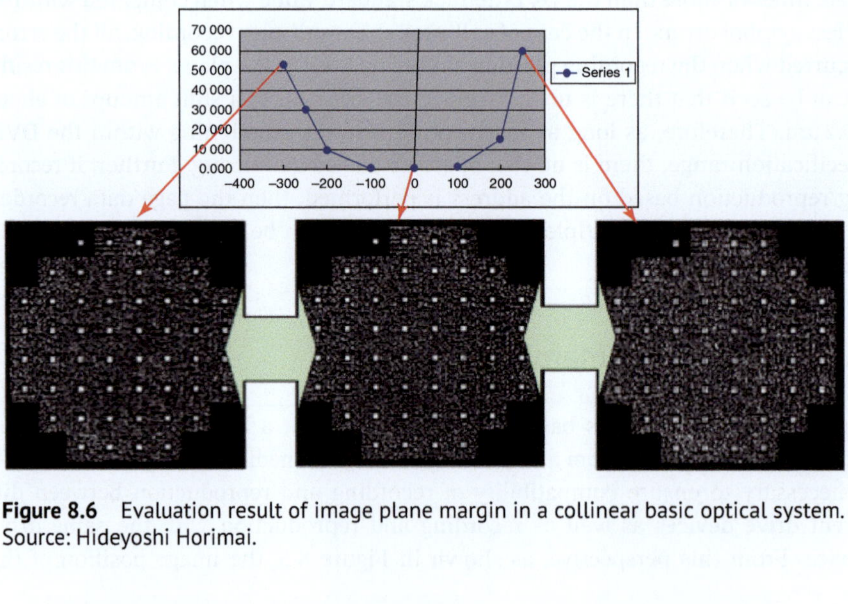


**Figure 8.6** Evaluation result of image plane margin in a collinear basic optical system. Source: Hideyoshi Horimai.

system, it can no longer be evaluated with a larger deviation. However, the linear way of following the surface eccentric lens of the optical disc is understood, and the optical design of the drive needs to be considered because of the effect of this merger.

## 8.4 Tilt Margin

As shown in Figure 8.7, the degree, to which the recording and reproduction are affected by the tilt of the disc, is a very important parameter for the optical disc system containing the rotating recording medium and is called a tilt margin. On the other hand, in the hologram, it is necessary to evaluate the tilt margin from the following two viewpoints:

(1) Absolute value tilt margin: This indicates whether or not the recording or reproduction is possible with respect to a tilted disk.
(2) Relative value tilt margin: This indicates by how much the disc tilt may change after the recording.

The results of experimental evaluation for each of these cases are discussed in the following sections.

### 8.4.1 Absolute Value Tilt Margin

Figure 8.8 shows the absolute tilt margin evaluation results. The angle of the disc with the reflective film was changed, and recording was performed with respect to the angle, and the disc was reproduced as it was. In this evaluation, the vertical axis represents the signal-to-noise ratio (SNR) and not the symbol error. As the disc is tilted, the SNR slowly decreases, and the page playback image shifts in-plane from the CMOS screen, but even when tilted by 2.25°, the symbol error due to recording and reproduction remains zero. This is because in a normal optical disc, aberration occurs at the focused spot due to tilt, and the pit cannot be reproduced, resulting in an increase in errors. In the case of the collinear method, interference fringes are produced on a volumetric recording medium. Because interference fringes are recorded in a form that includes aberration even if there is a tilt, the wavefront matches if it is in the same tilt state during the playback. This indicates that the information is played back properly. Further, as the recording medium is provided with a reflective

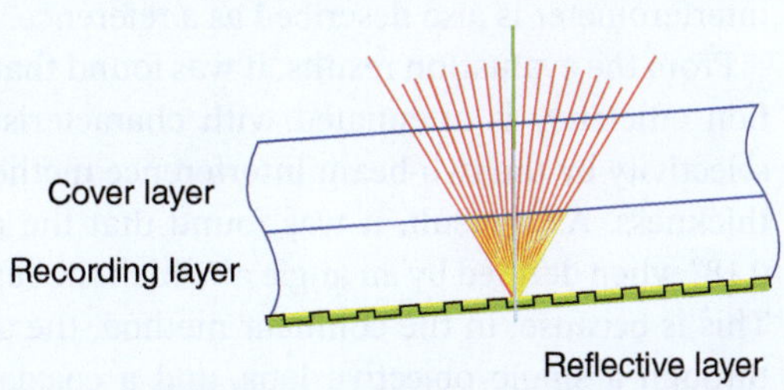


**Figure 8.7** Schematic of the tilt occurrence in the collinear method. Source: Hideyoshi Horimai.

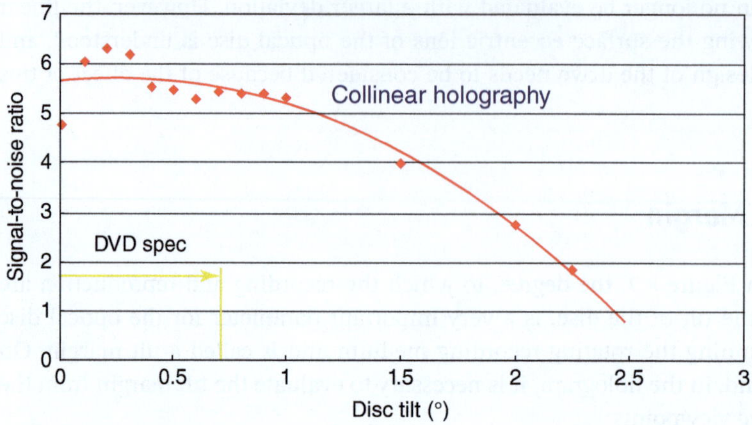


**Figure 8.8** Evaluation of the absolute value tilt margin by the collinear method. Source: Hideyoshi Horimai.

film, the traveling direction of the reproduced image is bent at a reflection angle twice that of the tilt and received by the CMOS sensor, but no significant error occurs.

The aforementioned results demonstrate that the absolute value margin is extremely large in the collinear method as well, as long as the shape of the recording medium does not change between recording and reproduction as in the conventional hologram. However, in an actual system, it is expected that the shape of the recording medium changes with time, and in the case of replacing the driving device, it is expected that the tilt will relatively change even within the range of the mechanical specifications. Therefore, in the collinear method, it is important to evaluate the next relative value tilt margin.

### 8.4.2 Relative Value Tilt Margin

Figure 8.9 shows the relative tilt margin evaluation results. Here, the characteristic is shown not by symbol error but by the change in the diffraction efficiency (reproduced signal intensity) calibrated by CMOS sensor. This relative value tilt margin corresponds to the angle selectivity of a so-called angle multiplexing method, in which multiplexing is performed by changing the angle of the reference light using the conventional two-beam interference method. Because the thickness of the recording material used to evaluate the collinear relative value tilt margin is 500 μm, the corresponding theoretical curve of the diffraction efficiency of the two-beam interferometer is also described as a reference.

From the evaluation results, it was found that in the collinear method, the diffraction efficiency is attenuated with characteristics that are poorer than the angle selectivity by the two-beam interference method with the same recording material thickness. As a result, it was found that the relative value tilt margin was about 0.18° when defined by an angle at which the reproduction signal intensity is halved. This is because, in the collinear method, the axis-shaped page data pattern passes through a single objective lens, and a coaxial hologram is formed. Because the

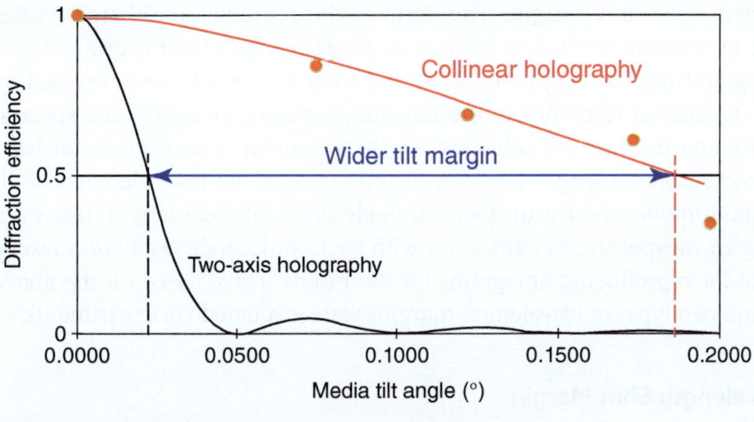


**Figure 8.9** Evaluation of the relative value tilt margin by collinear method. Source: Hideyoshi Horimai.

interference fringes are formed on the surface, it is thought that the wavefronts are matched even if the disk is slightly inclined.

In optical disks for data storage, such as magneto-optical disks, the tilt of the disk is about 5 mrad (0.28°), and the standard tilt value of a commercial DVD is 0.7°. The graphic optical discs are manufactured with a narrow flatness. Therefore, it can be said that the holographic optical disc using the collinear method has a margin value eligible for practical applications. However, in the future, we believe that it will be necessary to perform dynamic corrections, for example, by improving the margin by examining the page data or by adding a tilt servo mechanism compatible with the collinear method.

## 8.5 Wavelength Margin

Miniaturization and cost reduction of the apparatus are essential for the widespread use of holographic data storage (HDS), and thus, the optical system must be small and inexpensive. When considering the cost of the drive, it is particularly necessary to use a small and inexpensive semiconductor laser as the light source. The recording wavelength can be combined with the design of the recording material, and a considerably-high-output semiconductor laser can be developed at a specific wavelength. However, there are other two margins from the viewpoint of hologram recording/reproduction.

(1) Wavelength shift margin: This indicates the reproduction characteristics of a laser, whose oscillation wavelength is shifted from the recording wavelength. For a semiconductor laser, the oscillation wavelength varies significantly with different laser elements. In particular, it is assumed that the wavelength is different between the drive device at the time of recording and that at the time of reproduction. Thus, it is desired that this wavelength shift margin is wide for compatibility of the media device.

(2) Wavelength spectrum margin: This implies the reproduction characteristics due to a multimode oscillation laser. A single-wavelength laser is desirable for recording and reproducing holograms, and such a laser can be developed by adding an external resonator to a semiconductor laser to narrow the spectral line width and increase the coherent length. In general, a semiconductor laser has a short resonator length between the end faces of the laser element, oscillates in multimode, and features an extremely short coherent length. However, realizing an inexpensive system, even with such multimode oscillation lasers, is crucial for reproducing holograms for the future. Thus, based on the above viewpoint, two types of wavelength margins were evaluated via experiments.

### 8.5.1 Wavelength Shift Margin

The wavelength shift margin in the collinear method was evaluated by changing the wavelength during recording and reproduction. Since the laser incorporated in the S-VRD is a solid-state laser, the wavelength was fixed at 532 nm, and it is difficult to modify it to perform a reproduction experiment at a wavelength different from the recorded wavelength. For this reason, a collinear recording/reproducing optical system was constructed on the same optical plate using a dye laser capable of changing the wavelength, and an experiment for evaluating the wavelength shift margin was conducted.

Figure 8.10 shows an overview of the tunable laser and dye used in the experiment and the collinear recording/reproducing optical system developed on the optical surface plate. While selecting a dye, it is necessary to be able to oscillate at a wavelength with which the wavelength change width can be set large and the recording medium has sensitivity. Thus, the oscillation wavelength of the laser beam during recording was set to 495 nm. After recording the collinear page data at this wavelength, the collinear reproduction was performed by changing the wavelength to the longer wavelength side.

When reproducing while changing the wavelength, there is a possibility that the reproduced image is deteriorated by repeated irradiation of the reproduced image. To assess this irradiation-induced image deterioration, an experiment was repeated, in which the recording/reproduction was always performed at a new position; first recording at 495 nm and then changing the wavelength. Figure 8.11 shows a representative of the obtained reproduced images, and Figure 8.12 shows the diffraction efficiency obtained from the reproduced image and the wavelength dependency of the reproduced image. For comparison, data from two-beam interferometry with the same recording medium thickness (500 μm) are included as a reference [5].

Comparing with the relative diffraction efficiency of 50%, it is observed that the collinear method has a wavelength margin that is three times or more wider than that of the two-beam interference method. Further, when evaluated using the symbol error, as shown in Figure 8.13, even if the reproduction wavelength deviates by 2.7 nm from the recording wavelength, the reproduction page has no more than 100 symbol errors. The wavelength range shorter than the recording wavelength could not be evaluated because of the limitation of the experimental equipment.

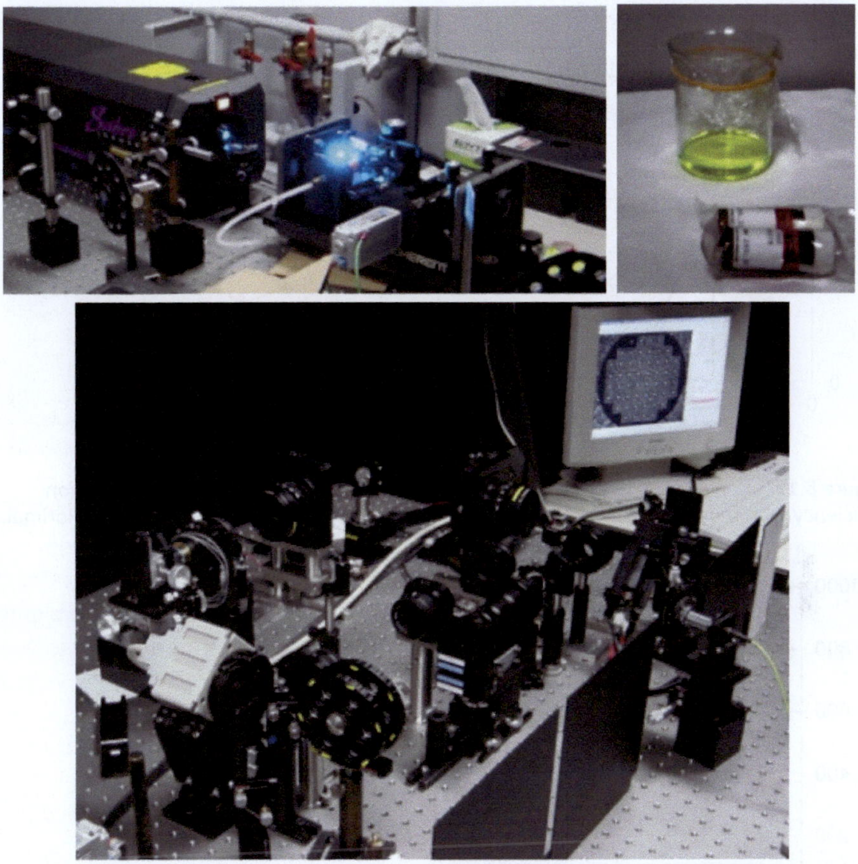

**Figure 8.10** Overview of the wavelength margin evaluation experiment system using the collinear method with a dye laser. Source: Hideyoshi Horimai.

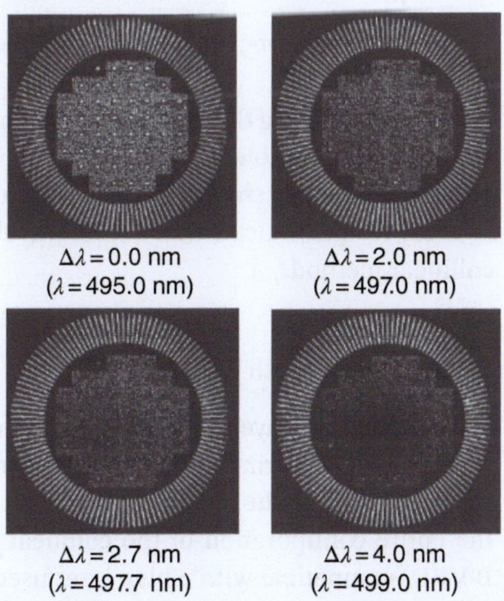


**Figure 8.11** Image reconstruction via wavelength change during the recording and reproduction. Source: Hideyoshi Horimai.

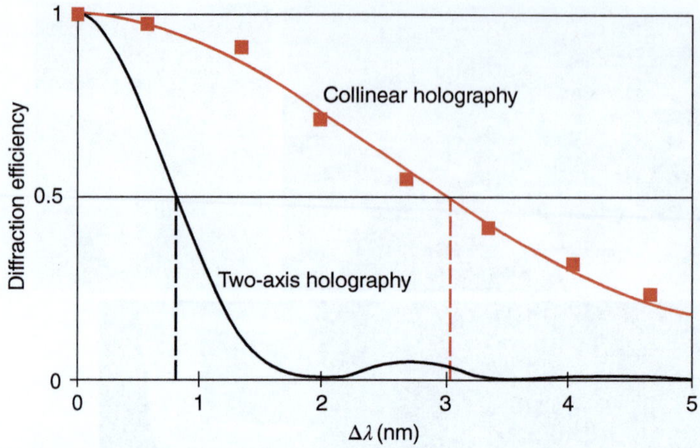


**Figure 8.12** Comparison of the wavelength margin evaluated based on the diffraction efficiency with the theoretical value of two-beam interferometry. Source: Hideyoshi Horimai.

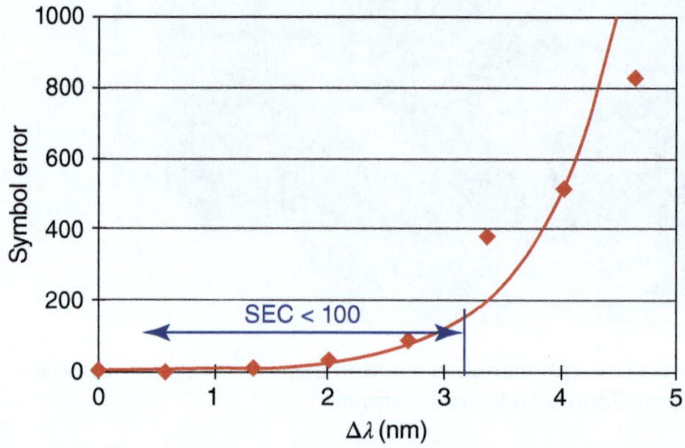


**Figure 8.13** Wavelength margin evaluation using symbol error. Source: Hideyoshi Horimai.

However, assuming that similar results could be obtained on the shorter wavelength side, the reproducible wavelength range for the recording wavelength of 497.7 nm, i.e. the wavelength shift margin, was estimated to be about ±2.7 nm.

These results indicate that a wide wavelength shift margin can be obtained in the collinear method.

### 8.5.2 Wavelength Spectrum Margin

To evaluate the wavelength spectrum margin, we performed a collinear page data reproduction experiment using a blue semiconductor laser (wavelength: ~405 nm) actually used in Blu-ray disc technology. Figure 8.14 shows a perspective view of the entire configuration of the collinear recording/reproducing apparatus (name: B-VRD) compatible with a blue laser used in the experiment. The optical system of

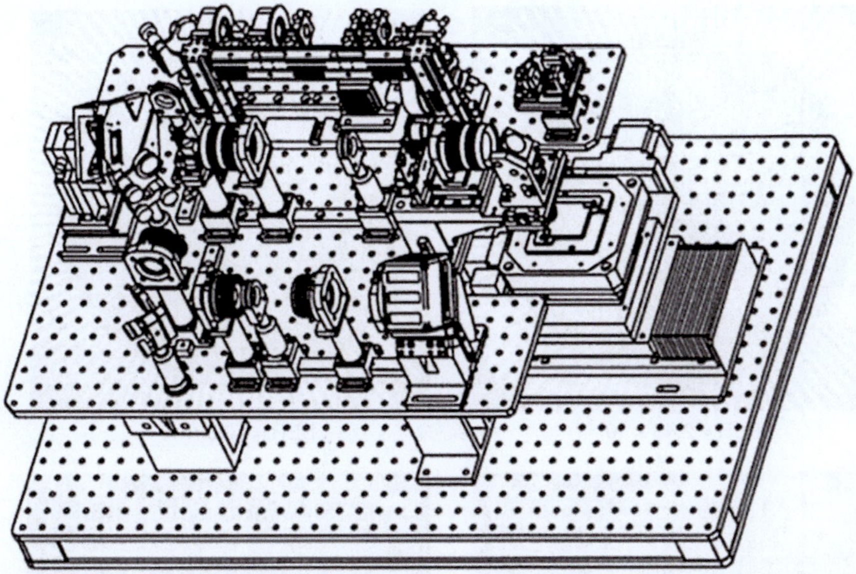

**Figure 8.14** Perspective view of the overall configuration of the collinear recording/reproducing device (B-VRD) using a blue laser. Source: Hideyoshi Horimai.

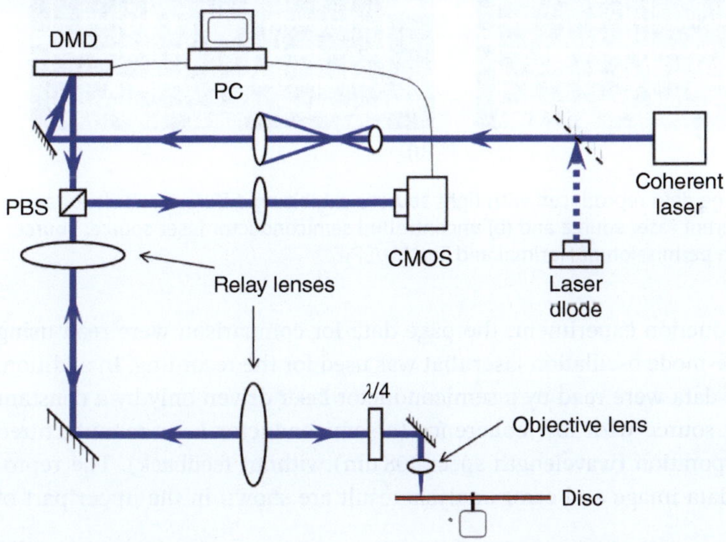


**Figure 8.15** B-VRD optical system configuration, incorporating a semiconductor laser light source. Source: Reproduced with permission of Horimai and Tan [6]/SPIE.

the experimental apparatus using this B-VRD is shown in Figure 8.15. In this experimental apparatus, two blue laser light sources were incorporated, and these were selected by a tiltable mirror. For the hologram recording of page data, which was the basis of the evaluation, a laser (wavelength of 407.9 nm measured), oscillated in a single mode by an external resonator, was used [7].

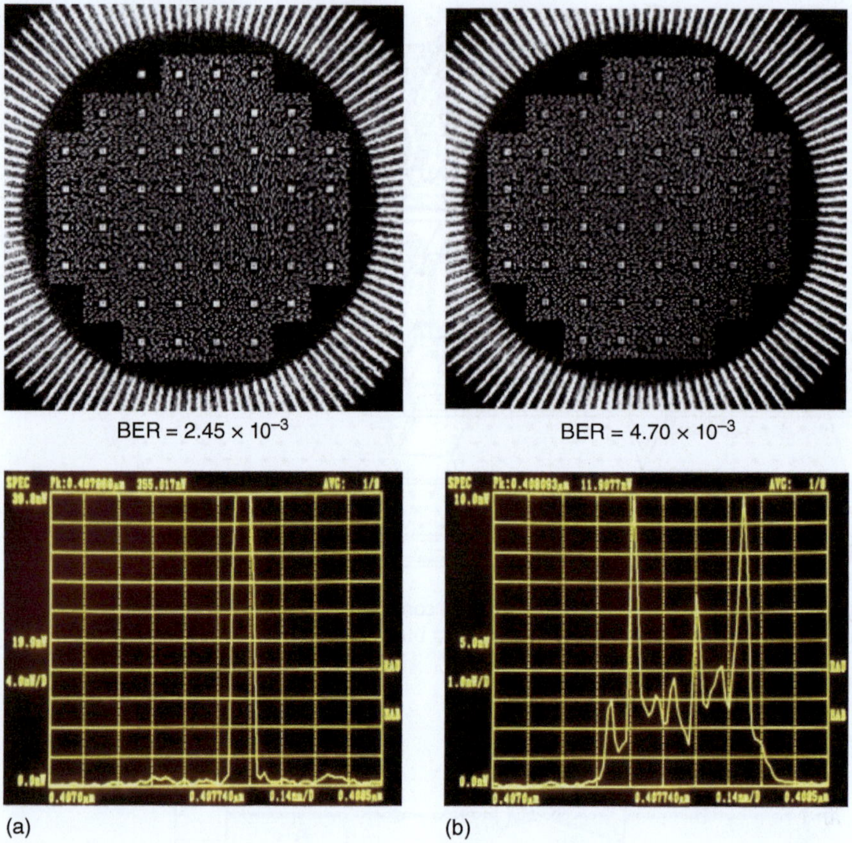


**Figure 8.16** Page data reproduced with light sources exhibiting different coherency and spectra. (a) Coherent laser source and (b) uncontrolled semiconductor laser source. Source: Reproduced with permission of Horimai and Tan [6]/SPIE.

In the reproduction experiment, the page data for comparison were read using the same single-mode oscillation laser that was used for the recording. In addition, the same page data were read by a semiconductor laser driven only by a constant current power source with low coherence (a semiconductor laser manufactured by Nichia Corporation (wavelength spec: 408 nm) without feedback). The reproduction page data image and error analysis result are shown in the upper part of Figure 8.16.

Both the evaluated BER of the reproduced page data were as low as $10^{-3}$. In contrast, when the oscillation wavelength of each semiconductor laser was measured with an optical spectrum analyzer, the wavelength spectrum shown in the lower part of Figure 8.16 was obtained. The screen display of the spectrum analyzer has a horizontal scale of 0.14 nm/div. Both the measurement screens display at approximately 407.0–408.5 nm. A laser oscillated in a single mode by an external resonator emits a single spectrum with a narrow line width of 407.9 nm, whereas a current-driven semiconductor laser exhibits a large number of spectral peaks, and the mode oscillation is visible. The spectral wavelength width of the multimode oscillation is about 0.9 nm as evident from the screen display.

Based on the above results, even when reproducing using a semiconductor laser that exhibits multiple oscillation spectra and oscillates in multiple modes, the collinear method can reproduce with a BER that is practically low and proved to have a wide wavelength spectrum margin.

## 8.6 Temperature Compensation for Collinear Holographic Data Storage

### 8.6.1 Introduction

For the consumer market, the HDS system must have a wide operating temperature range. However, when temperature is varied, the refractive index of the bulk and the period of the grating change. The individual grating components may also tilt if the expansion or contraction of the media is anisotropic. Because of these effects, the Bragg matching condition cannot be fulfilled, which may result in image recovery failure. Therefore, it is necessary to compensate for the thermal effects of HDS.

In this section, we discuss the temperature characteristics of a collinear holographic data storage system (CHDSS) and the compensation condition. Throughout this chapter, we use subscript $(\cdot)_0$ to denote the origin quantity of $(\cdot)$ for the origin temperature. We also assume the hologram is recorded at the origin temperature $T_0$ and read at temperature $T$; further, we consider that the system is in thermal equilibrium and there is no thermal grade inside the media.

### 8.6.2 Thermal Effects of the CHDSS

Due to the temperature change, the bulk index of refraction is changed from $n_0$ to $M_n n_0$, and the dimension of the media along the $x$, $y$, and $z$ directions becomes $M_x$, $M_y$, and $M_z$ times as that before. The other important parameters after the temperature change are:

(1) Wavelength in the storage media

$$\lambda = \lambda_0 / M_n \tag{8.1}$$

(2) Focal length inside the media

$$f = M_n f_0 \tag{8.2}$$

(3) Distribution of the grating (modulation of the dielectric constant)

$$\Delta\varepsilon(x_f, y_f, z_f) = \Delta\varepsilon_0(x_f/M_x, y_f/M_y, z_f/M_z) \tag{8.3}$$

We adopt the optical model for a transmission type of CHDSS. In this case, when the temperature is changed, the complex amplitudes of the reference beam can be written as

$$P(x_f, y_f, z_f) = \frac{\exp\left(2jk_0 f_0 M_n{}^2\right)}{j\lambda_0 f_0} \exp(jk_0 M_n z_f) \times \Im\left\{ RP(x_0, y_0) \exp\left( -j\frac{\pi z_f}{M_n \lambda_0 {f_0}^2} (x_0^2 + y_0^2) \right) \right\}_{f_x = \frac{x_f}{\lambda_0 f_0}, f_y = \frac{y_f}{\lambda_0 f_0}} \tag{8.4}$$

where $k_0 = 2\pi/\lambda_0$. The modulation of the dielectric constant of the medium for a planar hologram at a distance of $z$ can be expressed as:

$$\Delta\varepsilon_0(x_f, y_f, z_f) = R^*(x_f, y_f, z_f) S(x_f, y_f, z_f) \tag{8.5}$$

where $R$ and $S$ are the complex amplitudes of the reference and signal beams, respectively, and can be written as:

$$R(x_f, y_f, z_f) = \frac{\exp(2jk_0 f_0)}{j\lambda_0 f_0} \exp(jk_0 z_f) \times \Im\left\{ RP_0(x_0, y_0) \exp\left( -j\frac{\pi z_f}{\lambda_0 {f_0}^2} (x_0^2 + y_0^2) \right) \right\}_{f_x = \frac{x_f}{\lambda_0 f_0}, f_y = \frac{y_f}{\lambda_0 f_0}} \tag{8.6}$$

$$S(x_f, y_f, z_f) = \frac{\exp(2jk_0 f_0)}{j\lambda_0 f_0} \exp(jk_0 z_f) \times \Im\left\{ DP_0(x_0, y_0) \exp\left( -j\frac{\pi z_f}{\lambda_0 {f_0}^2} (x_0^2 + y_0^2) \right) \right\}_{f_x = \frac{x_f}{\lambda_0 f_0}, f_y = \frac{y_f}{\lambda_0 f_0}} \tag{8.7}$$

Here, $RP$ and $DP$ stand for the reference pattern and data pattern uploaded on the SLM, respectively. With the help of Eqs. (8.5)–(8.7), Eq. (8.3) can be readily transformed to

$$\Delta\varepsilon(x_f, y_f, z_f) = \frac{1}{(\lambda_0 f_0)^2} \times \Im^*\left\{ RP_0(x_0, y_0) \exp\left( -j\frac{\pi z_f}{M_z \lambda {f_0}^2} (x_0^2 + y_0^2) \right) \right\}_{f_x = \frac{x_f}{M_x \lambda_0 f_0}, f_y = \frac{y_f}{M_y \lambda_0 f_0}} \times \Im\left\{ DP_0(x_0, y_0) \exp\left( -j\frac{\pi z_f}{M_z \lambda {f_0}^2} (x_0^2 + y_0^2) \right) \right\}_{f_x = \frac{x_f}{M_x \lambda_0 f_0}, f_y = \frac{y_f}{M_x \lambda_0 f_0}} \tag{8.8}$$

During the reconstruction process, when the RP are used for the reading of this planar hologram, the diffraction signal can be written as:

$$U(x_f, y_f, z_f) = P(x_f, y_f, z_f) \Delta\varepsilon(x_f, y_f, z_f) = \frac{\exp\left(2jk_0 f_0 M_n{}^2\right)}{j(\lambda_0 f_0)^3} \exp(jk_0 M_n z_f)$$

$$
\begin{aligned}
&\times \Im\left\{RP(x_0,y_0)\exp\left(-j\frac{\pi z_f}{M_n\lambda_0 {f_0}^2}(x_0^2+y_0^2)\right)\right\}_{f_x=\frac{x_f}{\lambda_0 f_0},f_y=\frac{y_f}{\lambda_0 f_0}}\\
&\times \Im^*\left\{RP_0(x_0,y_0)\exp\left(-j\frac{\pi z_f}{M_z\lambda_0 {f_0}^2}(x_0^2+y_0^2)\right)\right\}_{f_x=\frac{x_f}{M_x\lambda_0 f_0},f_y=\frac{y_f}{M_y\lambda_0 f_0}}\\
&\times \Im\left\{DP_0(x_0,y_0)\exp\left(-j\frac{\pi z_f}{M_z\lambda_0 {f_0}^2}(x_0^2+y_0^2)\right)\right\}_{f_x=\frac{x_f}{M_x\lambda_0 f_0},f_y=\frac{y_f}{M_y\lambda_0 f_0}}
\end{aligned}
\tag{8.9}
$$

The optical field on the detector plane for this diffraction field can be expressed as:

$$
\begin{aligned}
E_z(x_d,y_d,z_f) \propto\ & \frac{\exp(2jk_0 f_0)}{j\lambda_0 f_0}\exp(-jM_n k_0 z_f)\\
&\times \exp\left(j\pi z_f\frac{x_d^2+y_d^2}{M_n\lambda_0 {f_0}^2}\right)\mathcal{F}\{U(x_f,y_f,z_f)\}_{f_x=\frac{x_d}{\lambda_0 f_0},f_y=\frac{y_d}{\lambda_0 f_0}}
\end{aligned}
\tag{8.10}
$$

Substituting Eq. (8.9) into Eq. (8.10) and ignoring the constant term, one gets

$$
\begin{aligned}
E_z(x_d,y_d,z_f) \propto\ & \exp\left(j\pi z_f\frac{x_d^2+y_d^2}{M_n\lambda_0 {f_0}^2}\right)\\
&\times\left[RP(-x_d,-y_d)\exp\left(-j\frac{\pi z_f}{M_n\lambda_0 {f_0}^2}(x_d^2+y_d^2)\right)\right]\\
&\star\left[RP_0(-M_x x_d,-M_y y_d)\exp\left(-j\frac{\pi z_f}{M_z\lambda_0 {f_0}^2}(M_x^2x_d^2+M_y^2y_d^2)\right)\right]\\
&\otimes\left[DP_0(-M_x x_d,-M_y y_d)\exp\left(-j\frac{\pi z_f}{M_z\lambda_0 {f_0}^2}(M_x^2x_d^2+M_y^2y_d^2)\right)\right]
\end{aligned}
\tag{8.11}
$$

Then, we can obtain the diffraction wave from the whole volume of the hologram with a thickness $t$ by integral the above expression along the $z$ direction as:

$$
\begin{aligned}
E(x_d,y_d) \propto\ & \int_{-t/2}^{t/2}\exp\left(j\pi z_f\frac{x_d^2+y_d^2}{M_n\lambda_0 {f_0}^2}\right)\\
&\times\left[RP(-x_d,-y_d)\exp\left(-j\frac{\pi z_f}{M_n\lambda_0 {f_0}^2}(x_d^2+y_d^2)\right)\right]\\
&\star\left[RP_0(-M_x x_d,-M_y y_d)\exp\left(-j\frac{\pi z_f}{M_z\lambda_0 {f_0}^2}(M_x^2x_d^2+M_y^2y_d^2)\right)\right]\\
&\otimes\left[DP_0(-M_x x_d,-M_y y_d)\exp\left(-j\frac{\pi z_f}{M_z\lambda_0 {f_0}^2}(M_x^2x_d^2+M_y^2y_d^2)\right)\right]\mathrm{d}z_f
\end{aligned}
\tag{8.12}
$$

### 8.6.3 Temperature Compensation Condition for the CHDSS

For simplicity, we only consider the point spread function, set the signal to be a point source located at the center of the SLM, and express the data pages as:

$$DP(x_d, y_d) = \delta(x_d, y_d) \tag{8.13}$$

Then, the reconstructed optical field for this point source signal on the detector plane can be written as:

$$\begin{aligned} PSF(x_d, y_d) \propto & \iint RP(-\varepsilon, -\eta) RP_0{}^*(-M_x(\varepsilon - x_d), -M_y(\eta - y_d)) \\ & \times t\,\mathrm{sinc}\left\{\frac{t}{2}\left[\left(\frac{1}{M_n\lambda_0 f_0{}^2} + \frac{M_x^2}{M_z\lambda_0 f_0{}^2}\right)x_d^2 + \left(\frac{M_x^2}{M_z\lambda_0 f_0{}^2} - \frac{1}{M_n\lambda_0 f_0{}^2}\right)\varepsilon^2\right.\right. \\ & - \frac{2M_x^2 x_d\varepsilon}{M_z\lambda_0 f_0{}^2} + \left(\frac{1}{M_n\lambda_0 f_0{}^2} + \frac{M_y^2}{M_z\lambda_0 f_0{}^2}\right)y_d^2 + \left(\frac{M_y^2}{M_z\lambda_0 f_0{}^2} - \frac{1}{M_n\lambda_0 f_0{}^2}\right)\eta^2 \\ & \left.\left. - \frac{2M_y^2 y_d\eta}{M_z\lambda_0 f_0{}^2}\right]\right\} \end{aligned} \tag{8.14}$$

when $x_d = 0, y_d = 0$, where corresponding to the image place of the point source, the amplitude of this field is

$$\begin{aligned} PSF(0,0) \propto & \iint RP(-\varepsilon, -\eta) RP_0{}^*(-M_x\varepsilon, -M_y\eta) \\ & \times t\,\mathrm{sinc}\left\{\frac{t}{2}\left[\left(\frac{M_x^2}{M_z\lambda_0 f_0{}^2} - \frac{1}{M_n\lambda_0 f_0{}^2}\right)\varepsilon^2\right.\right. \\ & \left.\left. + \left(\frac{M_y^2}{M_z\lambda_0 f_0{}^2} - \frac{1}{M_n\lambda_0 f_0{}^2}\right)\eta^2\right]\right\} \mathrm{d}\varepsilon\mathrm{d}\eta \end{aligned} \tag{8.15}$$

From Eq. (8.15), we can conclude that, when the temperature is changed, to keep the amplitude the same as before, the following two equations must be fulfilled:

$$RP(-\varepsilon, -\eta) = RP_0(-M_x\varepsilon, -M_y\eta) \tag{8.16}$$

$$\begin{cases} \dfrac{M_x^2}{M_z\lambda_0 f_0{}^2} - \dfrac{1}{M_n\lambda_0 f_0{}^2} = 0 \\ \dfrac{M_y^2}{M_z\lambda_0 f_0{}^2} - \dfrac{1}{M_n\lambda_0 f_0{}^2} = 0 \end{cases} \tag{8.17}$$

Since the pixels of SLM have a relatively large size, Eq. (8.16) can be well satisfied when the RP do not change. Equation (8.17) can be reduced to

$$\begin{cases} M_z = M_x^2 M_n \\ M_x = M_y \end{cases} \tag{8.18}$$

which impose requirement for the index and dimension change. The physical significant of Eq. (8.18) can interpreted by using the k-sphere method shown in Figure 8.17.

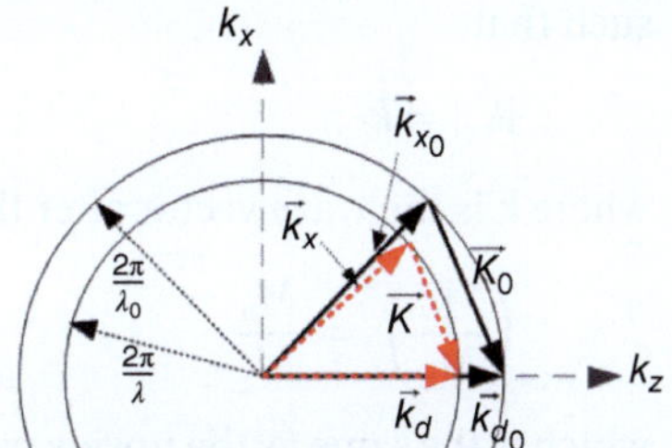


**Figure 8.17** $k$-space analysis for the temperature change. Source: Jianhua Li.

As shown in Figure 8.17, two wave vectors $\vec{k}_{x_0}$ and $\vec{k}_{d_0}$, which correspond to two pixels (one used as data and one as reference) on SLM with the coordinates of $(x_0, 0)$ and $(0, 0)$, form a grating vector $\vec{K}_0$, and this vector can be written as:

$$\vec{K}_0 = \vec{k}_{d_0} - \vec{k}_{x_0} \tag{8.19}$$

where

$$\vec{k}_{x_0} = k_0\left(\frac{x_0}{f_0}, 0, \sqrt{1-\left(\frac{x_0}{f_0}\right)^2}\right) \approx k_0\left(\frac{x_0}{f_0}, 0, 1-\frac{1}{2}\left(\frac{x_0}{f_0}\right)^2\right) \tag{8.20}$$

$$\vec{k}_{d_0} = k_0(0, 0, 1) \tag{8.21}$$

The substitution of Eqs. (8.20) and (8.21) into Eq. (8.19) gives

$$\vec{K}_0 = k_0\left(-\frac{x_0}{f_0}, 0, \frac{1}{2}\left(\frac{x_0}{f_0}\right)^2\right) \tag{8.22}$$

After the temperature change, the media expands along the $x$ and $z$ directions, and the grating vector $\vec{K}_0$ becomes

$$\vec{K} = k_0\left(-\frac{x_0}{f_0}\frac{1}{M_x}, 0, \frac{1}{2}\left(\frac{x_0}{f_0}\right)^2\frac{1}{M_z}\right) \tag{8.23}$$

To ensure the diffraction wave propagates in the same direction, the wave vector for the diffraction wave can be set as:

$$\vec{k}_d = M_n k_0(0, 0, 1) \tag{8.24}$$

Then, the wave vector of the reading beam, represented as $\vec{k}_x$, should be set as:

$$\vec{k}_x = \vec{k}_d - \vec{K} \tag{8.25}$$

to fulfill the Bragg condition. Equations (8.23) and (8.24) show that

$$\vec{k}_x = M_n k_0\left(\frac{x_0}{M_n f_0}\frac{1}{M_x}, 0, 1-\frac{1}{2}\left(\frac{x_0}{f_0}\right)^2\frac{1}{M_n M_z}\right) = k\left(\frac{x_0}{f}\frac{1}{M_x}, 0, 1-\frac{1}{2}\left(\frac{x_0}{f}\right)^2\frac{M_n}{M_z}\right) \tag{8.26}$$

such that

$$|\vec{k}_x| = k \tag{8.27}$$

where k is the wave vector after the temperature changes. Then, we have

$$\left(\frac{1}{M_x}\right)^2 = \frac{M_n}{M_z} \tag{8.28}$$

which is the same as the upside part of Eq. (8.18). By using the same method, we can derive the full part of Eq. (8.18). From Eq. (8.26), it is evident that this vector corresponds to the coordinate $(x_0/M_x, 0)$ on the SLM plane, in agreement with Eq. (8.16). To sum up, the two compensation conditions are such that the Bragg conditions are also fulfilled after the temperature changes. The substitution of Eqs. (8.16) and (8.17) into Eq. (8.12) yields the diffracted field after compensation, which can be written as

$$\begin{aligned} E(x_d, y_d) \propto \int_{-t/2}^{t/2} & \exp\left(j\pi z_f \frac{x_d^2 + y_d^2}{M_n \lambda_0 f_0{}^2}\right) \\ & \times \left[RP_0(-M_x x_d, -M_y y_d) \exp\left(-j\frac{\pi z_f}{M_n \lambda_0 f_0{}^2}(x_d^2 + y_d^2)\right)\right] \\ & \star \left[RP_0(-M_x x_d, -M_y y_d) \exp\left(-j\frac{\pi z_f}{M_n \lambda_0 f_0{}^2}(x_d^2 + y_d^2)\right)\right] \\ & \otimes \left[DP_0(-M_x x_d, -M_y y_d) \exp\left(-j\frac{\pi z_f}{M_n \lambda_0 f_0{}^2}(x_d^2 + y_d^2)\right)\right] \mathrm{d}z_f \end{aligned} \tag{8.29}$$

It should be noted that Eq. (8.29) is only valid for a transmit-type hologram (for a reflective-type hologram, the reader can refer to [8]).

### 8.6.4 Summary

A temperature compensation condition for the CHDSS is derived based on Born's approximation and the scalar diffraction theory. For temperature compensation, the coefficients of thermal expansion and the index change must fulfill a special relation, which imposes new limitation on the fabrication of the media. However, the result derived here is only valid for transmit-type holograms; reflective-type holograms are more sensitive to temperature variations, which may be harder to compensate.

## 8.7 Noise Margin

In an HDS, the quality of the reconstructed data pattern is decisive and directly affects the system performance. Although the HDS technology can achieve a high performance in theory, in practice, the performance is always deteriorated by the noise from optical systems, materials, and electrical components. Therefore, research on improving the SNR of HDS systems is widely conducted. However, despite the improvement in the SNR, noise still exists and is non-negligible.

Since an HDS requires the lowest possible error rate, the efforts are focused on the higher noise margin of the HDS system. It is obvious that the data error will be reduced if the decoding method has a better anti-noise ability [9].

### 8.7.1 Noise Margin Improvement in Binary Modulation HDS

In most cases, the encoding and decoding methods supplement each other. The 8:16 two-dimensional constant-weight sparse modulation code, which is also named 3/16 code, is popular in binary modulation owing to the balance between the code rate and raw BER [10]. It utilizes 16 pixels as a data block and picks 3 pixels as modulation pixels. Different combinations of the positions of the modulation pixels represent different data information. Partial data blocks of 8:16 two-dimensional constant-weight sparse modulation code are shown in Figure 8.18.

For this encoding method, a simple and popular decoded method is the threshold determination decoding. An intensity threshold value is settled to judge the modulation pixels; the pixels whose intensity is greater than the threshold value are considered as the modulation pixels. This decoding method is simple to carry out but not accurate enough. When the data page is polluted by noise, the threshold determination decoding easily results in misjudgment. A decrease in the intensity of the modulation pixels and an increase in that of the non-modulation pixels caused by noise might lead to wrong extra modulation pixel judgment and wrong missing modulation pixels; an example is shown in Figure 8.19.

To avoid this situation, the threshold determination decoding is improved by considering only three brightest pixels of the data block as the modulation pixels. Utilizing the three brightest pixels as the modulation pixels can realize a much higher noise margin compared to pure threshold determination [11]. The decoding error will not occur if the noise level is not larger than the intensity gap between the bright pixel and the dark pixel. However, this decoding method has its own defect; it is suitable for binary modulation but not multilevel modulation, which is considered as the future of this technology owing to its high code rate. Thus, this limitation of the decoding method hinders the practical applications of this technology.

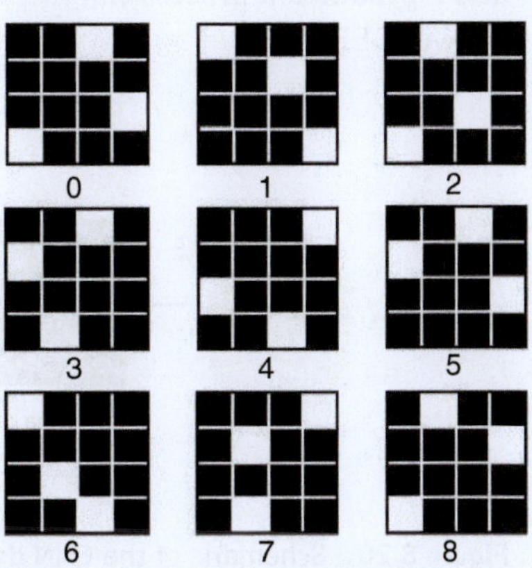


**Figure 8.18** Partial data blocks of the two-dimensional constant-weight sparse modulation code. Source: Xiao Lin.

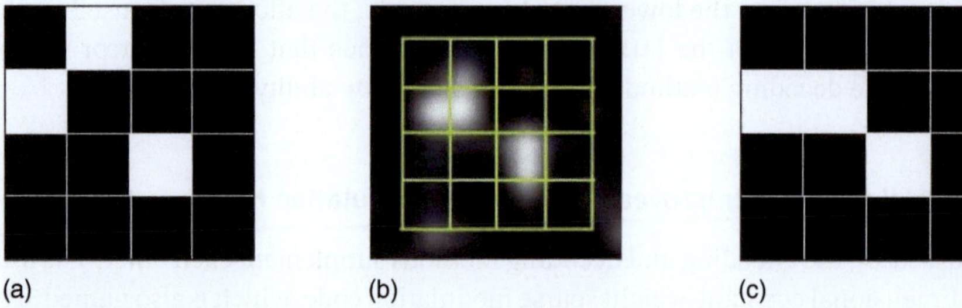


**Figure 8.19** Wrong judgment of data block by threshold determination decoding. (a) Correct data block, (b) image captured in the HDS system, and (c) wrong data block after threshold determination decoding. Source: Xiao Lin.

## 8.7.2 Noise Margin Improvement in Multilevel Modulation HDS

### 8.7.2.1 Convolutional Neural Network Decoding

To overcome the multilevel modulation decoding drawback, a data decoding method using a convolutional neural network (CNN) has been studied. The CNN learns a decoding rule based on the optical properties of the HDS, and the decoding errors decrease significantly without requiring any techniques to improve the SNR [12, 13]. The CNN is applied for data decoding, and the whole processing can be considered as image classification. The training of CNN must consider as many kinds of noise as possible, such as aberration of lens, dust pollution, resample mismatch, and so on, to achieve the highest possible noise margin. The schematic of the CNN decoding in HDS is shown in Figure 8.20.

In CNN decoding, the recording and reproduction noise are superimposed onto the reproduced data pages, and the corresponding original data streams are used for the training. Modulation blocks in the data pages are repeatedly input into the CNN for learning the rule of decoding, taking into consideration the noise. After training, the CNN accurately classifies each type of modulation block into the corresponding data block. Compared with the threshold determination decoding method, a CNN that learns the decoding rules based on the optical characteristics in the recording and reproduction process can realize decoding with noise-added modulation data at a lower BER.

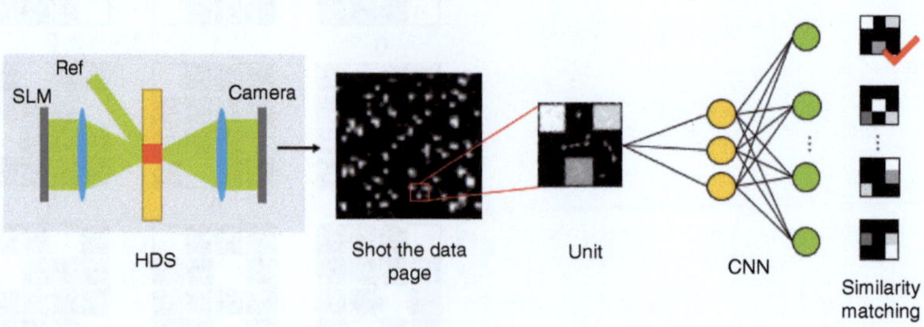


**Figure 8.20** Schematic of the CNN decoding in HDS. Source: Xiao Lin.

#### 8.7.2.2 Compressed Sensing Decoding

In addition to CNN decoding, there is another decoding method, called compressed sensing (CS) decoding, for improving the noise margin of HDS systems [14]. This decoding method developed from the CS technology is proposed to reduce the impact of noise in the decoding process. Compared with the threshold determination decoding method, the method is more robust to noise and more suitable for multilevel modulation. CS goes against the conventional wisdom in data acquisition, showing that signals can be recovered from fewer samples or measurements than those required by the Nyquist–Shannon sampling theorem for a sparse signal. The sparsity requirement results in noise depression, which is used by the CS decoding method to achieve a high noise margin. In the CS decoding method, the noise depression levels are determined by the choice of the sparse domain. Utilizing the sparse dictionary training method and setting the data symbols as the training set, a more targeted and suitable sparse domain of the data page can be found. The training flow is shown in Figure 8.21.

After finishing the sparse dictionary training, the reconstructed data page of the HDS in the reading process can be decoded using the CS technology with the trained sparse dictionary and orthogonal matching pursuit retrieval algorithm. By searching

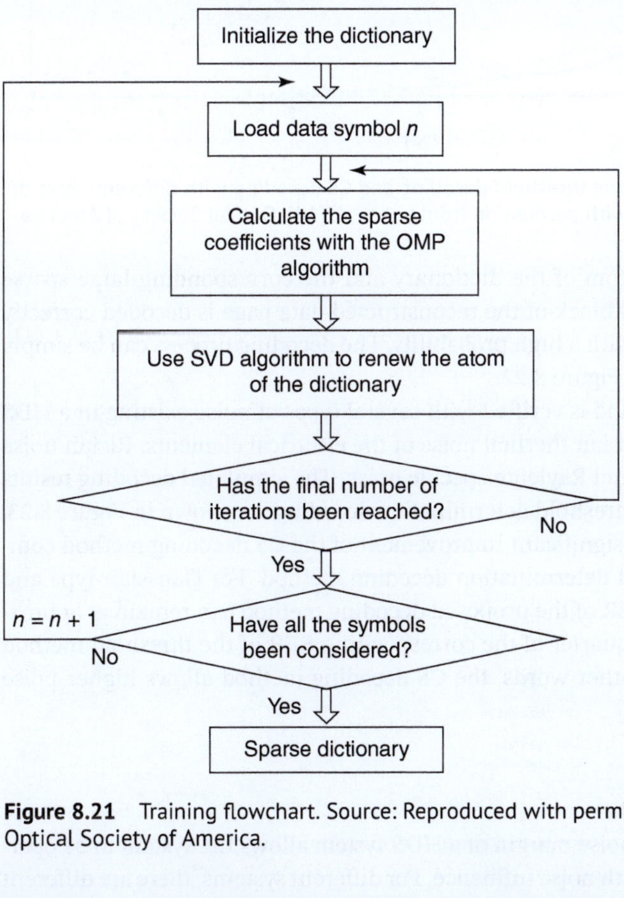


**Figure 8.21** Training flowchart. Source: Reproduced with permission from Liu et al. [14], Optical Society of America.

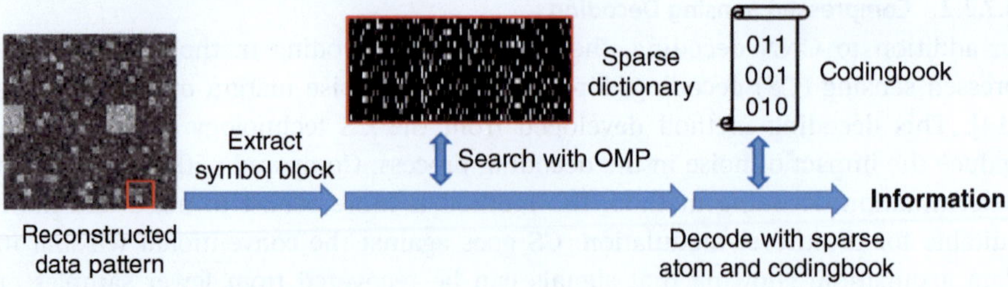


**Figure 8.22** CS decoding. Source: Liu et al. [14], Optical Society of America.

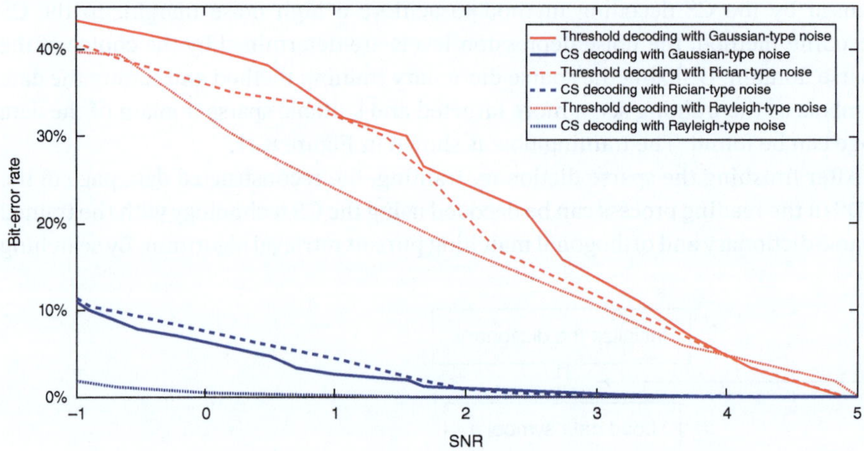


**Figure 8.23** BER curve of the threshold decoding and CS decoding with different types of noise. Source: Reproduced with permission from Liu et al. [14], Optical Society of America.

the most likely sparse atom of the dictionary and the corresponding large sparse coefficient, every symbol block of the reconstructed data page is decoded correctly to the data information with a high probability. The decoding process can be simply represented as shown in Figure 8.22.

The CS decoding method is verified with several types of noise existing in a HDS system, such as the Gaussian thermal noise of the electrical elements, Rician noise of the optical elements, and Rayleigh speckle noise. The simulated decoding results of the CS decoding and threshold determination decoding are shown in Figure 8.23.

Figure 8.23 shows the significant improvement of the CS decoding method compared with the threshold determination decoding method. For Gaussian-type and Rician-type noise, the BER of the proposed decoding method can remain at approximately 10%, which is a quarter of the corresponding BER of the threshold method with an SNR of −1. In other words, the CS decoding method allows higher noise margin in HDS system.

### 8.7.3 Summary

In conclusion, the high noise margin of a HDS system allows the system to be operated at a low error rate with noise influence. For different systems, there are different ways to achieve higher noise margins. Picking three brightest pixels as the modulation pixels in a binary system is a simple and an efficient method to improve the noise margin, but it is not suitable for multilevel modulation. By applying information processing technology, the CNN and CS decoding methods can achieve a high noise margin for multilevel HDS systems. Further, the current discussion on improving the noise margin of HDS systems is limited to amplitude modulation. The phase and polarization modulations need to be evaluated as well.

## References


1 Hopkins, H.H. (1979). Diffraction theory of laser read-out systems for optical video discs. *J. Opt. Soc. Am.* 69: 4–24.

2 米澤 実 (2005). システムの機構. サーホ技術. 東芝レヒュ゛ー, 60(1): 21–24.

3 Yanagisawa, T. and Nomoto, T. (2000). Reproducing margin simulation of next generation optical disk system. *PIONEER R&D* 12 (2): 3–12.

4 Yokogawa, F., Miyanabe, S., Ogasawara, M. et al. (2000). Read/write technology for 50 GB/layer at 100 Mbps. *Jpn. J. Appl. Phys.* 39: 819.

5 Li, J., Cao, L., Gu, H., Tan, X. et al. (2010). Wavelength and defocus margins of the collinear holographic storage system. In: *Proc. SPIE 7851, Information Optics and Optical Data Storage*, 785115.

6 Horimai, H. and Tan, X. (2006). Read-only holographic versatile disc system using laser diode. In: *International Conference on Holography, Optical Recording and Processing.*

7 Tanaka, T. Takahashi, K. Watanabe, K. et al. (2004). *Technical Digest of Optical Data Storage Topical Meeting 2004*, pp. 311-313, Monterey, California, USA.

8 Yu, Y.-W. (2009). The characteristic study, improvement and temperature compensation of the collinear holographic data storage system. PhD thesis. Taiwan: National Central University (In Chinese).

9 Hwang, E., Kim, K., Kim, J. et al. (2002). A new efficient error correctible modulation code for holographic data storage. *Jpn. J. Appl. Phys.* 41 (3S): 1763.

10 Gu, H., Cao, L., He, Q., and Jin, G. (2011). A two-dimensional constant-weight sparse modulation code for volume holographic data storage. *J. Zhejiang Univ. Sci. C* 12 (5): 430–435.

11 Heanue, J.F., Bashaw, M.C., and Hesselink, L. (1995). Channel codes for digital holographic data storage. *JOSA A* 12 (11): 2432–2439.

12 Katano, Y., Muroi, T., Kinoshita, N. et al. (2018). Data demodulation using convolutional neural networks for holographic data storage. *Jpn. J. Appl. Phys.* 57 (9S1): 09SC01.

13 Katano Y, Muroi T, Kinoshita N and Ishii N. (2018). Demodulation of multi-level data using convolutional neural network in holographic Ddata storage. *2018 Digital Image Computing: Techniques and Applications (DICTA)*. IEEE, pp. 1–5.

14 Liu, J., Zhang, L., Wu, A. et al. (2020). High noise margin decoding of holographic data page based on compressed sensing. *Opt. Express* 28 (5): 7139–7151.

# 9
# Applications

## 9.1 High-Speed Optical Correlation System Based on Collinear Holography

### 9.1.1 Introduction

Optical correlators capable of performing high-speed correlation calculations have been studied since the 1960s [1, 2], and can be broadly classified into Vander Lugt correlators (VLCs) [1] and joint transform correlators (JTCs) [2]. The VLC is an optical implementation of matched filtering and was proposed by Vander Lugt in 1964. This method uses a Fourier transform hologram optical system to create a spatial frequency filter with a complex amplitude distribution and then use it for correlation. VLC-based methods can be broadly classified into two types: methods that use spatial light modulators (SLMs) for filters [3–10] and those that use an analog hologram for filters [11–23]. Because filters are easier to rewrite, the SLM-based filters used in VLCs can be flexibly operated; for example, the compact and versatile optical correlator developed by Boulder Nonlinear Systems, Inc. [24]. Furthermore, an optical correlator containing only a single SLM was proposed by introducing a lens function in the SLM, followed by numerous other studies focused on the miniaturization of these devices [5], and in recent years, the Cambridge correlator [25] has been used in many practical applications. However, as the matched filter is saved in a computer and transferred to the SLM whenever required, the correlation speed is limited to the image display frame rate of the SLM and the rate at which the data is transferred to the SLM. From a practical viewpoint based on an enormous database, the high speed of light in optical correlation has not been fully utilized. In contrast, the latter method, which uses an analog hologram as the filter, allows recording in advance in the form of an analog hologram. Moreover, the associated machinery restrictions are fewer than those of the SLM filter-based method, and high-speed reading of the database is possible. Apart from these two methods, several other methods, including crystal-based hologram recording [11, 12, 14–17] and methods based on disk-type photopolymer holographic storage [13, 18–23], are also widely used. In principle, compared with the former method in which the switching speed of the filter depends on the transfer rate of the SLM, these methods can realize optical correlation computation at a higher speed.

*Collinear Holography: Devices, Materials, Data Storage*, First Edition.
Xiaodi Tan, Hideyoshi Horimai, Tsutomu Shimura, and Xiao Lin.

In this chapter, we describe an optical correlation system, in which the matched filter of a VLC is replaced with a volume analog hologram based on collinear holography. In Section 9.2, the concept of an optical correlator based on collinear holography is described. Owing to the collinear holography, the optical system can be easily implemented, and the stability of the system against wavelength drift, disturbance, and tilt of the disk could be improved. An optical correlation experiment demonstrating the possibility of high-speed optical correlation with a large dataset is also presented in Section 9.2. In Section 9.3, the basic configuration of an optical correlator system based on collinear holography is described. Further, the development of an autoencoder-based conversion module for face recognition and image retrieval as well as cross-domain image retrieval is elucidated. The optical correlator is particularly effective when used with a large-capacity database that exceeds the physical storage capacity of primary storage devices currently used in electronic systems.

### 9.1.2 Optical Correlator Based on Collinear Holography

#### 9.1.2.1 Basic Concept of Optical Correlator

The working of a holographic optical correlation calculation system based on a volume hologram can be divided into two processes: hologram recording, which is a database of correlation calculations, and optical correlation calculations. During the hologram recording process, an image for database recording, in which information and reference patterns are arranged in the same plane at the front focal point of the lens, is displayed on the SLM and irradiated with a light beam parallel to the SLM. As a result, the information and reference lights are Fourier transformed by the lens, and their interference fringes are recorded on the holographic medium. This hologram serves as a filter in the optical correlation calculation. During the optical correlation calculation process, only the information light is irradiated on the holographic medium as the correlation input. Because the intensity of the diffracted light from the hologram changes based on the similarity between the data recorded on the hologram and the input data, the light intensity is detected as a correlation signal. Figure 9.1 shows a schematic of the collinear holographic optical correlation system.

The holographic records of the interference fringes between the information and reference lights are formed during the recording, implying that the holographic data storage and optical correlation systems are similar to each other. However, a significant difference exists as well between these two systems, that is, an optical correlation system is only necessary when detecting the total light intensity of the reference light, whereas a holographic data storage system obtains an image of the information light.

#### 9.1.2.2 Optical Setup

The optical system in a collinear holography system is simple and easy to align because of the coaxiality of the reference and information lights. The system is highly stable against wavelength drift of the light source, disturbances (vibration), and inclination of the disk. Moreover, a facile diversion and active use of the servo-control mechanism of conventional optical disks, such as CD and DVD, can

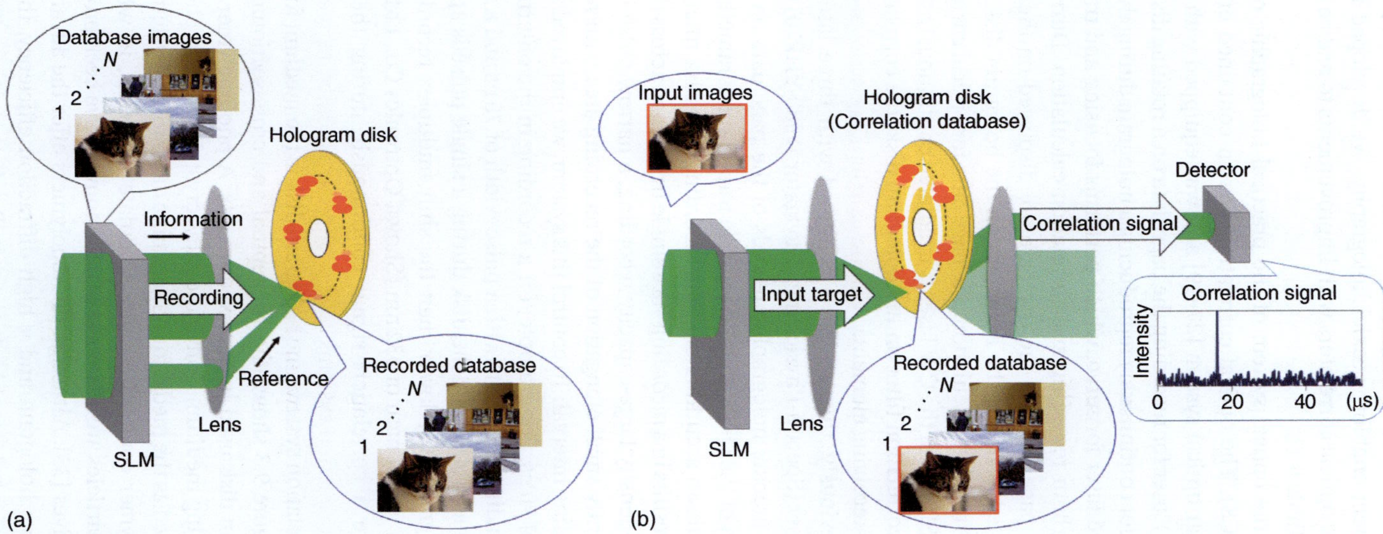


**Figure 9.1** Conceptual illustration of an optical correlator based on collinear holography. (a) Optical correlation filter recording, and (b) optical correlation process. Source: Kanami Ikeda.

be implemented [26, 27]. In this study, we generated a hologram, to be used as a correlation filter, on a disk-type holographic medium via shift multiplex recording using an optical system enabling coaxial holography. We developed four types of collinear holographic optical correlators with improvements to realize a high-speed optical correlation (Table 9.1).

Figure 9.2 shows the optical system of the practical holographic optical correlation system (PHOCS). The optical correlation system contained only one SLM based on the collinear optical system [27–29] and was equipped with three lasers: a Q-switched (Q-SW) laser for recording the page data on a rotating disk, a continuous wave (CW) laser for continuously reproducing a hologram during the correlation calculation, and a red laser for servo, which performs focusing and tracking of the rotating disk during both recording and correlation calculation. During the hologram recording, the page data for the database are displayed on the SLM, and a pulse laser is irradiated to record the hologram on the hologram disk. In contrast, during the optical correlation calculation, the page data for optical correlation input are displayed on the SLM, and the CW laser is irradiated continuously. After passing through the optical correlation filter on the hologram disk, the correlation signal is detected by the light-receiving element.

As mentioned previously, this system is equipped with three lasers. First, we employed a Q-SW laser (Spectra-Physics Inc., Navigator II J80-YHP70) with a wavelength of 532 nm to prepare holographic records of the page data on the rotating disk. The pulse width of the Q-SW pulse laser is an important parameter and should be controlled to enable an accurate hologram recording on a disk rotating at a high speed. A wide pulse results in a recording via a single irradiation across the wide disk, and thus, for a high-density large-capacity recording, a narrow pulse is imperative along with a sufficiently small elongation of the recording area with respect to the shift multiplex recording interval. To control this system, we employed a Q-SW laser with a pulse width of 70 ns. For example, for a recording in the outermost circumference of the disk (radius: 55.0 mm) with a pulse width of 70 ns and a disk rotation of 900 rpm, the distance rotated by the disk during a single pulse is approximately 0.36 μm, which is significantly lower than the shift multiplex recording interval. The CW laser with a wavelength of 532 nm (Showa Optronics Co., Ltd., H6780-01) was used to acquire a correlation signal continuously during the correlation calculation.

In this optical correlation system, the hologram disk is the medium for storing the correlation filter. Figure 9.3 shows the conceptual and cross-sectional schematics of the main hologram disk used in our experiments. A photopolymer was used as the hologram recording medium. Conventionally, crystals [30] and photopolymers [31, 32] have been used as the media for holographic data storage systems. However, in recent years, polymer composite nanoparticulate materials, in which organic and inorganic nanoparticles are dispersed in a monomer, have been proposed as the other viable alternatives [33]. Although photopolymers offer the advantage of easy recording of a volume hologram and a high diffraction efficiency, their practical application as media for holographic storage is still a distant dream. The suitability of a medium for holographic data storage is governed by the requirements of a high

**Table 9.1** Optical correlators based on collinear holography.

| | FRACO 2.0 [13] | FARCO 3.0 | R-PHOCS | W-PHCOS |
|---|---|---|---|---|
| **Year** | **2007** | **2008** | **2010** | **2012~** |
| Photograph |  | 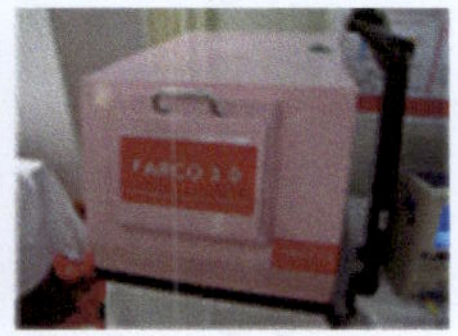 |  | 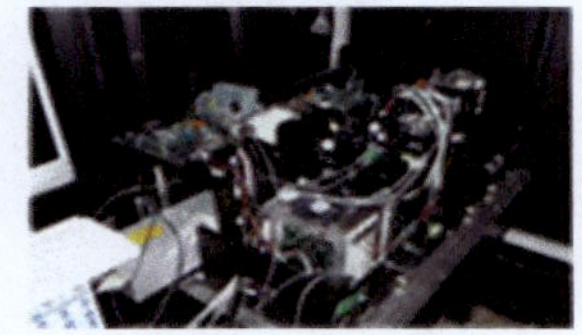 |
| Feature | World's first system of the optical correlator based on collinear holography using disk type hologram | For evaluation of the hologram materials | Compact only for correlation process | High speed system (more than 100 Gbps was achieved) |
| Light source (for hologram recording) | Q-SW laser | — | — | Q-SW laser (5 W) |
| Light source (for correlation) | CW laser (0.2 W) | CW laser (0.2 W) | CW laser (10 mW) | CW laser (10 mW) |
| SLM | DMD (768 × 1024 pixel) | DMD (768 × 1024 pixel) pixel size: 13.7 mm | FLC-SLM (1218 × 1218 pixel, pixel size: 10.7 μm) | FLC-SLM (1218 × 1218 pixel, pixel size: 10.7 μm) |
| Disk rotation speed | 60~300 rpm | 300~1500 rpm without servo control | ~1200 rpm | ~1200 rpm |
| Correlation speed | 16 Gbps | — | A few Gbps~150 Gbps | A few Gbps~150 Gbps |

Source: Kanami Ikeda.

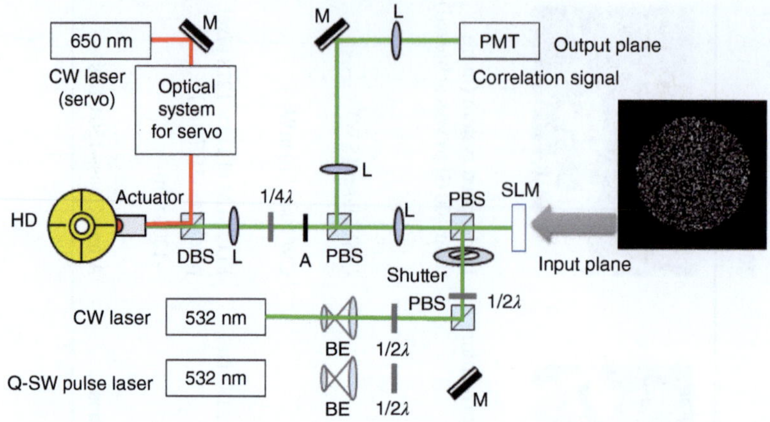


**Figure 9.2** Optical configuration of PHOCS, M, mirror; A, aperture; BE, beam expander, PBS, polarizing beam splitter; DBS, dichroic beam splitter; HD, holographic disk; SLM, spatial light modulator; PMT, photomultiplier tube. Source: Kanami Ikeda.

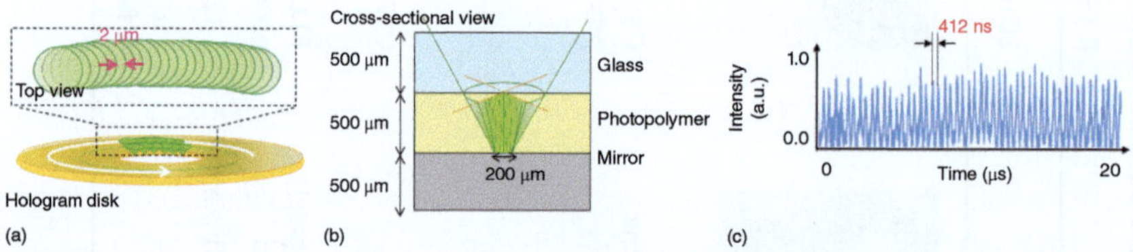


**Figure 9.3** (a) Conceptual illustration of shift-multiplex recording, (b) cross-sectional view of a spot on the hologram disk, and (c) obtained 100 auto-correlation signals in the high-speed optical correlation experiment. Source: Kanami Ikeda.

recording sensitivity, high refractive index modulation, low volume shrinkage, and good preservability. Currently, various research groups are working on identifying the essential requirements for materials to be used as the media for holographic storage [34]. The photopolymer used in our study exhibited a strong sensitivity to green light, and thus, we used a laser with a wavelength of 532 nm for the hologram recording and reproduction. However, as the photopolymer was insensitive to red light, a red laser with a wavelength of 650 nm was used for the servo modulation without producing any reaction with the recording material. This red laser allowed a high-precision recording and correlation while enabling focusing and tracking of the rotating disk during the recording and the correlation calculation. Using this servo laser, we could detect the eccentricity and surface blurring of the rotating disk, and a focused state was maintained. Moreover, as this system adopts a collinear holography optical system, changing the structure of the disk substrate allowed both focusing and tracking control. Furthermore, as mentioned earlier, while the holographic data storage system used a two-dimensional (2D) light-receiving element to read the recorded data as an image, only the integrated light intensity on the output surface needed to be detected without resolving the image itself in the holographic optical correlation system. Therefore, the holographic optical correlation system used only a single-pixel light-receiving element. The light-intensity detection speed of this element was higher than that of the 2D image-capturing element. This is also one of the reasons why the holographic optical correlation system could execute correlation calculations at high speeds.

#### 9.1.2.3 Correlation Speed

Figure 9.3a,b illustrate the shift multiplex recording and cross-section of a spot on the hologram disk, respectively. In the database recording process using the shift multiplex recording method, the recording spots overlap, as shown in Figure 9.3a, because one spot is recorded in terms of volume, as depicted in Figure 9.3b. The data transfer and correlation speed (in units of bits per second, i.e. bps) can be expressed as follows [19]:

$$V = C \cdot \frac{2\pi r}{\delta_x} \cdot \frac{R}{60} \tag{9.1}$$

where $C$ (bit), $R$ (rpm), $d$ (mm), and $r$ (mm) denote the data size of the page data, disk rotation speed, shift pitch of the shift multiplex recording, and diameter of the position on the disk, respectively. To determine the correlation with large-scale data in a general computer, first, data transfer is required from the secondary storage device, such as an HDD, to a primary storage memory, such as RAM, before the calculation. However, in the optical correlator, the database was recorded on the holographic disk, and the correlation calculation could be performed only by inputting the data without any data transfer. Therefore, we defined the speed of the optical correlator as the data transfer and correlation speed, which depends on the circumferential shift pitch of the shift multiplex recording and the rotational speed of the holographic optical disk. Figure 9.3c shows the optical correlation signal with a data size $C$ of $280 \times 210$ pixels, radius of the recording position on the disk ($r$) of 55 mm, disk rotation speed ($R$) of 900 rpm, and shift pitch of the shift multiplex recording ($dx$) of 2 μm. The interval of the autocorrelation signals was 0.4 μs, i.e. the page data correlation speed was $2.4 \times 10^6$ fps. Thus, a high-speed optical correlation was demonstrated with 142.9 Gbps of data transfer and optical computation speed [20].

### 9.1.3 Applications of the Optical Correlator Based on Collinear Holography

In this section, the basic configuration of the optical correlator system based on collinear holography is described. Based on this basic configuration, the face recognition, image retrieval, and cross-domain image retrieval applications of the optical correlator are presented with the corresponding feature extractors and conversion modules.

#### 9.1.3.1 Basic Configuration

Figure 9.4 shows the basic configuration of the optical correlator system based on collinear holography, comprising a feature extractor, a conversion module for the optical correlator, and an optical correlator. When using the optical correlator, the input data features are first extracted and later optimized for each application.

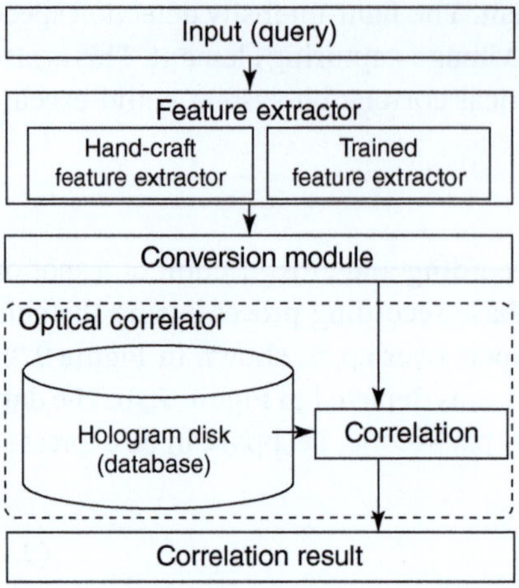


**Figure 9.4** Basic configuration of optical correlator system based on collinear holography. Source: Reproduced with permission from [23]/OSA.

Subsequently, each feature is converted to page data using a conversion module. Finally, the page data are sent as the input to the optical correlator, and the correlation result is obtained.

#### 9.1.3.2 Face Recognition Application

Based on the basic configuration, here we demonstrate face recognition using the optical correlator. Figure 9.5 shows the feature extractor and conversion module used for face recognition. In this process, first, the input data are captured using a commercial camera, and a pair of eyes is extracted. The face is then cropped, and the size of the face is normalized to $128 \times 128$ pixels. The colored face image is converted to grayscale and edge-enhanced using a Sobel filter. The edge-enhanced image is finally binarized via thresholding such that the number of white pixels is 20% of the total number of pixels. The optical energy of the input data is normalized during the binarization process.

We performed a correlation experiment using face images, which were feature extracted and converted for the optical-correlator-assisted face recognition. The intensities of the correlation peaks were compared with the threshold for verification. Figure 9.6 shows the dependences of the recognition error rates on the threshold; here, (a) and (b) represent the false-acceptance and false-rejection rates,

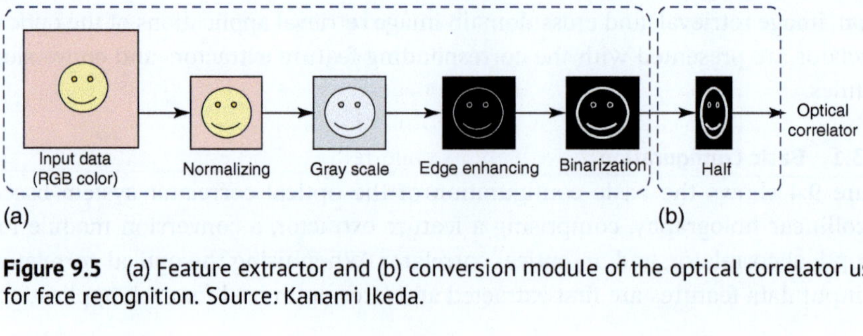


**Figure 9.5** (a) Feature extractor and (b) conversion module of the optical correlator used for face recognition. Source: Kanami Ikeda.

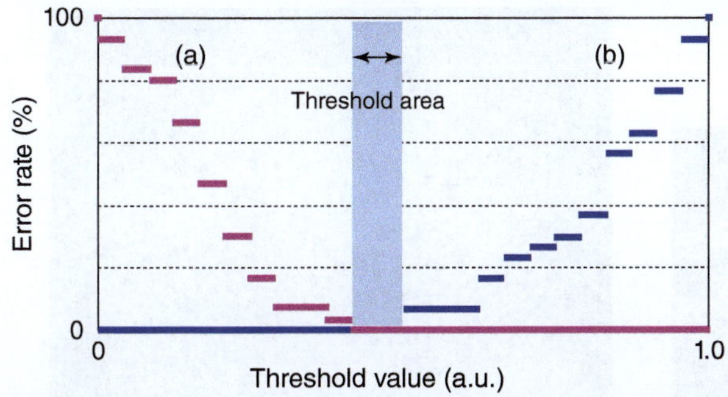


**Figure 9.6** Dependence of face recognition error rate on the threshold in the experiment, (a) false-acceptance rate, and (b) false-rejection rate. Source: Kanami Ikeda.

respectively, and the intersection of the lines represents the equal error rate (EER). An EER of 0% was achieved in this experiment.

#### 9.1.3.3 Image Retrieval Function

In this section, we describe the image retrieval function of an optical correlator system comprising an edge extraction-based feature extractor. Grayscaling, edge extraction, and binarization were performed on the input color image as the pre-processing steps for face recognition as described in the previous section; the same feature extractor was used for the image retrieval. However, for this application, the intensity distribution on the Fourier plane exhibits a greater difference depending on the input image, compared to that between the face images; this is a problem in high-density database recording. Therefore, a random address-rearrangement was introduced as a conversion step for the optical correlator. Figure 9.7a shows an example of the converted data for the optical correlator used in this application. The extracted feature was divided into multiple blocks, each containing $2 \times 2$ pixels, which were then randomly address-rearranged.

Because of the random address-rearrangement during the conversion of the input image, the intensity distribution on the Fourier plane is uniform between any type of input image. We performed an optical correlation experiment using the feature extraction and conversion algorithm with the random address-rearrangement, a shift multiplex recording pitch of 2.45 μm, a disk rotation speed of 900 rpm, and a disk radius 48.0 mm. Figures 9.7b shows an example of the recorded input page data used in the experiment. Figure 9.8 shows the correlation signal when data #30 are sent as the input to the database, in which data #1 to #100 are shift-multiplex recorded. A higher correlation is obtained when the database and input data are matched. Thus, we confirmed that image retrieval can be performed using thresholding for the obtained time-series correlation signal.

Based on the optical-correlator-based image retrieval system, we developed an optical–digital hybrid image searching system [22] consisting of a web server, a database server, a digital correlation server, a control server for the optical correlator, an analog-to-digital (AD) converter, and the optical correlator as shown in Figure 9.9a. Evidently, the network design is distributed, and the input and output

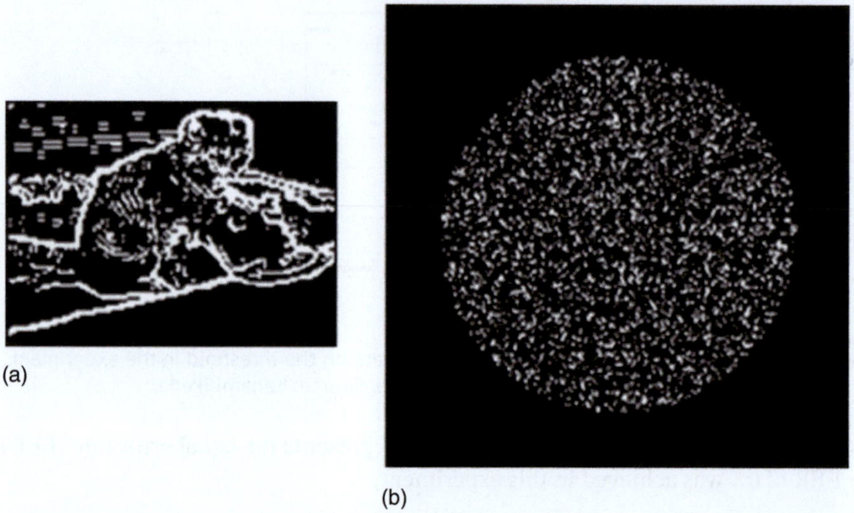


**Figure 9.7** Conversion module for an optical correlator applied in image retrieval. The extracted feature is divided into multiple blocks (each block consists of $2 \times 2$ pixels), and the blocks are then randomly address-rearranged. (a) the converted data, and (b) an example of the recorded input page data. Source: Kanami Ikeda.

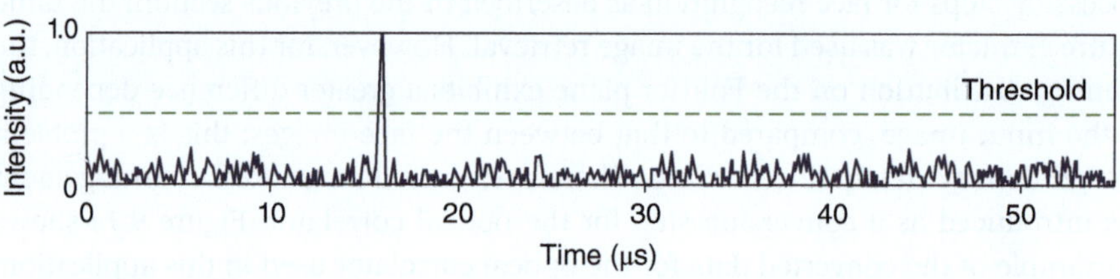


**Figure 9.8** Experimentally obtained correlation signal in the image retrieval system. Source: Kanami Ikeda.

data formats of the optical correlation system are the same as those of the digital correlation server. Therefore, the optical correlation system could be completely replaced or combined with a digital correlation server. Figure 9.9b,c show the interface designs of the system on a web browser; Figure 9.9b shows the page for setting the correlation input. We could select the hardware (either optical or digital), set the query data from the local folders, and send a request to start the correlation calculation. After the correlation calculation, the page containing the input image, result image (that is obtained with the highest correlation value), and graph of the correlation values, as shown in Figure 9.9c [35] were obtained. In an optical–digital hybrid image searching system, the optical correlator performs image searches in the cloud environment and can be controlled via a web browser. This hybrid system represents a basic technology for high-speed correlation of various types of data. The optical correlator of this system is more versatile and more widely used than the previous optical correlators, which require users to have particular skills. The reliability of the optical correlation experiment results can be improved using this system.

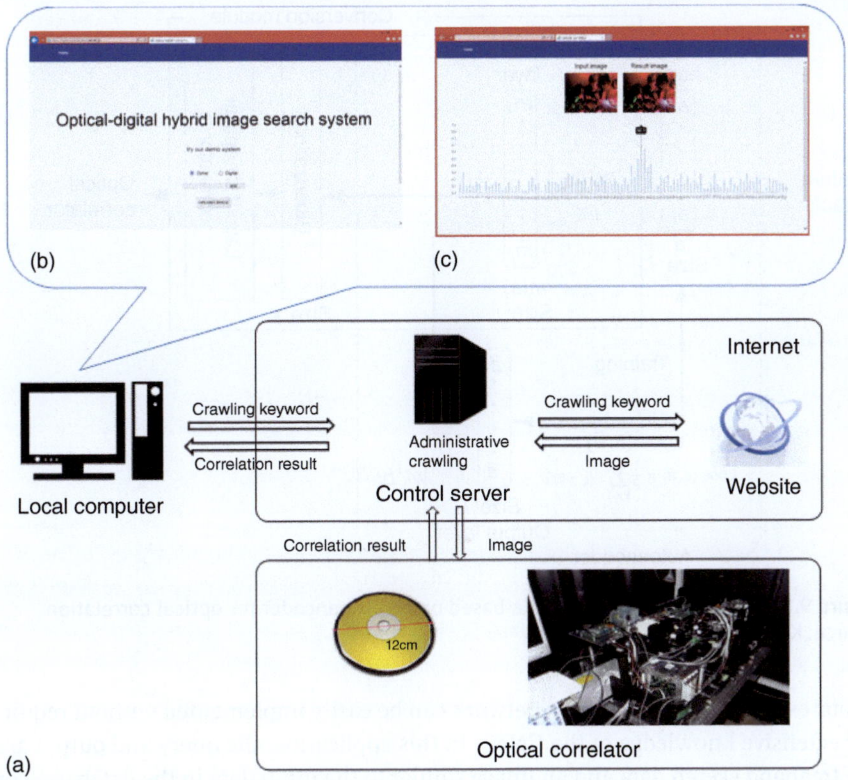


**Figure 9.9** Schematic of a hybrid optical–digital image searching system. (a) Optical correlator system, (b) interface designs of the system on the web browser, and (c) page of the correlation input and output. Source: Kanami Ikeda.

#### 9.1.3.4 Sketch-Based Image Retrieval Application

In this section, we introduce an optical correlator system facilitating cross-domain image retrieval using a deep neural network as the feature extractor. Here also, the input color image is preprocessed via grayscaling, edge extraction, and binarization for face recognition and image retrieval as described in the previous section. However, in this case, only verification is carried out through a search for an exact match of an image with a specific size. To extend the application of the optical correlator, we introduced the deep learning approach to extract features from the data. However, the conventional conversion module used in the optical correlator is incompatible with the deep-learning-based feature extractor. To overcome this problem, we developed a conversion module to convert the extracted feature to page data for the optical correlator and optimized it using a training scheme based on an autoencoder [36]. Figure 9.10 shows the data conversion module based on an autoencoder for optical correlation. By utilizing the conversion module, we introduced an existing model for deep learning as a feature extractor in the optical correlator system.

Sketch-a-Net [37], which is a convolutional neural network (CNN) used for feature extraction, was utilized for sketch-based image retrieval using the optical correlator

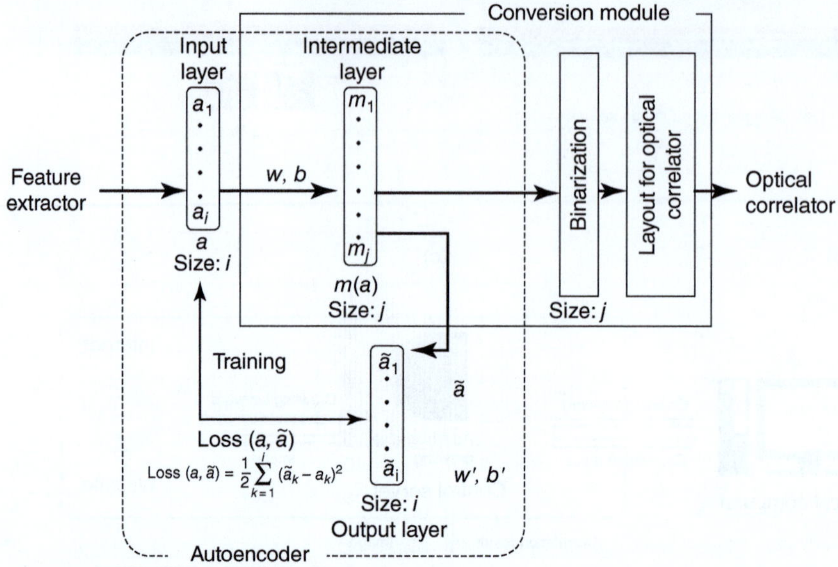


**Figure 9.10** Data conversion module based on an autoencoder for optical correlation. Source: Kanami Ikeda.

in our experiment, because this network can be easily implemented without requiring extensive knowledge of the CNNs. In this application, the query and output are the freehand sketch data and an image similar to the input data in the database. For the optical correlation experiment, we used the database previously used by Yu et al. [37]. This database contained 297 pairs of color images collected from an online shopping website as well as sketch data that were drawn from the color images by a volunteer. A conversion module that can convert various data to binary page data for the optical correlator was developed based on the autoencoder. The 200 sketch-photo pairs in the database were used in the same manner as that described in [38], wherein the CNN was trained for the feature extractor, and the autoencoder was trained for the conversion module. For the experimental evaluation of this system, 97 sketch-photo pairs, different from the data in the training process, were used to determine the accuracy of the system. The image retrieval system offers unique advantages; for instance, it allows the conversion of various types of data to page data and can be combined with the existing deep learning models as a feature extractor via the developed conversion module. In the experiment, after the features were extracted, binary matrices for optical correlation were generated with a $j$ value of 40 000 elements using the proposed module. The experiments were conducted using the optical correlation database, and the data for the converted database were recorded on a holographic optical disk with a 10-μm shift pitch of the shift multiplex recording. We obtained correlation peaks (Figure 9.11b) when the freehand human sketch was included in the input query (Figure 9.11a). Figure 9.11b shows the experimental result of the sketch-based cross-domain image retrieval process and example images from the database used in the experiment; similar images were obtained based on

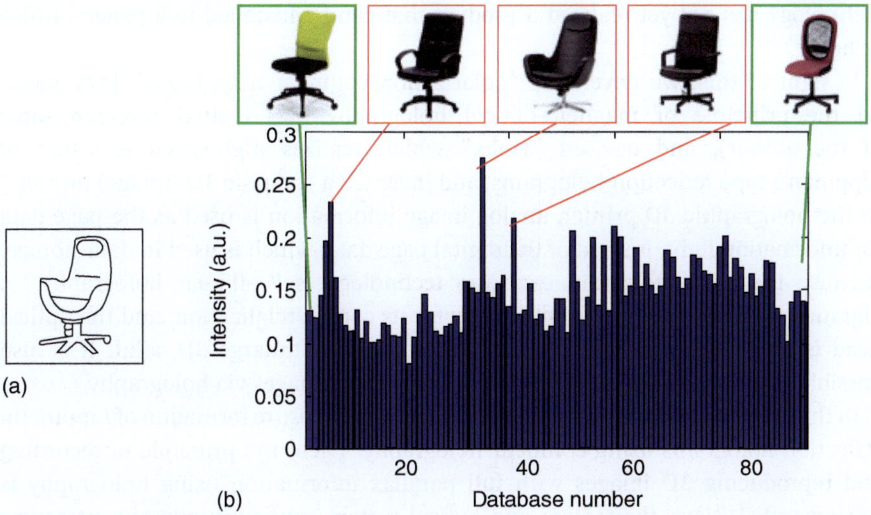


**Figure 9.11** Experimental result of the sketch-based cross-domain image retrieval process: (a) input sketch and (b) correlation values between the input sketch and database images. Source: Kanami Ikeda.

the optical correlation intensity. These results demonstrate the effectiveness of the optical correlation-based cross-domain image retrieval technique, via querying free-hand human sketches, in discovering similar images. With the conversion module and optical correlator, cross-domain image retrieval and the effectiveness of the image retrieval system were experimentally demonstrated.

## 9.2 3D Holographic Printer

### 9.2.1 Introduction

In recent years, 3D printers that precisely model 3D objects have been widely used in the field of prototype designing and medicine. Despite their names, they cannot print 3D stereoscopic images on thin sheets, such as computer peripherals, in general offices and homes.

However, in 1992, Yamaguchi et al. proposed a system that prints 3D stereoscopic images using holographic stereogram technology by exposing a thin hologram recording material such that the partial holograms are spread [39–43]. In addition, devices that are relatively small and have a high exposure speed have been developed for special business and research applications [44, 45]. In these systems, a Lippmann-type hologram is formed to reproduce the color of a 3D stereoscopic image with high fidelity and to raise the image via reflection, which results from the positional relationship between the illumination and the hologram [46]. However, exposing a Lippmann-type partial hologram on a hologram recording medium over a large optical system arrangement presents several drawbacks. Moreover, measurements against vibration remain unsolved. Therefore, the present technology has not yet realized a product that can be installed in a general office or home.

In light of this, we developed "polarization collinear holography" [47], based on the principle of magneto-optical holography [48] (initial research topic of the author), and devised "Holo," which realizes high-speed exposure of Lippmann-type reflection holograms, and invented a "graphic 3D (image) printer." In the holographic 3D printer, analog image information is used as the page data for information light, instead of the digital page data, which is used in data storage. Because the basic hologram exposure technology is "collinear holography," a vibration-isolation table is not necessary for its commercialization, and the optical head is reciprocated at a high speed to create an ultralarge 3D solid. It is also possible to perform high-speed exposure printing of images via holography.

In this section, first, we elucidate the principle of exposure formation of Lippmann reflection holograms using collinear holography. Then, the principle of recording and reproducing 3D images with full parallax information using holography is demonstrated. Next, the system and optical system configurations of a prototype collinear 3D printer system, viz. a collinear multidimensional printer system #1 (CDP-1), is presented with details. Finally, an example of a holographic 3D stereoscopic image sample actually exposed and printed using this CDP-1 is shown, which indicates that collinear holography can be utilized in applications other than holographic optical memory.

### 9.2.2 Principle of Exposure Formation of Lippmann Hologram by Collinear Method

In collinear holography, when the reference light and the information light travel in the same direction along the same optical axis, they interfere with each other in the focal plane of the objective lens, and a transmission hologram is formed. On the other hand, when the interference fringes are formed but travel in the opposite direction in the same optical axis, a reflection hologram, the so-called Lippmann hologram is formed. In the case of a holographic 3D printer, formation of a Lippmann hologram is desired to improve the color reproducibility and resolution of a reproduced image using wavelength selectivity. Therefore, it is necessary to realize reflection-type hologram recording with a simple collinear optical system configuration.

Therefore, an optical arrangement of a polarization collinear method in which the polarization directions of the reference light and the information light are orthogonal to each other is shown in Figure 9.12. Unlike applications in holographic memory, shift selectivity is not required, and there is no need to detect a reproduced image with a CMOS sensor. Therefore, the reference light is characterized by a single plane wave. In addition, since the reference light and the information light may not necessarily be spatially separated, the light is emitted from a position overlapping the information light in Figure 9.12. The quarter-wave plate (QWP) is placed in front of the objective lens as before, but the recording media and the reflective film part are separated, and the reflective film part is separated from the QWP and the reflective polarizing beam splitter. It is composed of a wire grid and an absorber, and has

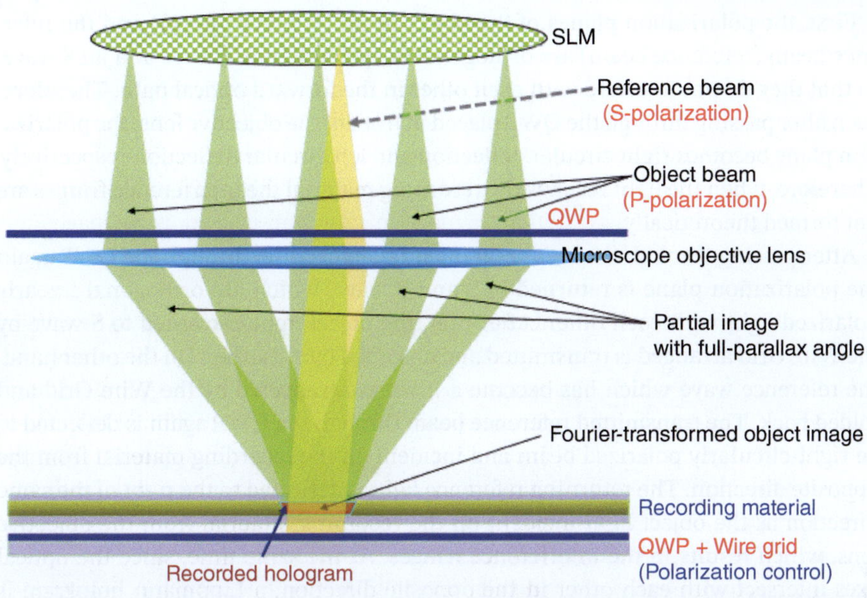


**Figure 9.12** Polarization collinear optical arrangement in which the polarization directions of the reference light and information light are orthogonal. Source: Hideyoshi Horimai.

a structure in which only light of a specific polarization plane is reflected, and a component of the polarization plane orthogonal thereto is then absorbed.

Figure 9.13 schematically shows how a Lippmann hologram by collinear holography is formed by using such a polarization collinear optical system. In the following, description will be made while following the polarization of each illustrated ray.

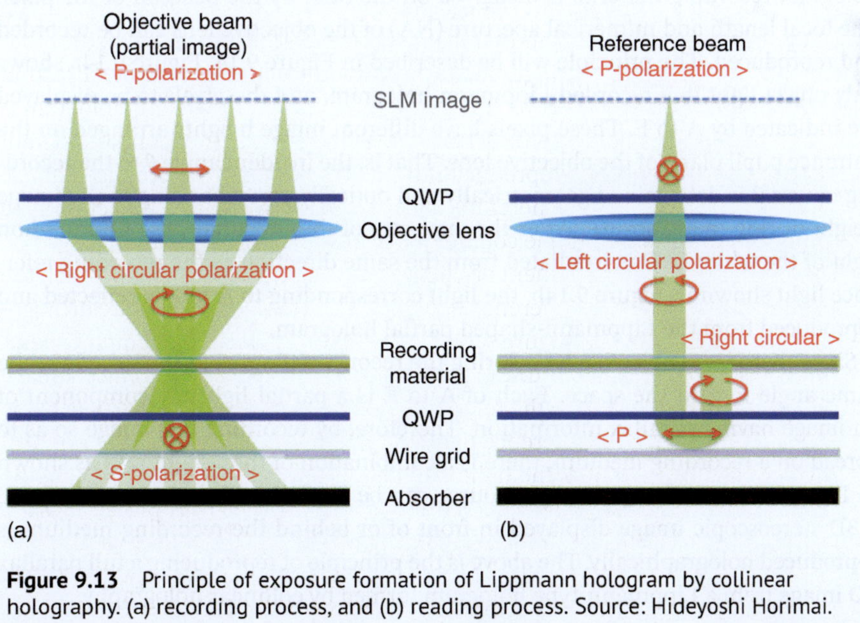


**Figure 9.13** Principle of exposure formation of Lippmann hologram by collinear holography. (a) recording process, and (b) reading process. Source: Hideyoshi Horimai.

First, the polarization planes of the object beam (reference beam) and the reference beam (reference beam) are orthogonal to each other as P wave and an S wave, so that they do not interfere with each other in the forward optical path. Therefore, even after passing through the QWP placed in front of the objective lens, the polarization plane becomes right circular deflection and left circular deflection respectively. Therefore, when the light reaches the recording material the interference fringes are not formed theoretically.

After passing through the recording material, by passing through the QWP again the polarization plane is returned to S and P wave, which are orthogonal linearly polarized light with each other. After that, the object light converted to S wave by the Wire Grid arranged is transmitted and absorbed by Absorber. On the other hand, the reference wave which has become a P wave, is reflected by the Wire Grid and folded back. The transmitted reference beam through the QWP again is deflected to be right-circularly polarized beam and incident on the recording material from the opposite direction. The returning reference light is deflected to the right of the same direction as the object light incident on the recording material from the objective lens, which results in the interference fringes. At the same time, since the optical axes intersect with each other in the opposite direction, a Lippmann hologram is formed. In addition, since the plane of polarization is orthogonal to the reference light incident on the recording material from the objective lens, no interference fringes are formed in principle.

### 9.2.3 Hologram Recording/Reproduction Principle of Full Parallax 3D Image

Full parallax including all vertical and horizontal information which parallax to hologram recording material is displayed on the SLM by the position of the pixel. The focal length and numerical aperture (NA) of the objective lens can be recorded and reproduced. The principle will be described in Figure 9.14. Figure 9.14a shows only object light in a recorded Lippmann hologram, and the pixels to be displayed are indicated by A to E. These pixels have different image heights arranged on the entrance pupil plane of the objective lens. That is, the incident angle $\theta$ to the recording material is determined geometrically and optically corresponding to the image height during the exposure. After the exposure of the hologram, the illumination light of the white LED is irradiated from the same direction as the recording reference light shown in Figure 9.14b, the light corresponding to A to E is reflected and reproduced from the Lippmann-shaped partial hologram.

Since there is no objective lens during the reconstruction, A to E proceeds at the same angle toward the space. Each of A to E is a partial light ray component of an image having parallax information. Therefore, by recording this image so as to spread on a recording medium, there is a combination of those light rays as shown in Figure 9.15. Thus, a point light source can be formed in the space. As a result, a 3D stereoscopic image displayed in front of or behind the recording medium is reproduced holographically. The above is the principle of reproducing a full parallax 3D image from a Lippmann-type hologram formed by collinear holography.

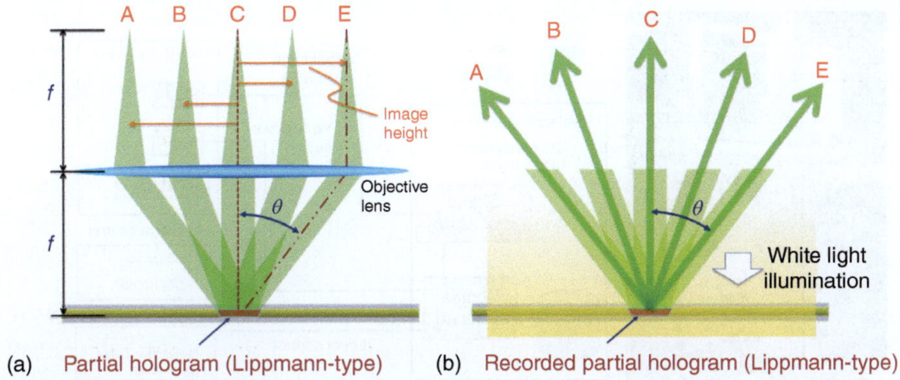


**Figure 9.14** Hologram formation and reproduction principle of full parallax 3D image information. Source: Hideyoshi Horimai.

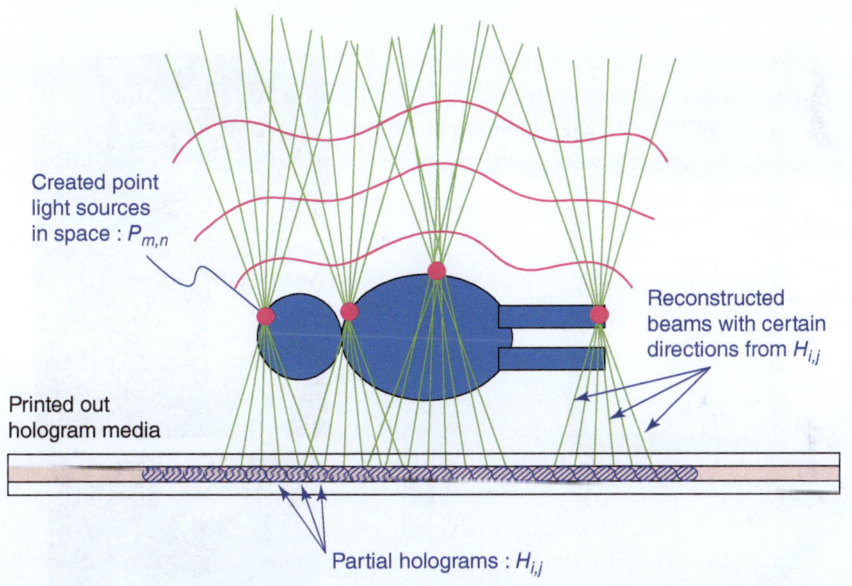


**Figure 9.15** Principle of spatial point light source generation using a set of partial holograms. Source: Hideyoshi Horimai.

### 9.2.4 Holographic 3D Printer System: CDP-1

Based on the above principle, we have developed a prototype 3D image printer system that uses the collinear method: CDP-1. Figure 9.16 shows a block diagram of the appearance and system configuration of the system. Figure 9.17 shows a schematic diagram of the configuration of the optical system, and the appearance of the optical system and a photograph of the main part. A Q-SW pulse laser with a wavelength of 532 nm is applied for recording. Since the excessive interference distance is extremely short, after the object light and the reference light are separated by PBS1, the optical path length of the reference light is corrected using a pair of 45° mirrors

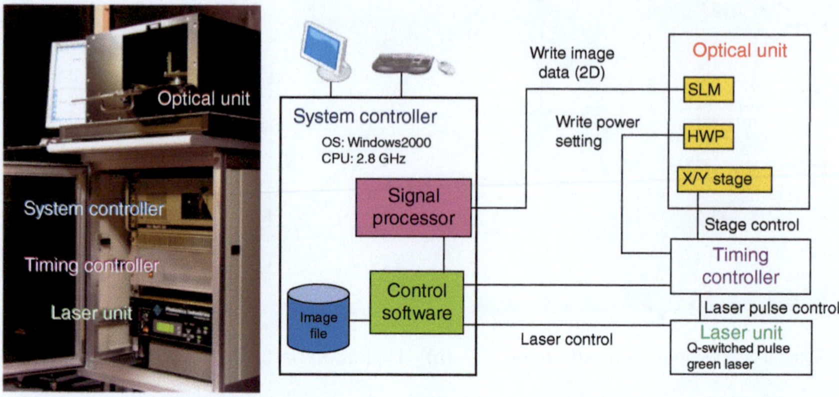


**Figure 9.16** Collinear 3D printer system: CDP-1 appearance and system configuration block diagram. Source: Hideyoshi Horimai.

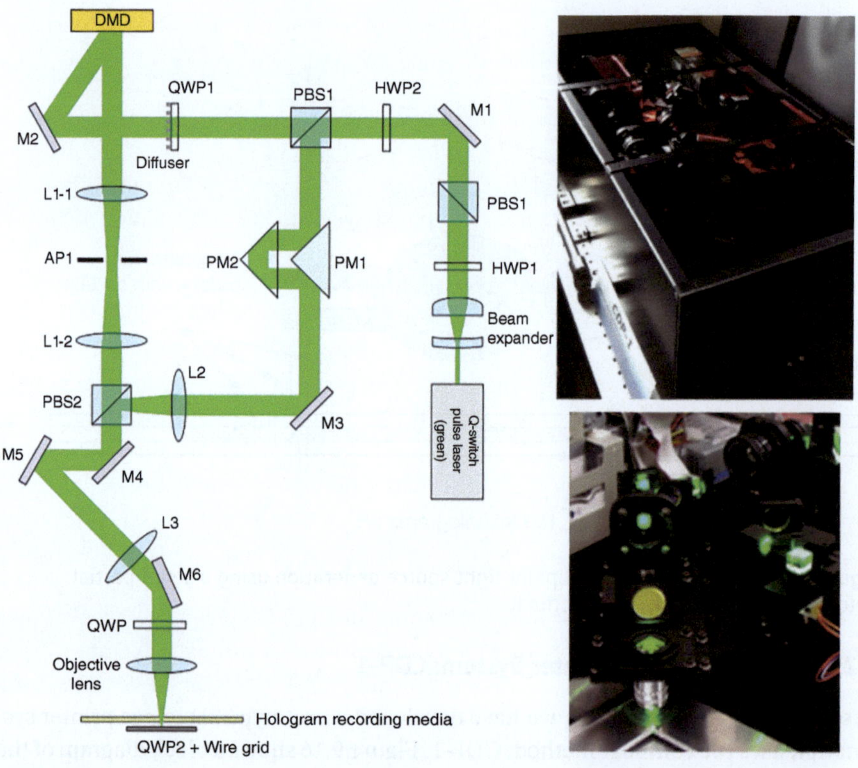


**Figure 9.17** Schematic of the optical system configuration of CDP-1, and appearance of the optical unit and main part of the optical system. Source: Hideyoshi Horimai.

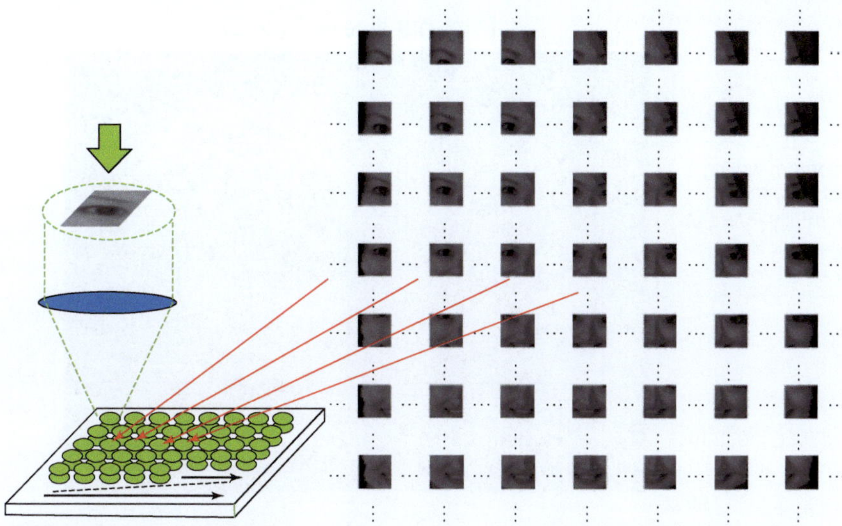

**Figure 9.18** Spread exposure of partial holograms carrying full parallax information (schematic). Source: Hideyoshi Horimai.

PM1 and PM2. The digital micromirror device (DMD) is used as the SLM for information light display. The QWP1 placed in front of the DMD is inserted for correcting the birefringence inherent in the DMD glass and improving the light utilization efficiency in the polarization collinear optical system. The information light and the reference light incidented on the PBS2 resulting in the coaxial reference light and information light whose polarization planes are orthogonal to each other, then guided to the objective lens to form an interference fringe.

The page data carrying full parallax information to be recorded for each partial hologram are pre-rendered, sliced and dice processed. Then they are stored in the hard disk (Image File) in the system controller and displayed on the DMD as an XGA-size image. They are read out in synchronism with the X/Y stage and recorded so as to be spread over a recording medium as schematically shown in Figure 9.18.

### 9.2.5 Exposure Printing of 3D Images Using CDP-1

Using CDP-1, exposure printing of holographic 3D stereoscopic images was actually performed. Figure 9.19 is a photograph obtained by observing a hologram formed by exposure with a microscope. The size of the partial hologram is about 500 μm in diameter, but the exposure pitch of the hologram is 300 μm. This is a kind of multiplex recording utilizing the multiple exposure performance of a photopolymer so that the boundary line for each partial hologram does not appear in the reproduced image. However, since the reference light is one parallel light generated from one pixel, it can be said that the reference light is a unique collinear shift multiplex recording having no shift selectivity.

Figure 9.20 shows an example of the observation result of the hologram that is actually exposed and printed; overall, $64 \times 64$ holograms are recorded. This is

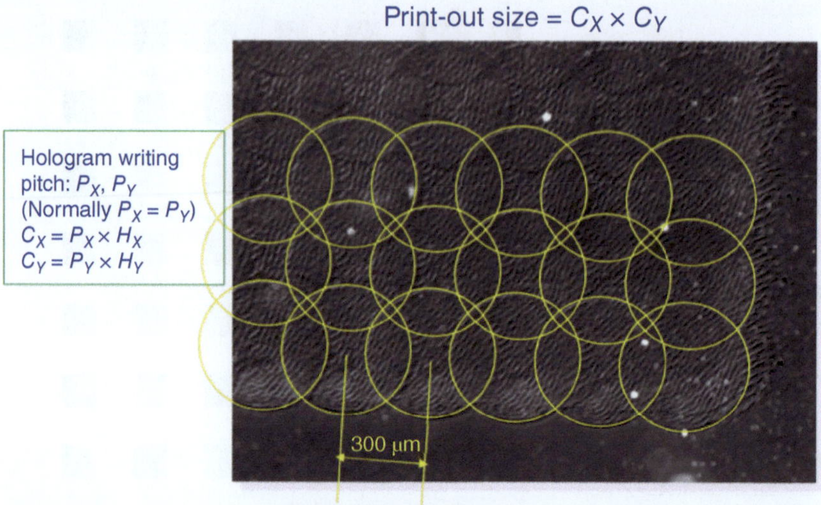


**Figure 9.19** Micrograph of a partial hologram formed by exposure, using the collinear shift multiplexing method. Source: Hideyoshi Horimai.

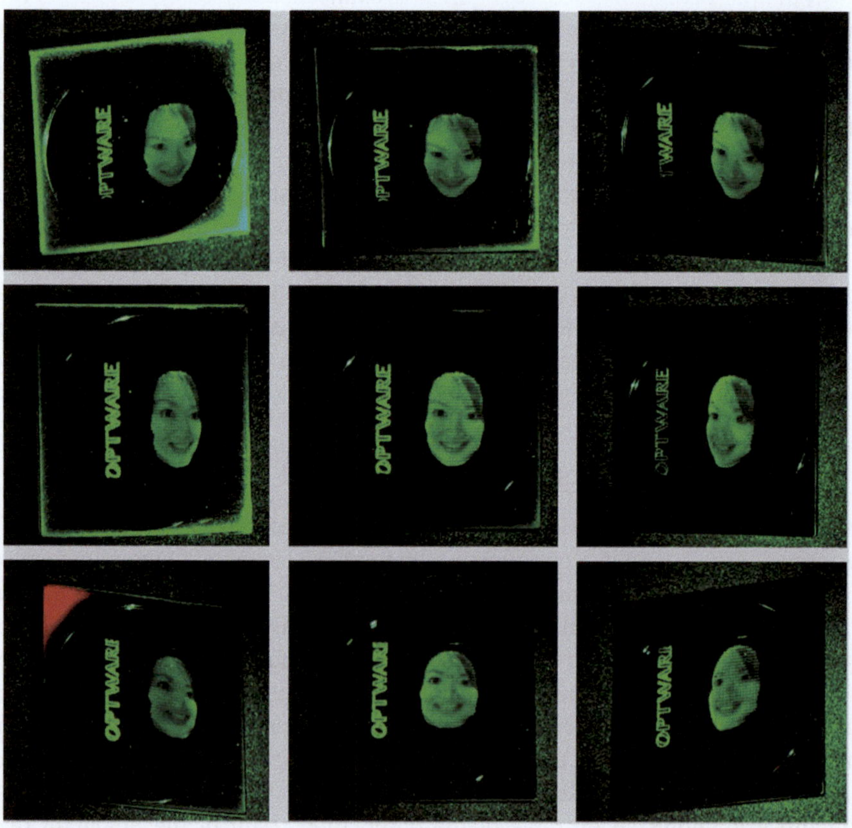

**Figure 9.20** Observation results of exposure-printed hologram (photographed from three horizontal directions × three vertical directions). Source: Hideyoshi Horimai.

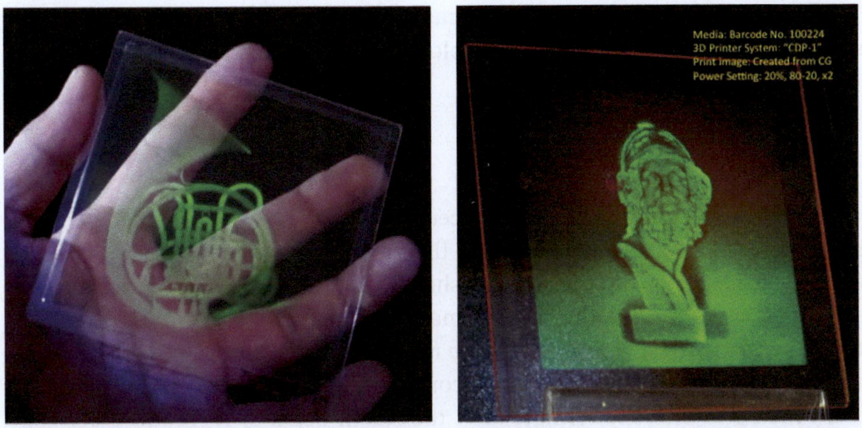


**Figure 9.21** Example of high-definition 3D hologram exposure-printing sample using collinear shift multiplexing. Source: Hideyoshi Horimai.

a picture captured by changing the parallax from three horizontal directions and three vertical directions, and the captured images are shown side by side. Since the displayed image changes when the parallax is changed, it was confirmed that a full parallax 3D stereoscopic image was formed by exposure.

Furthermore, Figure 9.21 shows an example of image observation of a sample that is exposed and printed with 512 × 512 pixels horizontally with a hologram pitch of 200 μm or less. The exposure size was 90 mm × 90 mm, and a partial hologram was dynamically exposed and printed at a speed of 20 mm/s on a continuously moving hologram recording medium. From the observation results, it was suggested that even in the case of exposure printing of holographic 3D images, high-density 3D reproduction images can be obtained as a result of high-density multiple exposure by collinear shift multiplexing.

For the practical application, improvement of viewing angle and realization of color exposure printing should be considered in the future issues.

### 9.2.6 Summary

The principle of exposure formation of Lippmann holograms using polarization collinear holography technology, in which the polarization directions of the object and reference beams are orthogonalized, and the control of forward and backward polarizations are described here. We also developed CDP-1, a holographic 3D (image) printer that uses this technology to expose and print 3D images with full parallax information via holography. A partial hologram exposed and printed at a pitch of <200 μm on a recording medium, continuously moving at a relative speed of 20 mm/s, facilitated high-resolution 3D stereoscopic image exposure printing. Because the size of the partial hologram is 500 μm, it is suggested that the collinear shift multiplexing recording is effective for recording and reproducing high-resolution images even in a holographic 3D printer.

The aforementioned results demonstrate the flexible applicability of collinear holography in applications other than holographic optical memory.

## 9.3 Encryption

In Section 3.4 (Chapter 3), we introduced the simulation of interference fringes formed by the collinear method. Using this method, we can derive the intensity distribution in the media. Figure 9.22 shows that two different reference beams (radial and random) and the same signal beam as the input produce different interference fringes. Obviously, when the reference beam changes, the interference fringe also changes. If only the same recording reference beam is used to read the corresponding interference fringe, then the signal beam is reproduced correctly. Thus, based on this characteristic, encryption can be realized by using the reference beam as the key. Moreover, the reference beam can be both amplitude and phase modulated, thereby guaranteeing the strength of the encryption.

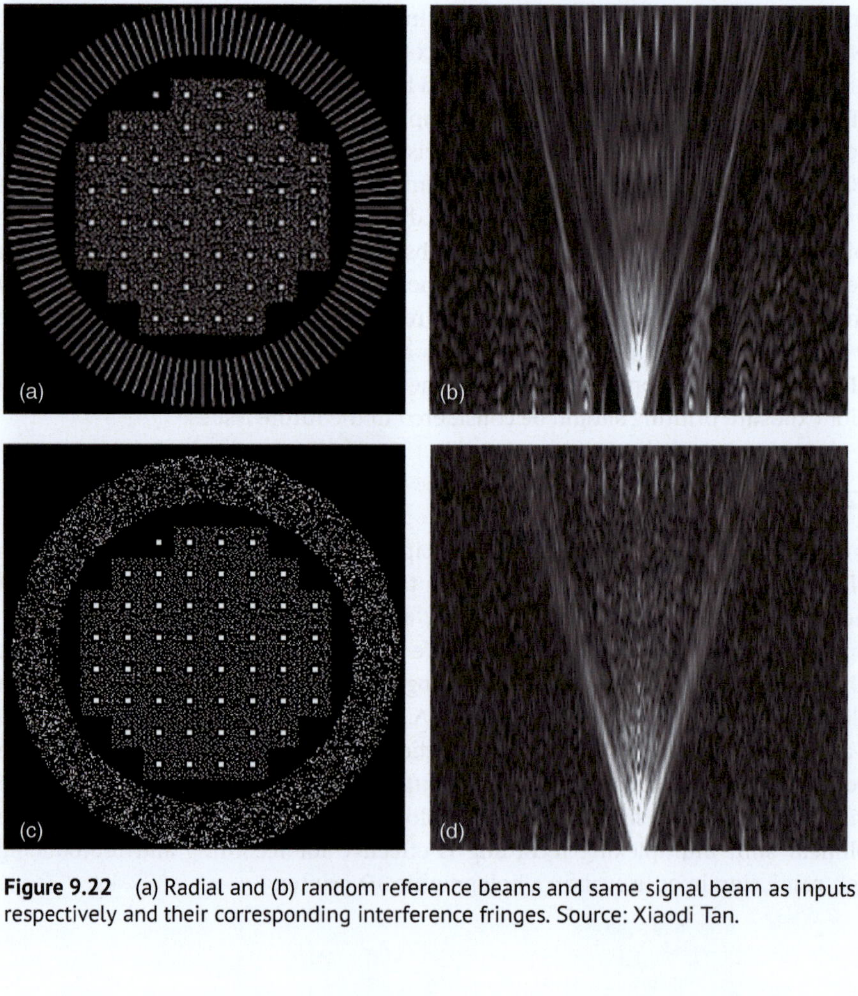


**Figure 9.22** (a) Radial and (b) random reference beams and same signal beam as inputs respectively and their corresponding interference fringes. Source: Xiaodi Tan.

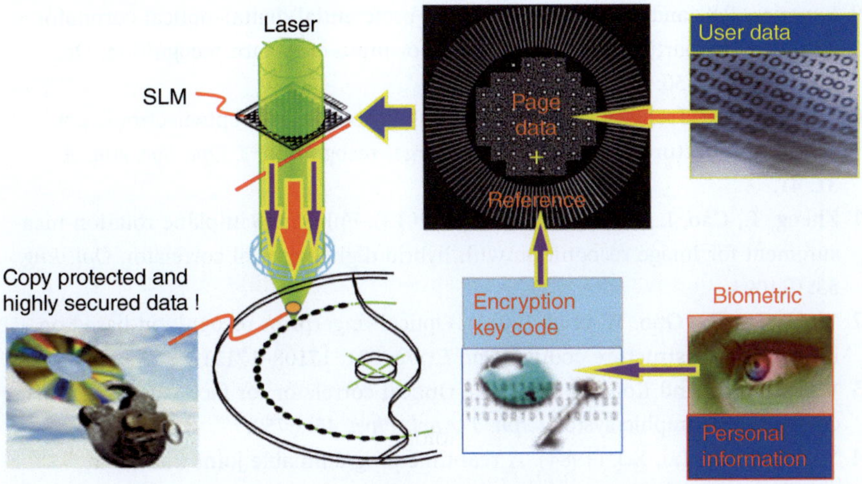


**Figure 9.23** Schematic of the encryption process. Source: Xiaodi Tan.

Figure 9.23 shows the schematic of the encryption process. Personal information, such as biometric information, is the unique encryption key code, which is related to the unique reference pattern. Furthermore, the user data are encoded to many page data and stored using the unique reference pattern with high security and copy protection.

## References


**1** Lugt, A.V. (1964). Signal detection by complex spatial filtering. *IEEE Trans. Inf. Theory* 10: 139–145.

**2** Weaver, C.S. and Goodman, J.W. (1966). A technique for optically convolving two functions. *Appl. Opt.* 5: 1248–1249.

**3** Watanabe, E. and Kodate, K. (2005). Implementation of a high-speed face recognition system that uses an optical parallel correlator. *Appl. Opt.* 44: 666–676.

**4** Xu, P., Hong, C., Cheng, G. et al. (2015). Planar optical correlators integrated with binary optical lens. *Opt. Express* 23: 6773–6779.

**5** Zeng, X., Bai, J., Hou, C., and Yang, G. (2011). Compact optical correlator based on one phase-only spatial light modulator. *Opt. Lett.* 36: 1383.

**6** Alfalou, A. and Brosseau, C. (2011). Robust and discriminating method for face recognition based on correlation technique and independent component analysis model. *Opt. Lett.* 36: 645.

**7** Goyal, S., Nishchal, N.K., Beri, V.K., and Gupta, A.K. (2006). Wavelet-modified maximum average correlation height filter for rotation invariance that uses chirp encoding in a hybrid digital-optical correlator. *Appl. Opt.* 45: 4850.

**8** Fukuchi, N., Inoue, T., Toyoda, H., and Hara, T. (2009). Lensless Vanderlugt optical correlator using two phase-only spatial light modulators. *Chin. Opt. Lett.* 7: 1–3.

9 Banerjee, P.K. and Datta, A.K. (2013). A preferential digital–optical correlator optimized by particles warm technique for multi-class face recognition. *Opt. Laser Technol.* 50: 33.

10 Monjur, M.S., Tseng, S., Tripathi, R. et al. (2014). Hybrid optoelectronic correlator architecture for shift-invariant target recognition. *J. Opt. Soc. Am. A* 31: 41.

11 Zheng, T., Cao, L., He, Q., and Jin, G. (2013). Full-range in-plane rotation measurement for image recognition with hybrid digital–optical correlator. *Opt. Eng.* 53: 011003.

12 Yi, Y., Cao, L., Guo, W. et al. (2013). Optical fingerprint recognition based on local minutiae structure coding. *Opt. Express* 21: 17108–17121.

13 Watanabe, E. and Kodate, K. (2006). Optical correlator for face recognition using collinear holographic system. *Jpn. J. Appl. Phys.* 45: 6759.

14 Yu, F.T.S. and Lu, X.J. (1984). A real-time programmable joint transform correlator. *Opt. Commun.* 52: 10.

15 Heifetz, A., Shen, J.T., Lee, J.K. et al. (2006). Translation-invariant object recognition system using an optical correlator and a superparallel holographic random access memory. *Opt. Eng.* 45: 1.

16 Gu, C., Fu, H., and Lien, J.R. (1995). Correlation patterns and cross-talk noise in volume holographic optical correlators. *J. Opt. Soc. Am. A* 12: 861.

17 Pu, A., Denkewalter, R., and Psaltis, D. (1997). Real-time vehicle navigation using a holographic memory. *Opt. Eng.* 36: 2737.

18 Watanabe, E., Ichikawa, Y., Akiyama, R., and Kodate, K. (2008). Ultrahigh-speed optical correlation system using holographic disc. *Jpn. J. Appl. Phys.* 47: 5964.

19 Ikeda, K. and Watanabe, E. (2015). High-speed optical correlator with coaxial holographic system. *Jpn. J. Appl. Phys.* 54: 09ME02.

20 Ikeda, K. and Watanabe, E. (2016). High-speed image matching with coaxial holographic optical correlator. *Jpn. J. Appl. Phys.* 55: 09SC01.

21 Ikeda, K. and Watanabe, E. (2015). Reducing input data via image categorization to improve the speed of copyright content management systems. *Opt. Rev.* 22: 93–98.

22 Ikeda, K., Kodate, K., and Watanabe, E. (2016). Optical–digital hybrid image search system in cloud environment. *Jpn. J. Appl. Phys.* 55: 09SC03.

23 Ikeda, K., Suzuki, H., and Watanabe, E. (2017). Optical-correlation based cross-domain image retrieval system. *Opt. Lett.* 42 (13): 2603–2606.

24 Simon, D.M. and Serati, S.A. (1994). Optical correlator using ferroelectric liquid crystal spatial light modulators and Fourier transform lenses. US Patent US5418380 A, 1994 April 12.

25 Wilkinson, T.D. (1997). Optical correlator. UK Patent Application Number WO2004029746, 9726386.7, 1997 December 12.

26 Horimai, H., Tan, X., Li, J., and Suzuki, K. (2005). Wavelength margin analysis in advanced collinear holography. *Jpn. J. Appl. Phys.* 44: 3493–3494.

27 Tan, X., Horimai, H., Shimura, T. et al. (2006). Collinear technology for holographic versatile disc (HVDTM) system. *Proc. SPIE* 6343: 63432W.

**28** Horimai, H., Tan, X., and Li, J. (2005). Collinear holography. *Appl. Opt.* 44: 2575–2579.

**29** Nobukawa, T., Wani, Y., and Nomura, T. (2015). Multiplexed recording with uncorrelated computer-generated reference patterns in coaxial holographic data storage. *Opt. Lett.* 40: 2161–2164.

**30** Ashkin, A., Boyd, G.D., Dziedzic, J.M. et al. (1966). Optically-induced refractive index inhomogeneities in $LiNbO_3$ and $LiTaO_3$. *Appl. Phys. Lett.* 9: 72.

**31** Martin, S., Feely, C.A., and Toal, V. (1997). Holographic recording characteristics of an acrylamide-based photopolymer. *Appl. Opt.* 36: 5757.

**32** Kwon, J.H., Hwang, H.C., and Woo, K.C. (1999). Analysis of temporal behavior of beams diffracted by volume gratings formed in photopolymers. *J. Opt. Soc. Am. B* 16: 1651.

**33** Momose, K., Takayama, S., Hata, E., and Tomita, Y. (2012). Shift-multiplexed holographic digital data page storage in a nanoparticle-(thiol-ene) polymer composite film. *Opt. Lett.* 37: 2250.

**34** Hesselink, L., Orlov, S.S., and Bashaw, M.C. (2004). Holographic data storage systems. *Proc. IEEE* 92: 1231.

**35** Ikeda, K., Taniguchi, Y., Maskawa, S. et al. (2015). Optical–digital hybrid image search system operable from web browsers. *International Symposium on Optical Memory 2015 Technical Digest*, October 2015.

**36** Hinton, G.E. and Salakhutdinov, R.R. (2006). Reducing the dimensionality of data with neural networks. *Science* 313: 504–507.

**37** Yu, Q., Liu, F., Song, Y.Z. et al. (2016). Sketch me that shoe. *Proceedings of IEEE Conference on Computer Vision and Pattern Recognition (IEEE 2016)*, pp. 799–807.

**38** Wang, F., Kang, L., and Li, Y. (2015). Sketch-based 3D shape retrieval using Convolutional Neural Networks. *Proceedings of IEEE Conference on Computer Vision and Pattern Recognition (IEEE 2015)*, pp. 1875–1883.

**39** Yamaguchi, M., Ohyama, N., and Honda, T. (1992). Holographic three-dimensional printer new method. *Appl. Opt.* 31 (2): 217.

**40** Yamaguchi, M., Endoh, H., Honda, T., and Ohyama, N. (1994). High-quality recording of a lull-parallax holographic sterogram with a digital diffuser. *Opt. Lett.* 19 (2): 135.

**41** Mamyama, S., Ono, Y., and Yamaguchi, M. (2008). High-density recording of lull-parallax holographic stereogram. *Proc. SPIE* 6912, 69120N–10.

**42** Yang, F., Murakami, Y., and Yamaguchi, M. (2012). Digital color management in full-color holographic three-dimensional printer. *Appl. Opt.* 51 (19): 4343–4352.

**43** Utsugi, T. and Yamaguchi, M. (2013). Reduction of the recorded speckle noise in holographic 3D printer. *Opt. Express* 21 (1): 662–674.

**44** 木原信宏 (1998). 白倉明, 馬場茂幸, "高速ホログラムポートレイトプリントシステム," 3 次元画像コンファレンス (1998).

**45** 白倉明, 木原信宏, 馬場茂幸, "即時ホログラフィック 3D プリント技術," ホログラフィックディスプレイ研究会会報, 18–3 (1998).

**46** Lippmann, G. (1894). Sur la théorie de la photographie des couleurs simples et composées par la méthode intérférentielle. *J. Phys. Theor. Appl.* 3 (3): 97–107.

**47** Horimai, H., Lim, P.B., Hesselink, L., and Inoue, M. (2002). Volumetric optical disk storage with collinear polarized holography. Technical Digest of ISOM/ODS'02, TuP26, pp. 43–44.

**48** 堀米秀嘉, 井上明, 見山友裕, 藤井壽崇, 鈴木新一, 草鹿履一郎, 徳島忠夫, 松本誠二 (1984). TbFe 非晶質薄膜を用いた高密度磁気ホログラム記録. *電子通信学会研究会* CPM83-85: 55–60.

# Index



*Collinear Holography: Devices, Materials, Data Storage*, First Edition.
Xiaodi Tan, Hideyoshi Horimai, Tsutomu Shimura, and Xiao Lin.